S0-AHQ-288

INTRODUCTION
TO INTERVAL COMPUTATIONS

This is a volume in
COMPUTER SCIENCE AND APPLIED MATHEMATICS
A Series of Monographs and Textbooks

Editor: WERNER RHEINBOLDT

A complete list of titles in this series is available from the Publisher upon request.

INTRODUCTION TO INTERVAL COMPUTATIONS

Götz Alefeld

FAKULTÄT FÜR MATHEMATIK
UNIVERSITÄT KARLSRUHE
KARLSRUHE, FEDERAL REPUBLIC OF GERMANY

Jürgen Herzberger

FAKULTÄT FÜR MATHEMATIK
UNIVERSITÄT OLDENBURG
OLDENBURG, FEDERAL REPUBLIC OF GERMANY

Translated by *Jon Rokne*

DEPARTMENT OF COMPUTER SCIENCE
THE UNIVERSITY OF CALGARY
CALGARY, ALBERTA, CANADA

1983

ACADEMIC PRESS
A Subsidiary of Harcourt Brace Jovanovich, Publishers
New York London
Paris San Diego San Francisco São Paulo Sydney Tokyo Toronto

UNIVERSITY LIBRARY
GOVERNORS STATE UNIVERSITY
PARK FOREST SOUTH, ILL.

COPYRIGHT © 1983, BY ACADEMIC PRESS, INC.
ALL RIGHTS RESERVED.
NO PART OF THIS PUBLICATION MAY BE REPRODUCED OR TRANSMITTED IN ANY FORM OR BY ANY MEANS, ELECTRONIC OR MECHANICAL, INCLUDING PHOTOCOPY, RECORDING, OR ANY INFORMATION STORAGE AND RETRIEVAL SYSTEM, WITHOUT PERMISSION IN WRITING FROM THE PUBLISHER.

ACADEMIC PRESS, INC.
111 Fifth Avenue, New York, New York 10003

United Kingdom Edition published by
ACADEMIC PRESS, INC. (LONDON) LTD.
24/28 Oval Road, London NW1 7DX

Library of Congress Cataloging in Publication Data

Alefeld, G., Date.
Introduction to interval computations.

(Computer science and applied mathematics)
Translation of: Einführung in die Intervallrechnung.
Bibliography: p.
1. Interval analysis (Mathematics) I. Herzberger, Jürgen. II. Title. III. Series.
QA297.75.A3613 519.4 82-6715
ISBN 0-12-049820-0

PRINTED IN THE UNITED STATES OF AMERICA

83 84 85 86 9 8 7 6 5 4 3 2 1

This is a revised and expanded, English-language edition of *Einführung in die Intervallrechnung,* Bibliographisches Institut AG, Mannheim, 1974.

7
77.75
3613
783
1

To Josef Heinhold
who pointed the way for us

CONTENTS

PREFACE TO THE ENGLISH EDITION

This book is a revised and expanded version of the original German text. The arrangement of the material and the structure are essentially unchanged. All remarks in the Preface to the German Edition regarding naming conventions for formulas, theorems, lemmas, and definitions are still valid as are those concerning the arrangement and choice of material.

In almost all sections of the book there are minor changes, some rearrangements, and some additional material. More extensive additions of new material have taken place in Chapters 3, 4, 7, 12, 14, 15, 18, and 19 (interval arithmetic evaluation, rounded interval calculations, symmetric single step method, feasibility of the Gaussian algorithm, solution of nonlinear equations, and systems of equations).

As in the original version, we could not present a comprehensive survey of the applications of interval computations, which are now even more numerous. Such a survey would have made the text too large. Instead we concentrated on the principal tools and results of interval arithmetic. The interested reader is therefore referred to the new expanded Bibliography in the English edition.

The English manuscript was typed by Mrs. B. Trajanovic, Berlin, and Mrs. F. Knight, Calgary. Dr. H. Cornelius, Dr. L. Platzöder, Dr. H. Schwandt, and Dr. G. Mayer read various versions of the English manuscript.

We are thankful to Professor W. Rheinboldt and Academic Press for publishing the book in the series "Computer Science and Applied Mathematics." Our main thanks are directed to our colleague Professor Jon Rokne, Calgary, who translated the book in a very careful and correct manner during the tenure of Killam Resident Fellowship held at the University of Calgary.

Karlsruhe and Oldenburg G. ALEFELD AND J. HERZBERGER

PREFACE TO THE GERMAN EDITION

This book originated from seminars and lectures that we have assisted with and conducted at the University of Karlsruhe. It is intended to be both a systematic introduction to the tools of interval analysis as well as a unifying presentation of the interval analytic methods developed over the past years using these tools. For the latter, concern is with methods for the solution of algebraic problems in numerical mathematics developed using interval analysis. The book is therefore not a complete representation and description of what has been developed recently under the unifying concept of interval analysis, nor does it contain a complete summary of all possible applications of interval analysis.

Because of the aim of the book, the first chapters contain an elementary and complete description of the properties of the four basic operations on real intervals and of the expressions formed from these. The aim is to develop here a system of rules for calculating with intervals sufficient for further applications. This system depends on only a few concepts, such as absolute value, width, and distance. Concepts from real vector spaces are extended to interval spaces whenever appropriate. Because of the nonlinear structure of the spaces occurring in interval analysis, it follows that the properties of the operations in interval spaces will differ from those of linear spaces. Two realizations of interval arithmetics over the complex numbers are also introduced within the same framework. These are the so-called rectangular and circular complex interval arithmetics. By restricting the interval operations, it is possible to demonstrate analogies among the three interval arithmetics and to exploit these in the applications. Moreover, the reader is helped by the inclusion of rules common to vector spaces and is also made aware of other more specialized interval operations and of other specialized concepts by references to the literature. These specialized concepts will, however, not be treated beyond that in this book. The tools of interval arithmetic over the real and complex numbers, respectively, vectors or matrices, are applied to a number of selected problems. The problems that are treated may collectively be characterized as problems relating to the calculation and improvement of inclusions of solutions of mathematical problems. The basic problems may be divided into two classes: problems with "inexact" and with "exact" initial data. The first class generally contains the problem of solving an equation whose data are allowed to vary over an interval by computing a best possible inclusion set. An example of this is a set of linear equations whose coefficients are allowed to vary over an interval. In the class of problems where the exact initial data are given, the interval analysis is mainly used to develop methods that generate convergent sequences of bounds converging to the solution under comparatively weak conditions.

The starting point for the application of interval analysis was, in retrospect, the desire in numerical mathematics to be able to execute algorithms on digital computers capturing all the roundoff errors automatically and therefore to calculate strict error bounds automatically. This rather naïve outlook, occasionally found even today, that is, the replacement of real values in the algorithms that are developed in the real field by intervals as well as the execution of these algorithms using a machine interval arithmetic, seldom delivers error bounds of practical utility. The effects of doing this are not properly investigated and only a few works are concerned with this. This investigation is carried the furthest in one paper [B268] in the Bibliography.

The problems treated in this book are of the types described above with either "inexact" or "exact" data. The calculations of the inclusions, mostly through the use of an iteration procedure, are executed in such a manner that such basic properties of the solution as monotonicity of the sequence of inclusions or convergence are preserved. When these calculations are performed in an appropriate machine interval arithmetic, the same is assumed to hold true. The treatment of machine interval arithmetic is therefore limited to the description of conditions for appropriate realizations of the interval operations on a computer. Interval procedures written in ALGOL are found in Appendix B, and they are applied in the algorithms in Appendix C. We have chosen this form for a realization of a machine interval arithmetic since most computers have an ALGOL translator wherein such computations may be executed. The cost of programming the interval operations as procedures is therefore small, as seen in the programming examples in Appendix C. The introduction of higher programming languages with the possibility of introducing nonstandard data types and operations in higher programming languages is not considered. It would, however, be desirable in the future to introduce machine intervals and machine interval operations as part of these languages. It is already possible to do this, for example, in ALGOL 68 or in SIMULA, since these are extensible languages. They are not, however, in universal use. Until universal languages are defined with this ability, it seems that the method suggested above is a reasonable compromise in order that the programs have a general utility.

After this short description of the motivation and content of the book, the individual sections will be described by their key points.

The book consists of 22 chapters and three appendixes. Most of the chapters conclude with remarks on the material in the chapter and references to results in further works.

In Chapter 1 the real interval operations are introduced in an elementary manner and the most important relations between real intervals are given. Chapter 2 treats the metric properties. In Chapter 3 the concept of the interval evaluation of a real function is explained and some qualitative statements are proved. This chapter is not described in full generality for reasons given in the Remarks. It does contain, however, all properties of interval evaluations needed in the sequel. In Chapter 4 the realization of real interval computations on digital computers is introduced and discussed. The criteria for a machine interval arithmetic are given, guaranteeing that the most important properties of the real interval arithmetic remain valid when approximated by a machine interval arithmetic. In Chapter 5 two possible interval concepts are introduced for the complex plane and the metric properties for these are treated in Chapter 6. Chapter 7 brings the first applications. Different methods are given for the improvement of an inclusion of the solution of a real ("exact") equation and convergence statements and statements about the order of convergence are proved. The two following chapters contain methods for the simultaneous calculation of all the roots of a polynomial with exact coefficients. Of utmost interest is that the method given in Chapter 8 for polynomials with all real zeros lying in pairwise disjoint intervals always converges to the zeros at least quadratically. In Chapter 9 methods are given for the general case of a polynomial with complex zeros. The zeros are continuously included in circular domains. These methods are locally convergent and have convergence of order 3. Chapter 10 introduces an arithmetic for interval matrices as well as some metric properties for these. In Chapter 11 some statements that are in part elementary are proved for the determination of a fixed point of a system of equations with interval coefficients. An interpretation is also given for such fixed points. Of particular interest in this chapter are the possibilities that are displayed for the computation of Lipschitz constants for a system of equations with interval coefficients. Chapters 12–14 concern themselves with the particular problems of a linear system of equations in a form amenable to iterations. In Chapters 12 and 13 the corresponding generalizations of the total step, single step, and relaxation methods are given for the application of systems of linear equations with interval coefficients. Necessary and sufficient conditions are given for the convergence. It is remarkable that the total step method and the single step method are either both convergent or both divergent. The relaxation method does not offer any advantage. In Chapter 13 it is furthermore shown that the methods for solving a linear system of equations using the theory of monotonically decomposable operators are special cases of the methods developed using interval analysis. In Chapter 14 some modifications of the total step and the single step methods are discussed and an optimal method is presented. Chapters 15–17 are mostly concerned with carrying over the formulas of the Gaussian algorithm for computing an including set for the solution of a system of equations having interval coefficients. In Chapter 15 two classes of interval matrices are given for which it is shown that the Gaussian algorithm always computes an inclusion. In Chapter 16 one of the first suggestions for "solving" a general linear system of equations is presented and some qualitative statements are given. In Chapter 17 we consider a possibility for the improvement of inclusions for system of equations with interval coefficients. In Chapter 18 methods of arbitrarily high order of convergence are presented for the iterative inclusion of the inverse of a real matrix and convergence statements are given for these. Chapters 19–22 are concerned with the solution of "exact" nonlinear systems of equations. In these chapters interval analysis is used—as was the case in Chapters 8 and 9—exclusively to guarantee the convergence of various modifications to methods from traditional numerical analysis. By using interval analysis it is possible to prove far-reaching convergence statements that can be proved only locally or using restricting conditions as, for example, convexity of the system, if interval analytic techniques are not applied. In Chapter 19 some modifications of Newton's method for the solution of a nonlinear system of equations by repeated inclusions are considered. In this chapter we also bring the application of the Gaussian algorithm to the solution of a system of linear equations with interval coefficients that was mentioned in Chapter 15. Chapter 20 gives methods not requiring the solution of linear systems with interval coefficients still having at least second order of convergence. In Chapter 21 special methods for nonlinear systems, suitable, for example, for systems arising from the discretization of nonlinear boundary value problems, are considered. In this chapter it is demonstrated most clearly that the clever use of certain properties of interval analysis may lead to very useful methods. Chapter 22 treats methods useful for linear systems of equations with interval coefficients that arise when one uses Newton's method for the solution of nonlinear systems of equations. The solution set of this linear system of equations with interval coefficients contains the solution of the nonlinear system and it is approximately solved by the use of one or more steps of the total step or the single step method. Such methods are, as is well known, appropriate for the solution of the kind of large nonlinear systems of equations that arise, for example, from the discretization of nonlinear partial differential equations. It turns out that one may prove the convergence of these methods under relatively general conditions. In Appendix A we introduce a definition for the order of convergence of an interval arithmetic iteration method. This definition is used in the formulation of statements of the order of convergence in the rest of the manuscript. In Appendix B we present a realization of a machine interval arithmetic in ALGOL 60. Appendix C contains three ALGOL procedures for algorithms previously discussed.

At the end of each chapter there are references that list the papers from which the material treated in the chapter originates. Additional references are given for papers treating similar problems. At the end of the book there is a general list of references. We do not claim that this Bibliography is complete. It refers to an extensive collection of publications that we have assembled and built up in a systematic manner over some years as material for our own work, lectures, and seminars. We believe that it does contain the most important papers on the subject of interval mathematics as well as the other papers used for the book. To this belong, in particular, the numerous publications that are concerned with the solution of continuous problems through discretization capturing the discretization errors, especially in the treatment of initial and boundary value problems. If the results from these papers had been included, then the bounds of this book would certainly have been exceeded. The references in the text are written in square brackets. If the reference is to the general list of references at the end of the book, then the letter B, standing for Bibliography, is written in front of the number. Otherwise the references always refer to the corresponding number in the References at the end of the chapter.

Definitions, theorems, lemmas, and corollaries are numbered sequentially throughout each chapter, as are the formulas, which are quoted by chapter and number. Therefore, formula (6.8) means formula (8) of Chapter 6. If a formula is quoted in the same chapter, then the chapter reference is omitted. The same holds true for quoting theorems, lemmas, corollaries, and definitions. Proofs of theorems, lemmas, and corollaries are concluded by ■.

In most chapters that treat numerical methods, there are numerical examples inserted. Except for a few examples, these were programmed by Mr. Peter T. Speck on the X8 computer at the Computing Center of the University of Karlsruhe. We are grateful to him for his efforts.

Large parts of this text were treated in the Angewandt Mathematisches Seminar, led by Professor U. Kulisch and Professor J. Weissinger at the University of Karlsruhe. We are grateful to both of them for making this possible, as well as to all the participants of the seminar for the numerous discussions. Through these discussions we were often able to simplify the representation as well as to clarify some of the concepts. In addition, we learned a great deal from the lecture series by Professor Kulisch.

Dr. Christian Ullrich of the Institut für Angewandte Mathematik of the University of Karlsruhe read the complete manuscript. Many small errors were eliminated in this manner.

The second author would like to thank the Stiftung Volkswagenwerk at this point for support in completing some parts of the book.

We should also like to again thank Mrs. Lioba Schindele for typing the manuscript very carefully.

Our main thanks, however, go to Professor Kulisch, who introduced us to the problems treated here and who strongly supported the development of this book.

Karlsruhe and Berkeley G. ALEFELD AND J. HERZBERGER

Chapter 1 / REAL INTERVAL ARITHMETIC

In the following sections the field of real numbers is denoted by $\mathbb{R}$, and members of $\mathbb{R}$ are denoted by lowercase letters $a, b, c, \ldots, x, y, z$. A subset of $\mathbb{R}$ of the form

$$A = [a_1, a_2] = \{t \mid a_1 \leqslant t \leqslant a_2, a_1, a_2 \in \mathbb{R}\}$$

is called a closed real interval or an interval if no confusion arises. At several points later we shall denote the bounds of an interval A by

$$i(A) = a_1, \qquad s(A) = a_2$$

in order to avoid confusion. The set of all closed real intervals is denoted by $I(\mathbb{R})$ and the members of $I(\mathbb{R})$ by uppercase letters $A, B, C, \ldots, X, Y, Z$. Real numbers $x \in \mathbb{R}$ may be considered special members $[x, x]$ from $I(\mathbb{R})$, and they will generally be called point intervals.

Definition 1: Two intervals $A = [a_1, a_2]$ and $B = [b_1, b_2]$ are called equal, that is, $A = B$, if they are equal in the set theoretic sense. ■

From this definition it follows immediately that

$$A = B \quad \Leftrightarrow \quad a_1 = b_1, \quad a_2 = b_2.$$

The relation $=$ between two elements from $I(\mathbb{R})$ is reflexive, symmetric, and transitive.

We now generalize the arithmetic of real numbers by introducing operations on elements from $I(\mathbb{R})$.

Definition 2: Let $* \in \{+, -, \cdot, :\}$ be a binary operation on the set of real numbers $\mathbb{R}$. If $A, B \in I(\mathbb{R})$, then

$$A * B = \{z = a * b \mid a \in A, b \in B\} \tag{1}$$

defines a binary operation on $I(\mathbb{R})$. ■

It is asssumed that $0 \notin B$ in the case of division, and this will not be explicitly mentioned in the sequel. It should also be noted that the same symbols are used for operations in $I(\mathbb{R})$ and $\mathbb{R}$. This should cause no confusion since it will always be clear from the context whether it is an operation on reals or on intervals.

The operations on intervals $A = [a_1, a_2]$ and $B = [b_1, b_2]$ may be calculated explicitly as

$$\begin{cases} A + B = [a_1 + b_1, a_2 + b_2], \\ A - B = [a_1 - b_2, a_2 - b_1] = A + [-1, -1] \cdot B, \\ A \cdot B = [\min\{a_1b_1, a_1b_2, a_2b_1, a_2b_2\}, \max\{a_1b_1, a_1b_2, a_2b_1, a_2b_2\}], \\ A \div B = [a_1, a_2] \cdot [1/b_2, 1/b_1]. \end{cases} \tag{2}$$

This follows from the fact that $z = f(x, y)$ with $f(x, y) = x * y$, $* \in \{+, -, \cdot, :\}$, is a continuous function on a compact set. The function $f(x, y)$ therefore takes on a largest and a smallest value as well as all values in between. $A * B$ is therefore again a closed real interval. Clearly the formulas (2) then calculate the smallest (resp. largest) values of $f(x, y)$. The set $I(\mathbb{R})$ is therefore closed under these operations. Furthermore, it is immediately clear that the real numbers $x, y, \ldots$ are isomorphic to intervals of the form $[x, x], [y, y], \ldots$. We shall therefore frequently simplify the operation $[x, x] * A$ of a point interval $[x, x]$ with an interval A to $x * A$. The multiplication symbol $\cdot$ is often omitted also in the case of (2).

In addition to the operations (1) there are further, mostly unary, common operations on intervals.

Definition 3: If $r(x)$ is a continuous unary operation on $\mathbb{R}$, then

$$r(X) = \left[\min_{x \in X} r(x), \max_{x \in X} r(x)\right]$$

defines a (subordinate) unary operation on $I(\mathbb{R})$. ■

Examples of such unary operations on $I(\mathbb{R})$ are

$$X^k (k \in \mathbb{R}),\ e^X,\ \ln X,\ \sin X,\ \cos X, \text{ etc.}$$

We now collect together the most important rules for the operations on $I(\mathbb{R})$.

Theorem 4: A, B, and C are members of $I(\mathbb{R})$. Then it follows that

(3) $$A + B = B + A, \qquad A \cdot B = B \cdot A \qquad \text{(commutativity)},$$

(4) $$(A + B) + C = A + (B + C), \qquad (A \cdot B) \cdot C = A \cdot (B \cdot C)$$
$$\text{(associativity)},$$

(5) $X = [0, 0]$ and $Y = [1, 1]$ are the unique neutral elements with respect to addition and multiplication; that is,

$$A = X + A = A + X \quad \text{for all} \quad A \in I(\mathbb{R}) \Leftrightarrow X = [0, 0],$$
$$A = Y \cdot A = A \cdot Y \quad \text{for all} \quad A \in I(\mathbb{R}) \Leftrightarrow Y = [1, 1],$$

(6) $I(\mathbb{R})$ has no zero divisors,

(7) an arbitrary element $A = [a_1, a_2] \in I(\mathbb{R})$ with $a_1 \neq a_2$ has no inverse element with respect to $+$ and $\cdot$. Nevertheless it follows that

$$0 \in A - A \quad \text{and} \quad 1 \in A : A,$$

$$A(B + C) \subseteq AB + AC \quad \text{(subdistributivity)},$$

(8) $$a(B + C) = aB + aC, \quad a \in \mathbb{R},$$

$$A(B + C) = AB + AC \quad \text{if} \quad bc \geqslant 0 \text{ for all } b \in B \text{ and } c \in C.$$

Proof: Of (3): Let $* \in \{+, \cdot\}$. Then

$$A * B = \{z = a * b \mid a \in A, b \in B\}$$
$$= \{z = b * a \mid b \in B, a \in A\} = B * A.$$

Of (4): Let $* \in \{+, \cdot\}$. Then

$$(A * B) * C = \{z = y * c \mid y \in A * B, c \in C\}$$
$$= \{z = (a * b) * c \mid a \in A, b \in B, c \in C\}$$
$$= \{z = a * (b * c) \mid a \in A, b \in B, c \in C\}$$
$$= \{z = a * x \mid a \in A, x \in B * C\} = A * (B * C).$$

Of (5): The $\Leftarrow$ of the assertion is trivial. If N and $\bar{N}$ are two neutral elements with respect to the addition, then

$$N + \bar{N} = \bar{N} \quad \text{and} \quad \bar{N} + N = N.$$

From the commutativity (3) we get $N = \bar{N}$. The uniqueness of $Y = [1, 1]$, the neutral element under multiplication, is shown in a similar manner.

Of (6): Let $A \cdot B = 0$; that is,

$$A \cdot B = \{z = a \cdot b \mid a \in A, b \in B\} = [0, 0].$$

From this it follows that at least one of the intervals $A, B \in I(\mathbb{R})$ must be equal to $[0, 0]$.

Of (7): Both assertions are equivalent to

$$A - B = [0,0] \Rightarrow A = [a,a] = B,$$

$$A \cdot B = [1,1] \Rightarrow A = [a,a], B = [1/a, 1/a].$$

Let

$$A - B = \{z = a - b \mid a \in A, b \in B\} = [0,0].$$

Then it follows that $z = a - b = 0$ for all $a \in A, b \in B$. By fixing $b \in B$ it follows that $a = b$ for all $a \in A$; that is, $A = [b,b]$. Correspondingly one concludes that $B = [a,a]$ and therefore $a = b$. The second statement is proved in a similar manner.

Since

$$0 = a - a \in \{z = x - y \mid x \in A, y \in A\} \qquad \text{for} \quad a \in A,$$

it follows that $0 \in A - A$. Similarly we have $1 \in A : A$ for $0 \notin A$.

Of (8):

$$\begin{aligned} A(B + C) &= \{z = a(b + c) \mid a \in A, b \in B, c \in C\} \\ &\subseteq \{y = ab + \tilde{a}c \mid a, \tilde{a} \in A, b \in B, c \in C\} \\ &= AB + AC. \end{aligned}$$

In order to show that equality does not hold in general we consider the example

$$A = [0,1], \qquad B = [1,1], \qquad C = [-1, -1],$$

$$A(B + C) = [0,0] \subset [-1,1] = AB + AC.$$

Furthermore, we have that

$$\begin{aligned} a(B + C) &= \{z = a(b + c) \mid b \in B, c \in C\} \\ &= \{z = ab + ac \mid b \in B, c \in C\} \\ &= \{x = ab \mid b \in B\} + \{y = ac \mid c \in C\} \\ &= aB + aC. \end{aligned}$$

In order to prove the last equation we restrict ourselves to $b_1 \geqslant 0$, $c_1 \geqslant 0$ without loss of generality. If $a_1 \geqslant 0$, then we have

$$A(B + C) = [a_1(b_1 + c_1), a_2(b_2 + c_2)]$$

and

$$AB + AC = [a_1b_1, a_2b_2] + [a_1c_1, a_2c_2] = [a_1(b_1 + c_1), a_2(b_2 + c_2)],$$

that is the assertion holds for this case.

The case $a_2 \leqslant 0$ can be reduced to the case $a_1 \geqslant 0$ by considering $-A$ instead of A. If $a_1a_2 < 0$, then one gets

$$A(B + C) = [a_1(b_2 + c_2), a_2(b_2 + c_2)]$$

as well as

$$AB + AC = [a_1b_2, a_2b_2] + [a_1c_2, a_2c_2] = [a_1(b_2 + c_2), a_2(b_2 + c_2)],$$

which proves the final case. ■

A complete characterization of the cases for which the distributive law holds is found in Ratschek [16] and Spaniol [19].

We now want to touch upon the question of the solvability of the equation

$$AX = B$$

when $A \neq [0,0]$ and for an $X \in I(\mathbb{R})$. In order to answer this question an auxiliary function χ defined by

$$\chi A = \begin{cases} a_1/a_2 & \text{if} \quad |a_1| \leqslant |a_2| \\ a_2/a_1 & \text{otherwise,} \end{cases}$$

was introduced in Ratschek [17]. The following statement holds:

The equation $AX = B$ is satisfied by an $X \in I(\mathbb{R})$ iff

$$\chi A \geqslant \chi B.$$

The solution is not unique iff

$$\chi A = \chi B \leqslant 0.$$

The proof of this assertion is found in Ratschek [17]. (Compare also with Berti [4].)

We now want to illustrate the above statement using an example. Let

$$[1,2]X = [-1,3]$$

be given. Then this equation is satisfied only by $X = [-\frac{1}{2}, \frac{3}{2}]$, since

$$\chi[1,2] = \tfrac{1}{2} > \chi[-1,3] = -\tfrac{1}{3}.$$

If, on the other hand, one considers the set of solutions of the equations

$$ax = b, \qquad a \in [1,2], \qquad b \in [-1,3],$$

then one gets

$$\{x = b/a \mid a \in [1,2], b \in [-1,3]\} = [-1,3]/[1,2] = [-1,3] \supset X.$$

The solution set is here genuinely different from the interval X that satisfies the equation $AX = B$. We therefore do not call X a solution of the equation $AX = B$, but rather an "algebraic" solution.

One may in general prove the following:

Let the equation $AX = B$, $0 \notin A$, be satisfied by an $X \in I(\mathbb{R})$. Then

$$X \subseteq B : A.$$

This is shown by

$$x \in X \Rightarrow \text{there exists } a \in A, b \in B \text{ for which } ax = b \Rightarrow x = b/a \in B : A.$$

It should be noted that the equation $AX = B$ may also be satisfied when $B : A$ is not defined. This is shown by the example

$$[-\tfrac{1}{3}, 1]X = [-1, 2]$$

with $\chi[-\frac{1}{3}, 1] > \chi[-1, 2]$, where $X = [-1, 2]$ is uniquely determined.

A fundamental property of interval computations is inclusion monotonicity. The following theorem shows this property.

Theorem 5: Let $A^{(k)}, B^{(k)} \in I(\mathbb{R})$, $k = 1, 2$, and assume that

$$A^{(k)} \subseteq B^{(k)}, \qquad k = 1, 2.$$

Then for the operations $* \in \{+, -, \cdot, :\}$ it follows that

$$A^{(1)} * A^{(2)} \subseteq B^{(1)} * B^{(2)}. \tag{9}$$

Proof: Since $A^{(k)} \subseteq B^{(k)}$, $k = 1, 2$, it follows that

$$\begin{aligned} A^{(1)} * A^{(2)} &= \{z = x * y \mid x \in A^{(1)}, y \in A^{(2)}\} \\ &\subseteq \{w = u * v \mid u \in B^{(1)}, v \in B^{(2)}\} \\ &= B^{(1)} * B^{(2)}. \quad \blacksquare \end{aligned}$$

A special case of Theorem 5 is given by

Corollary 6: Let $A, B \in I(\mathbb{R})$ with $a \in A$, $b \in B$. Then it follows that

$$a * b \in A * B$$

for $* \in \{+, -, \cdot, :\}$. ■

The unary operations $r(X)$ of Definition 3 have the corresponding properties

$$\begin{aligned} &X \subseteq Y \Rightarrow r(X) \subseteq r(Y), \\ &x \in X \Rightarrow r(x) \in r(X). \end{aligned} \tag{10}$$

A direct generalization of these statements to interval expressions is given in Theorem 3.1.

Remarks: This elementary introduction to the real interval arithmetic on $I(\mathbb{R})$ corresponds to the description by Moore [11]. It is also possible to give simple descriptions of most of the unary operations of Definition 3 using the upper and lower bounds of the argument interval. This can, for example, easily be done in the case of the monotonic functions x^k and $\sqrt{x}$.

The four basic operations $+$, $-$, $\cdot$, and $:$ between general point sets were introduced by Young in [21]. Some of the elementary relations, for example, (3), (4), and (8), were also derived.

Kulisch [10] investigated which properties of operations on a set M are carried over to the power set $P(M)$. The interval operations (1) are contained in these results as a special case.

The representation of intervals by Sunaga in [20] corresponds to the notation for circular complex intervals in Chapter 5. In this case the pair of numbers (a, r) corresponds to the interval $[a - r, a + r]$. Using this representation the interval operations (1) were explicitly described and used in Sunaga [20]. They do not correspond, however, to the operations in Chapter 5 when these are restricted to real intervals.

In [13], Ortolf identified intervals $A = [a_1, a_2]$ with points $(a_1, a_2) \in \mathbb{R} \times \mathbb{R}$. Using this identification he succeeded in defining operations for all members of $\mathbb{R} \times \mathbb{R}$. If $a_1 \leqslant a_2$, then his definition reduces to the operations (2). In this manner the formal inverse with respect to $+$, and if $0 \notin A$, also the inverse with respect to $\cdot$, can be given as points of $\mathbb{R} \times \mathbb{R}$.

An extension of the interval operations (2) is described by Kahan [7]. There the real operations are completed by using the arguments $+\infty$ and $-\infty$. In certain cases the result of an operation is the "interval" Ω of all real numbers. In addition to Ω one also allows intervals $[a_1, a_2[$, $]a_1, a_2[$, $]a_1, a_2]$ with the possibility of having both $a_1 = \pm\infty$ or $a_2 = \pm\infty$ as well as $a_1 > a_2$. In order to interpret such interval representations one makes use of the oriented real circle of numbers. In this manner intervals may contain $\infty \equiv +\infty \equiv -\infty$ and also be half open or open. The arithmetic is defined by (1).

Interval computations in partially ordered spaces were described in a general manner by Apostolatos [1]. Again $I(\mathbb{R})$ appears as a particular case.

A three-valued set theory was constructed by Klaua [9]. He considers the so-called partial sets and partial cardinal numbers. The resulting cardinal number arithmetic corresponds in the finite case exactly to the interval arithmetic operations. Correspondingly one gets in this manner with the three-valued numbers an analog to the interval arithmetic $I(\mathbb{R})$. Together with the equality $=$ as in Definition 1 the weaker relation $=_\sharp$ is considered, where for $A = [a_1, a_2]$ and $B = [b_1, b_2]$ the definition is

$$A =_\sharp B \Leftrightarrow A \cap B \neq \varnothing \Leftrightarrow \max\{a_1, b_1\} \leqslant \min\{a_2, b_2\}.$$

The relation $=_\sharp$ is reflexive and symmetric and furthermore

$$A = B \Rightarrow A =_\sharp B.$$

This then means in the set theoretic sense that if $A \neq_\sharp B$, then for all $a \in A$ and all $b \in B$ we always have $a \neq b$. Correspondingly we also have that

$$A \neq_\sharp B \Rightarrow A \neq B.$$

In the sense of the relation $=_\sharp$ one may consider $I(\mathbb{R})$ as a kind of generalized field. One may for example prove the following properties:

$$X - X =_\sharp 0 \qquad \text{for } X \in I(\mathbb{R}),$$

$$AX =_\sharp B \Leftrightarrow X =_\sharp B : A \qquad \text{for } A, B, X \in I(\mathbb{R}) \text{ with } A \neq_\sharp 0,$$

$$X(Y + Z) =_\sharp XY + XZ \qquad \text{for } X, Y, Z \in I(\mathbb{R}).$$

Similar properties were used in the thesis and papers by Jahn [B137–B139].

The set $I(\mathbb{R})$ was extended to the set $\overline{I(\mathbb{R})}$ by Kaucher [8] by the addition of so-called nonregular intervals. These intervals are interpreted as intervals with negative width. The point intervals $[a, a]$ are then no longer minimal elements with respect to the ordering $\subseteq$. All the structures of $I(\mathbb{R})$ are carried over to $I(\mathbb{R}) \cup \overline{I(\mathbb{R})}$ and a completion through two improper elements p and $-p$ is achieved. In this manner the division by an interval $A = [a_1, a_2]$ with $a_1 \leqslant 0 \leqslant a_2$, $a_1 \neq a_2$, can also be defined.

REFERENCES

[1] N. Apostolatos, Allgemeine Intervallarithmetiken und Anwendungen. *Bull. Soc. Math. Grèce (N.S.)* **10**, 136–180 (1969).

[2] N. Apostolatos and U. Kulisch, Grundlagen einer Maschinenintervallarithmetik. *Computing* **2**, 89–104 (1967).

[3] N. Apostolatos and U. Kulisch, Approximation der erweiterten Intervallarithmetik durch einfache Maschinenintervallarithmetiken. *Computing* **2**, 181–194 (1967).

[4] S. Berti, The solution of an interval equation. *Mathematica* **11**, 189–194 (1969).

[5] S. Berti, Some relations between interval functions (I). *Mathematica* **14**, 9–26 (1972).

[6] S. Berti, Aritmetica si analiza intervalelor. *Rev. Anal. Numer. Teoria Aproximatiei* **1**, 21–39 (1972).

[7] W. Kahan, A more complete interval arithmetic. Report, Univ. of Toronto (1968).

[8] E. Kaucher, Über metrische und algebraische Eigenschaften einiger beim numerischen Rechnen auftretender Räume. Ph.D. Thesis, Univ. Karlsruhe (1973).

[9] D. Klaua, Partielle Mengen und Zahlen. *Mtber. Dt. Akad. Wiss.* **11**, 585–599 (1969).

[10] U. Kulisch, Grundzüge der Intervallrechnung. *In* "Überblicke Mathematik," Vol. 2, Bibliograph. Inst. Mannheim, 1969.

[11] R. E. Moore, "Interval Analysis." Prentice-Hall, Englewood Cliffs, New Jersey, 1966.

[12] E. Nuding, Intervallarithmetik. Lecture Notes, Rechenzentrum der Univ. Heidelberg (1972).

[13] H. J. Ortolf, Eine Verallgemeinerung der Intervallarithmetik. *Ber. Ges. Math. Dat. Nr.* **11** (1969).

[14] H. Ratschek, Über einige intervallarithmetische Grundbegriffe. *Computing* **4**, 43–55 (1969).

[15] H. Ratschek, Die binären Systeme der Intervallarithmetik. *Computing* **6**, 295–308 (1970).

[16] H. Ratschek, Die Subdistributivität in der Intervallarithmetik. *Z. Angew. Math. Mech.* **51**, 189–192 (1971).

[17] H. Ratschek, Teilbarkeitskriterien der Intervallarithmetik. *J. Reine Angew. Math.* **252**, 128–137 (1971).

[18] H. Ratschek, Intervallrechnung – mit Zirkel und Lineal. *Elem. Math.* 93–96 (1973).
[19] O. Spaniol, Die Distributivität in der Intervallarithmetik. *Computing* **5**, 6–16 (1970).
[20] T. Sunaga, Theory of an interval algebra and its application to numerical analysis. *RAAG Memoirs* **2**, 547–564 (1958).
[21] R. C. Young, The algebra of many-valued quantities. *Math. Ann.* **104**, 260–290 (1931).

Chapter 2 / FURTHER CONCEPTS AND PROPERTIES

We shall now introduce the idea of distance on the set of real intervals $I(\mathbb{R})$. This is done in

Definition 1: The distance between two intervals $A = [a_1, a_2]$, $B = [b_1, b_2] \in I(\mathbb{R})$ is defined as

$$q(A, B) = \max\{|a_1 - b_1|, |a_2 - b_2|\}. \quad \blacksquare$$

It is easy to show that the map q introduces a metric in $I(\mathbb{R})$ since q has the properties (see, for example, Hausdorff [1])

$$q(A, B) \geqslant 0 \qquad \text{and} \qquad q(A, B) = 0 \Leftrightarrow A = B,$$

$$q(A, B) \leqslant q(A, C) + q(B, C) \qquad \text{(triangle inequality).}$$

The triangle inequality may be verified as follows:

$$\begin{aligned} q(A, C) + q(B, C) &= \max\{|a_1 - c_1|, |a_2 - c_2|\} + \max\{|b_1 - c_1|, |b_2 - c_2|\} \\ &\geqslant \max\{|a_1 - c_1| + |b_1 - c_1|, |a_2 - c_2| + |b_2 - c_2|\} \\ &\geqslant \max\{|a_1 - b_1|, |a_2 - b_2|\} = q(A, B). \end{aligned}$$

This distance reduces to the usual distance for real numbers when applied to point intervals. That is, it follows that

$$q([a, a], [b, b]) = |a - b|.$$

The metric for real intervals that has been introduced here is the Hausdorff metric for $I(\mathbb{R})$ (see also Hausdorff [1]). The Hausdorff metric is a generalization of the distance between two points in a metric space – in this case $\mathbb{R}$ with $q(x, y) = |x - y|$ – to the space of all compact nonempty subsets of the space. If U and V are such compact, nonempty sets of real numbers, then the Hausdorff distance is defined by

$$q(U, V) = \max\left\{ \sup_{v \in V} \inf_{u \in U} q(u, v), \sup_{u \in U} \inf_{v \in V} q(u, v) \right\}.$$

There exist further useful characterizations of the Hausdorff metric (see again Hausdorff [1]). In the case of real intervals A and B it is easy to convince oneself that the Hausdorff metric is described by the expression given in Definition 1.

The introduction of a metric in $I(\mathbb{R})$ makes $I(\mathbb{R})$ into a topological space. The concepts of convergence and continuity may therefore be used in the sequel in the usual manner as for a metric space. In this connection we get that a sequence of intervals $\{A^{(k)}\}_{k=0}^{\infty}$ converges to an interval A iff the sequence of bounds of the individual members of the sequence converges to the corresponding bounds of $A = [a_1, a_2]$. We can therefore write

$$\text{(1)} \qquad \lim_{k \to \infty} A^{(k)} = A \Leftrightarrow \left(\lim_{k \to \infty} a_1^{(k)} = a_1 \text{ and } \lim_{k \to \infty} a_2^{(k)} = a_2 \right).$$

The proof of this assertion follows easily from the definition of the distance between two intervals, and it will therefore be omitted here.

The following theorem is valid for the above metric.

Theorem 2: The metric space $(I(\mathbb{R}), q)$ with the metric of Definition 1, is a complete metric space. ■

(This means that every Cauchy sequence of intervals converges to an interval.)

The following theorem discusses the convergence behavior of a useful class of interval sequences.

Theorem 3: Every sequence of intervals $\{A^{(k)}\}_{k=0}^{\infty}$ for which

$$A^{(0)} \supseteq A^{(1)} \supseteq A^{(2)} \supseteq \cdots$$

is valid converges to the interval $A = \bigcap_{k=0}^{\infty} A^{(k)}$.

Proof: Let us consider the sequences of bounds

$$a_1^{(0)} \leqslant a_1^{(1)} \leqslant a_1^{(2)} \leqslant a_1^{(3)} \leqslant \cdots \leqslant a_2^{(3)} \leqslant a_2^{(2)} \leqslant a_2^{(1)} \leqslant a_2^{(0)}.$$

The sequence of lower bounds of $\{A^{(k)}\}_{k=0}^{\infty}$ is therefore a monotonic nondecreasing sequence of real numbers that is bounded above by $a_2^{(0)}$. Such a sequence converges to a real number a_1. Similarly the monotonic nonincreasing sequence of real numbers $\{a_2^{(k)}\}_{k=0}^{\infty}$ converges to a real number a_2 for which $a_1 \leqslant a_2$. The equality $A = \bigcap_{k=0}^{\infty} A^{(k)}$ is verified in an equally simple manner. ■

As the proof shows, we also get that each sequence $\{A^{(k)}\}_{k=0}^{\infty}$ with

$$A^{(0)} \supseteq A^{(1)} \supseteq A^{(2)} \supseteq A^{(3)} \supseteq \cdots \supseteq B$$

converges to an interval A with $A \supseteq B$.

In connection with the operations $+$, $-$, $\cdot$, and $:$ for intervals as well as the further operations we have

Theorem 4: The operations $+$, $-$, $\cdot$, and $:$ introduced in Chapter 1 between intervals are continuous.

Proof: We carry through the proof only for the operation $+$. Let $\{A^{(k)}\}_{k=0}^{\infty}$ and $\{B^{(k)}\}_{k=0}^{\infty}$ be two sequences of intervals for which $\lim_{k\to\infty} A^{(k)} = A$ and $\lim_{k\to\infty} B^{(k)} = B$. The sequence of interval sums $\{A^{(k)} + B^{(k)}\}_{k=0}^{\infty}$ then satisfies

$$\begin{aligned} \lim_{k\to\infty} (A^{(k)} + B^{(k)}) &= \lim_{k\to\infty} [a_1^{(k)} + b_1^{(k)}, a_2^{(k)} + b_2^{(k)}] \\ &= \left[\lim_{k\to\infty} (a_1^{(k)} + b_1^{(k)}), \lim_{k\to\infty} (a_2^{(k)} + b_2^{(k)}) \right] \\ &= [a_1 + b_1, a_2 + b_2] = A + B \end{aligned}$$

from (1).

The proof of the continuity of the remaining operations can be carried out in a similar manner. ■

As an extension of Theorem 4 we have (see Definition 1.3)

Corollary 5: Let r be a continuous function and let $r(X) = [\min_{x\in X} r(x), \max_{x\in X} r(x)]$; then $r(X)$ is a continuous interval expression. ■

The elementary proof of this corollary follows immediately from the continuity of the function r, and it will not be given here. Corollary 5 guarantees the continuity of expressions like X^k, $\sin X$, and e^X.

Definition 6: The absolute value of an interval $A = [a_1, a_2] \in I(\mathbb{R})$ is defined as

$$|A| = q(A, [0, 0]) = \max\{|a_1|, |a_2|\}. \quad ■$$

The absolute value of an interval can also be written

$$(2) \qquad |A| = \max_{a\in A} |a|.$$

Clearly, if $A, B \in I(\mathbb{R})$, then

$$(3) \qquad A \subseteq B \Rightarrow |A| \leqslant |B|.$$

We now prove some properties related to the metric in $I(\mathbb{R})$.

Theorem 7: Let $A = [a_1, a_2]$, $B = [b_1, b_2]$, $C = [c_1, c_2]$, $D = [d_1, d_2] \in I(\mathbb{R})$. Then

$$q(A + B, A + C) = q(B, C), \tag{4}$$

$$q(A + B, C + D) \leqslant q(A, C) + q(B, D), \tag{5}$$

$$q(aB, aC) = |a| q(B, C), \qquad a \in \mathbb{R}, \tag{6}$$

$$q(AB, AC) \leqslant |A| q(B, C). \tag{7}$$

Proof: Of (4): From the definition of the metric q it follows that

$$\begin{aligned} q(A + B, A + C) &= \max\{|a_1 + b_1 - (a_1 + c_1)|, |a_2 + b_2 - (a_2 + c_2)|\} \\ &= \max\{|b_1 - c_1|, |b_2 - c_2|\} = q(B, C). \end{aligned}$$

Of (5): From the triangular inequality, the above property (4), as well as the symmetry of q, it follows that

$$\begin{aligned} q(A + B, C + D) &\leqslant q(A + B, B + C) + q(C + D, B + C) \\ &= q(A, C) + q(B, D). \end{aligned}$$

Of (6):

$$q(aB, aC) = \max\{|ab_1 - ac_1|, |ab_2 - ac_2|\} = |a| q(B, C).$$

Of (7): Let $A = [a_1, a_2]$. For brevity we use the notations $i(A) = a_1$ and $s(A) = a_2$. The assertion may then be written

$$\max\{|i(AB) - i(AC)|, |s(AB) - s(AC)|\} \leqslant |A| q(B, C).$$

We shall prove

$$|i(AB) - i(AC)| \leqslant |A| q(B, C).$$

The inequality

$$|s(AB) - s(AC)| \leqslant |A| q(B, C)$$

is proved in an analogous manner.

The previous relation (6) is rewritten

$$\max\{|i(aB) - i(aC)|, |s(aB) - s(aC)|\} = |a| q(B, C).$$

It is now assumed without loss of generality that

$$i(AB) \geqslant i(AC).$$

(The case $i(AB) < i(AC)$ is treated similarly.)

Since

$$AC = \{ac \mid a \in A, c \in C\},$$

there is an $a \in A$ such that

$$i(AC) = i(aC).$$

From inclusion monotonicity we get

$$aB \subseteq AB$$

and therefore

$$i(aB) \geqslant i(AB)$$

and furthermore that

$$i(aB) - i(aC) \geqslant i(AB) - i(AC) \geqslant 0.$$

Finally we get the result

$$\begin{aligned} |i(AB) - i(AC)| = i(AB) - i(AC) &\leqslant i(aB) - i(aC) \\ &= |i(aB) - i(aC)| \leqslant |a|q(B, C) \\ &\leqslant |A|q(B, C). \quad \blacksquare \end{aligned}$$

If we use $|A| = q(A, 0)$, then we have the following easily verified properties for the absolute value:

$$\begin{aligned} &|A| \geqslant 0 \quad \text{and} \quad |A| = 0 \Leftrightarrow A = [0, 0], \\ &|A + B| \leqslant |A| + |B|, \\ &|xA| = |x|\,|A| \quad \text{for} \quad x \in \mathbb{R}, \\ &|AB| = |A|\,|B|. \end{aligned} \tag{8}$$

The last relation is proved by

$$\begin{aligned} |AB| = \max_{c \in AB}|c| = \max_{a \in A, b \in B}|ab| &= \max_{a \in A, b \in B}(|a|\,|b|) \\ &= \max_{a \in A}|a| \max_{b \in B}|b| = |A|\,|B|. \end{aligned}$$

The remaining relations may be proven in a similar manner.

Definition 8: The width of an interval $A = [a_1, a_2]$ is defined to be

$$d(A) = a_2 - a_1 \geqslant 0. \quad \blacksquare$$

The set of point intervals may now be characterized as

$$\{A \in I(\mathbb{R}) \mid d(A) = 0\}.$$

From the above definition we immediately get the properties

$$A \subseteq B \Rightarrow d(A) \leqslant d(B), \tag{9}$$

$$d(A \pm B) = d(A) + d(B). \tag{10}$$

The proof of (9) is trivial, and it follows immediately from the equivalent form of the width

(11) $$d(A) = \max_{a,b \in A} |a - b|.$$

Property (10) is verified for the operation $+$ in the following manner:

$$\begin{aligned} d(A + B) &= d([a_1 + b_1, a_2 + b_2]) \\ &= a_2 + b_2 - (a_1 + b_1) \\ &= a_2 - a_1 + b_2 - b_1 = d(A) + d(B). \end{aligned}$$

The case $-$ is proved in a completely analogous manner.

Furthermore, we have

Theorem 9: Let $A, B \in I(\mathbb{R})$ be real intervals. Then

(12) $$d(AB) \leqslant d(A)|B| + |A|\, d(B),$$

(13) $$d(AB) \geqslant \max\{|A|\, d(B), |B|\, d(A)\},$$

(14) $$d(aB) = |a|\, d(B), \qquad a \in \mathbb{R},$$

(15) $$d(A^n) \leqslant n|A|^{n-1} d(A), \qquad n = 1, 2, \ldots,$$

$$(A^n := A \cdot A \cdot \cdots \cdot A,\ n \text{ times}),$$

(16) $$d((X - x)^n) \leqslant 2(d(X))^n \qquad \text{for} \quad x \in X, \qquad n = 1, 2, \ldots,$$

$$((X - x)^n := (X - x)(X - x) \cdots (X - x),\ n \text{ times}).$$

For an interval $C \in I(\mathbb{R})$ with $0 \in C$ it furthermore follows that

(17) $$|C| \leqslant d(C) \leqslant 2|C|.$$

Proof: Of (12): Using the equivalent form (11) we obtain

$$\begin{aligned} d(AB) &= \max_{a,a' \in A, b, b' \in B} |ab - a'b'| \\ &= \max_{a,a' \in A, b, b' \in B} |ab - ab' + ab' - a'b'| \\ &\leqslant \max_{a,a' \in A, b, b' \in B} \{|a(b - b')| + |(a - a')b'|\} \\ &\leqslant \max_{a \in A, b, b' \in B} |a|\,|b - b'| + \max_{a,a' \in A, b' \in B} |a - a'|\,|b'| \\ &= \left(\max_{a \in A} |a|\right)\left(\max_{b,b' \in B} |b - b'|\right) + \left(\max_{a,a' \in A} |a - a'|\right)\left(\max_{b' \in B} |b'|\right) \\ &= |A|\, d(B) + d(A)|B|. \end{aligned}$$

Of (13): We first prove

$$d(AB) = \max_{a,a' \in A, b,b' \in B} |ab - a'b'| \geq \max_{a \in A, b,b' \in B} |ab - ab'|$$
$$= \max_{a \in A, b,b' \in B} |a|\,|b - b'| = |A|\,d(B).$$

Similarly it can be shown that

$$d(AB) \geq |B|\,d(A)$$

and (13) follows immediately.

Of (14):

$$d(aB) = \max_{b,b' \in B} |ab - ab'| = \max_{b,b' \in B} \{|a|\,|b - b'|\}$$
$$= |a| \max_{b,b' \in B} |b - b'| = |a|\,d(B).$$

Of (15): Equality holds for $n = 1$. If the inequality is true for an $n \geq 1$, then it follows using (12) and the last relation of (8) that

$$d(A^{n+1}) = d(A^n A) \leq d(A^n)|A| + |A|^n d(A)$$
$$\leq n|A|^{n-1} d(A)|A| + |A|^n d(A)$$
$$= (n + 1)|A|^n d(A).$$

Of (16): Since $x \in X$ we have from (9) and inclusion monotonicity that

$$d((X - x)^n) \leq d((X - X)^n) = d([-d(X), d(X)]^n)$$
$$= d([-(d(X))^n, (d(X))^n]) = 2(d(X))^n.$$

Of (17): Since $0 \in C = [c_1, c_2]$, we have that $c_1 \leq 0 \leq c_2$, from which

$$d(C) = c_2 - c_1 = |c_2| + |c_1| \geq \max\{|c_1|, |c_2|\} = |C|.$$

Furthermore

$$d(C) = |c_1| + |c_2| \leq 2\max\{|c_1|, |c_2|\} = 2|C|. \quad \blacksquare$$

We now prove the following theorem.

Theorem 10: Let $A, B \in I(\mathbb{R})$ be intervals, and assume that A is a symmetric interval; i.e., $A = -A$. The following properties then hold:

(18) $$AB = |B|A,$$

(19) $$d(AB) = |B|\,d(A).$$

The second property is also valid for $0 \in A$ if either $b_1 \geq 0$ or $b_2 \leq 0$.

Proof: We assume first that $A = -A$ or equivalently $a_2 = a = -a_1$. Then

$$\begin{aligned} AB &= [\min\{ab_1, ab_2, -ab_1, -ab_2\}, \max\{ab_1, ab_2, -ab_1, -ab_2\}] \\ &= [a\min\{b_1, -b_1, b_2, -b_2\}, a\max\{b_1, -b_1, b_2, -b_2\}] \\ &= [a(-|B|), a|B|] = [-a, a]|B| = |B|A. \end{aligned}$$

This implies (19) because of Eq. (14). The other cases may be proved in an analogous manner. ■

In this connection we also prove

Theorem 11: The following properties are valid for intervals $A, B \in I(\mathbb{R})$:

(20) $$d(A) = |A - A|,$$

(21) $$A \subseteq B \Rightarrow \tfrac{1}{2}(d(B) - d(A)) \leqslant q(A, B) \leqslant d(B) - d(A).$$

Proof: Of (20):

$$d(A) = a_2 - a_1 = |A - A|.$$

Of (21): Let $A \subseteq B$. Then $b_1 \leqslant a_1 \leqslant a_2 \leqslant b_2$, and therefore

$$\begin{aligned} q(A, B) &= \max\{|a_1 - b_1|, |a_2 - b_2|\} = \max\{a_1 - b_1, b_2 - a_2\} \\ &\leqslant b_2 - a_2 + a_1 - b_1 = b_2 - b_1 - (a_2 - a_1) = d(B) - d(A) \end{aligned}$$

and furthermore

$$\begin{aligned} q(A, B) &= \max\{a_1 - b_1, b_2 - a_2\} \geqslant \tfrac{1}{2}(a_1 - b_1 + b_2 - a_2) \\ &= \tfrac{1}{2}(d(B) - d(A)). \quad ■ \end{aligned}$$

We now introduce another binary operation in $I(\mathbb{R})$. Suppose $A, B \in I(\mathbb{R})$ are two intervals. Then the relation

(22) $$A \cap B = \{c \mid c \in A,\ c \in B\}$$

denotes the set theoretic intersection of two intervals. The result of this operation is in $I(\mathbb{R})$ iff the set theoretic intersection is nonempty. For this case it follows that

(23) $$A \cap B = [\max\{a_1, b_1\}, \min\{a_2, b_2\}].$$

The important properties of the intersection operation are collected in the following:

Corollary 12: Let $A, B, C, D \in I(\mathbb{R})$. Then

(24) $A \subseteq C,\ B \subseteq D \Rightarrow A \cap B \subseteq C \cap D$ (inclusion monotonicity).

The intersection operation is a continuous operation if it can be performed in $I(\mathbb{R})$.

Proof: The inclusion monotonicity (24) follows from the definition (22). The proof of the continuity may be carried out with the help of (23). ■

Remarks: The Hausdorff metric in $I(\mathbb{R})$ corresponding to Definition 1 was used by Moore [7]. Both in Moore [7] and in Kulisch [4] one finds several of the rules for calculating with the absolute value $|A|$ and the width $d(A)$ of intervals A. The inequality (7), important in applications, was first proved by Mayer [5], and it may also be found in [B122].

Occasionally one may find the following definition based on Definition 1.3:

$$\mathrm{abs}(A) = \left[\min_{a \in A}|a|, \max_{a \in A}|a| \right]$$

for the absolute value. Since this definition is not widely used in the applications, it will not be used in the sequel.

According to S. M. Rump (personal communication) the factor of 2 in (16) can be eliminated through an improved estimation. For $x \in X$, we have $X - x = [a, b]$ with $a \leqslant 0, b \geqslant 0$. It can be assumed that $b \geqslant -a = |a|$ (if not, consider $x - X$). Then

$$(X - x)(X - x) = [ab, b^2],$$

and by complete induction we get

$$(X - x)^n = [ab^{n-1}, b^n].$$

Therefore

$$d((X - x)^n) = b^n - ab^{n-1} = b^{n-1}(b - a).$$

Now $b - a = d(X - x) = d(X)$ and because of $a \leqslant 0$, $b \geqslant 0$, we get $b \leqslant d(X - x) = d(X)$. Finally

$$d((X - x)^n) \leqslant d(X)^n.$$

REFERENCES

[1] F. Hausdorff, "Mengenlehre." de Gruyter, Leipzig, 1972.

[2] J. Herzberger, Metrische Eigenschaften von Mengensystemen und einige Anwendungen. Ph.D. Thesis, Univ. Karlsruhe (1969).

[3] W. Kahan, A More Complete Interval Arithmetic. Report, Univ. of Toronto (1968).

[4] U. Kulisch, Grundlagen der Intervallrechnung. In "Überblicke Mathematik 2." Bibliograph. Inst., Mannheim, 1969.

[5] O. Mayer, Über die in der Intervallrechnung auftretenden Räume und einige Anwendungen. Ph.D. Thesis, Univ. Karlsruhe (1969).

[6] O. Mayer, Algebraische und metrische Strukturen in der Intervallrechnung und einige Anwendungen. *Computing* **5**, 144–162 (1970).

[7] R. E. Moore, "Interval Analysis." Prentice-Hall, Englewood Cliffs, New Jersey, 1966.

[8] E. Nuding, Intervallarithmetik. Lecture Notes, Rechenzentrum der Univ. Heidelberg (1972).
[9] F. Ris, Interval Analysis and Applications to Linear Algebra. Ph.D. Thesis, Oxford Univ. (1972).
[10] W. J. Thron, "Topological Structures." Holt, New York, 1966.

Chapter 3 / INTERVAL EVALUATION AND RANGE OF REAL FUNCTIONS

In this chapter we consider continuous real functions f. An expression $f(x)$ belonging to f is a calculating procedure that will determine a value of the function f for every argument x. We then assume that all occurring expressions are composed of finitely many operations and operands for which the corresponding interval operations are defined in accordance with Definitions 1.2 and 1.3. If an expression belonging to f also contains constants $a^{(0)}, \ldots, a^{(m)}$, then this will be clarified by writing $f(x; a^{(0)}, \ldots, a^{(m)})$. For the purpose of simplifying matters later, we shall always assume that each constant $a^{(k)}$, $0 \leqslant k \leqslant m$, occurs only once in an expression. If this is not the case, then by introducing new constants that are made equivalent for the multiple occurrences of these new constants, one may transform the expression into the required form.

Example: Two expressions for the function g are

$$g^{(1)}(x; a) = \frac{ax}{1 - x}, \qquad x \neq 1, \qquad x \neq 0,$$

and

$$g^{(2)}(x; a) = \frac{a}{1/x - 1}, \qquad x \neq 1, \qquad x \neq 0.$$

The expression

$$W(f, X; A^{(0)}, \ldots, A^{(m)})$$
$$= \{f(x; a^{(0)}, \ldots, a^{(m)}) \mid x \in X, a^{(k)} \in A^{(k)}, 0 \leqslant k \leqslant m\}$$
$$= \left[\min_{\substack{x \in X \\ a^{(k)} \in A^{(k)},\, 0 \leqslant k \leqslant m}} f(x; a^{(0)}, \ldots, a^{(m)}), \max_{\substack{x \in X \\ a^{(k)} \in A^{(k)},\, 0 \leqslant k \leqslant m}} f(x; a^{(0)}, \ldots, a^{(m)}) \right],$$

will in the sequel denote the interval of all values of the function f when $x \in X$, and $a^{(k)} \in A^{(k)}$, $0 \leqslant k \leqslant m$, are considered independent of each other. This definition is independent of the expression for f.

Example: With g as in the previous example and with

$$A = [0, 1] \qquad \text{and} \qquad X = [2, 3]$$

we get

$$W(g, [2, 3]; [0, 1]) = \left\{ \frac{ax}{1 - x} \,\middle|\, 2 \leqslant x \leqslant 3, 0 \leqslant a \leqslant 1 \right\} = [-2, 0].$$

We define an interval evaluation of a real function f as follows.

Let an expression be given for the function f. In this expression all operands are replaced by intervals and all operations by interval operations resulting in the expression $f(X; A^{(0)}, \ldots, A^{(m)})$. If all operands are within the domain of definition of the operations defined in Definitions 1.2 and 1.3 then this is called the *interval evaluation* or *interval arithmetic evaluation* for f.

The above replacement is always possible for the functions that are considered here. The constants $a^{(0)}, \ldots, a^{(m)}$ as well as the variable x are replaced by intervals. Clearly the interval evaluation of a function f is dependent on the choice of expression for f. We shall make use of this fact later. Here we give a simple example.

Example: Let g be the function of the previous two examples. Using $A = [0, 1]$ and $X = [2, 3]$, we get two possible evaluations:

$$g^{(1)}([2, 3]; [0, 1]) = \frac{[0, 1][2, 3]}{1 - [2, 3]} = [-3, 0],$$

$$g^{(2)}([2, 3]; [0, 1]) = \frac{[0, 1]}{1/[2, 3] - 1} = [-2, 0] \neq g^{(1)}([2, 3]; [0, 1]).$$

The notation introduced above may also be applied to functions in several variables. The expression $f(x^{(1)}, \ldots, x^{(n)}; a^{(0)}, \ldots, a^{(m)})$ then has the interval range of values $W(f, X^{(1)}, \ldots, X^{(n)}; A^{(0)}, \ldots, A^{(m)})$ when both $x^{(k)} \in X^{(k)}$, $1 \leqslant k \leqslant n$, and $a^{(j)} \in A^{(j)}$, $0 \leqslant j \leqslant m$, are considered independent. The interval evaluation $f(X^{(1)}, \ldots, X^{(n)}; A^{(0)}, \ldots, A^{(m)})$ is defined in a similar manner.

We now give an example of an expression that does not lead to a well-defined interval expression when operands and operations are replaced by their interval analogues. The real function

$$f(x) = 1/(x^2 + \tfrac{1}{2})$$

is defined for all $x \in \mathbb{R}$. An expression for f is, for example,

$$\mathfrak{f}(x) = 1/(x \cdot x + \tfrac{1}{2}).$$

The independent variable x is now replaced by the interval $X = [-1, 1]$ clearly contained in the domain of definition of f. Replacing all operations by interval operations leads to the interval expression

$$\mathfrak{f}([-1,1]) = \frac{1}{[-1,1][-1,1] + \frac{1}{2}} = \frac{1}{[-1,1] + \frac{1}{2}} = \frac{1}{[-\frac{1}{2}, \frac{3}{2}]}.$$

This expression is not defined.

We now introduce a series of properties that hold for interval evaluations. The following two properties, important in further considerations, are easy consequences of Theorem 1.5 and Corollary 1.6.

Theorem 1: Let f be a continuous function of the real variables $x^{(1)}, \ldots, x^{(n)}$, and let $\mathfrak{f}(x^{(1)}, \ldots, x^{(n)}; a^{(0)}, \ldots, a^{(m)})$ be an expression for f. Furthermore, assume that the interval evaluation $\mathfrak{f}(Y^{(1)}, \ldots, Y^{(n)}; B^{(0)}, \ldots, B^{(m)})$ is defined for the intervals $Y^{(1)}, \ldots, Y^{(n)}, B^{(0)}, \ldots, B^{(m)}$.

It then follows that

(a) for all

$$X^{(k)} \subseteq Y^{(k)}, \qquad A^{(j)} \subseteq B^{(j)}, \qquad 1 \leqslant k \leqslant n, \qquad 0 \leqslant j \leqslant m,$$

it holds that

(1) $W(f, X^{(1)}, \ldots, X^{(n)}; A^{(0)}, \ldots, A^{(m)})$

$\subseteq \mathfrak{f}(X^{(1)}, \ldots, X^{(n)}; A^{(0)}, \ldots, A^{(m)})$ (inclusion property);

(b) for all

$$X^{(k)} \subseteq Z^{(k)} \subseteq Y^{(k)}, \qquad A^{(j)} \subseteq C^{(j)} \subseteq B^{(j)}, \qquad 1 \leqslant k \leqslant n, \qquad 0 \leqslant j \leqslant m,$$

it holds that

(1′) $\mathfrak{f}(X^{(1)}, \ldots, X^{(n)}; A^{(0)}, \ldots, A^{(m)})$

$\subseteq \mathfrak{f}(Z^{(1)}, \ldots, Z^{(n)}; C^{(0)}, \ldots, C^{(m)})$ (inclusion monotonicity). ■

Example: The function f is defined by the expression

$$\mathfrak{f}(x; a) = a - x/(1 + x), \qquad x \neq -1.$$

By choosing

$$X = [-\tfrac{1}{2}, 1], \qquad Z = [-\tfrac{1}{2}, 2], \qquad A = C = [2, 3],$$

one obtains the relations

$$W(f, [-\tfrac{1}{2}, 1]; [2, 3]) = [\tfrac{3}{2}, 4] \subset f([-\tfrac{1}{2}, 1]; [2, 3]) = [0, 4],$$
$$f([-\tfrac{1}{2}, 1]; [2, 3]) = [0, 4] \subset f([-\tfrac{1}{2}, 2]; [2, 3]) = [-2, 4].$$

The inclusion property (1) gives a connection between the range of a real function and its interval evaluation. In this section we shall also give formulas that give a qualitative estimate of the interval evaluation compared to the range.

It is possible to give examples where one always gets equality in (1). This is clearly the case if each of the quantities $x^{(1)}, \ldots, x^{(n)}, a^{(0)}, \ldots, a^{(m)}$ only occurs once in the expression $f(x^{(1)}, \ldots, x^{(n)}; a^{(0)}, \ldots, a^{(m)})$. Other classes of functions with this property are described in Apostolatos and Kulisch [3, 4].

Theorem 2: Let p be a polynomial in the real variable x defined by the expression

$$p(x; a^{(0)}, \ldots, a^{(m)}) = (\cdots((a^{(m)}x + a^{(m-1)})^{n_{m-1}} + a^{(m-2)})^{n_{m-2}} + \cdots + a^{(1)})^{n_1} + a^{(0)},$$

where $n_\nu \geqslant 2, 1 \leqslant \nu \leqslant m - 1$.

If the powers appearing in the expression are evaluated as

$$X^k = \left[\min_{x \in X} x^k, \max_{x \in X} x^k\right]$$

(see Definition 1.3), then

$$W(p, X; a^{(0)}, \ldots, a^{(m)}) = p(X; a^{(0)}, \ldots, a^{(m)}).$$

Proof: For the case $m = 2$ with $p(x; a^{(0)}, a^{(1)}, a^{(2)}) = (a^{(2)}x + a^{(1)})^{n_1} + a^{(0)}$ the proof is immediate. The remainder of the proof follows by complete induction. ■

It is not possible in general to transform a polynomial to the form required by Theorem 2. A polynomial of second degree

$$p(x; b^{(0)}, b^{(1)}) = x^2 + b^{(1)}x + b^{(0)}$$

may, however, be transformed to

$$p(x; a^{(0)}, a^{(1)}) = (x + a^{(1)})^2 + a^{(0)},$$

where

$$a^{(1)} = b^{(1)}/2, \qquad a^{(0)} = b^{(0)} - (b^{(1)})^2/4.$$

See [3, 4] for further details.

Together with the general statement of Theorem 1 and the special cases mentioned above one is also interested in a qualitative statement of the approximation of the range of a function f by an interval evaluation. In the case of a function of one real variable we formulate

Theorem 3: Let f be a real function of the real variable x, and let $f(x; a^{(0)}, \ldots, a^{(m)})$ be an expression for f. A new expression $\tilde{f}(x^{(1)}, \ldots, x^{(n)}; a^{(0)}, \ldots, a^{(m)})$ is defined by replacing each occurrence of the real variable x by a new variable $x^{(k)}$, $1 \leqslant k \leqslant n$. Let the interval evaluation $f(Y; A^{(0)}, \ldots, A^{(m)})$ exist for Y, $A^{(0)}, \ldots, A^{(m)} \in I(\mathbb{R})$. Furthermore, let the expression $\tilde{f}(x^{(1)}, \ldots, x^{(n)}; a^{(0)}, \ldots, a^{(m)})$ satisfy, for each variable $x^{(k)}, 1 \leqslant k \leqslant n$, in the interval Y, a Lipschitz condition for an arbitrary choice of $x^{(j)} \in Y$, $1 \leqslant j \leqslant n$, $j \neq k$, and $a^{(j)} \in A^{(j)}$, $0 \leqslant j \leqslant m$. Otherwise the notation is as in Theorem 1. Then for $X \subseteq Y$ it follows that

$$(2) \quad q(W(f, X; A^{(0)}, \ldots, A^{(m)}), f(X; A^{(0)}, \ldots, A^{(m)})) \leqslant \gamma\, d(X), \qquad \gamma \geqslant 0.$$

Proof: First, we have

$$\tilde{f}(x, \ldots, x; a^{(0)}, \ldots, a^{(m)}) = f(x; a^{(0)}, \ldots, a^{(m)}), \qquad x \in Y.$$

We may now write the interval evaluation of f as

$$f(X; A^{(0)}, \ldots, A^{(m)}) = W(\tilde{f}, X, \ldots, X; A^{(0)}, \ldots, A^{(m)}), \qquad X \subseteq Y.$$

It remains to show that

$$q(W(f, X; A^{(0)}, \ldots, A^{(m)}), W(\tilde{f}, X, \ldots, X; A^{(0)}, \ldots, A^{(m)})) \leqslant \gamma\, d(X), \qquad X \subseteq Y.$$

If we now write for $X \subseteq Y$

$$W(f, X; A^{(0)}, \ldots, A^{(m)}) = [f(u; a^{(0)}, \ldots, a^{(m)}), f(v; b^{(0)}, \ldots, b^{(m)})],$$
$$u, v \in X, \qquad a^{(j)}, b^{(j)} \in A^{(j)}, \qquad 0 \leqslant j \leqslant m,$$

$$\begin{aligned} &W(\tilde{f}, X, \ldots, X; A^{(0)}, \ldots, A^{(m)}) \\ &\quad = [\tilde{f}(x^{(1)}, \ldots, x^{(n)}; c^{(0)}, \ldots, c^{(m)}), \tilde{f}(y^{(1)}, \ldots, y^{(n)}; e^{(0)}, \ldots, e^{(m)})], \\ &x^{(k)}, y^{(k)} \in X, \qquad 1 \leqslant k \leqslant n, \qquad c^{(j)}, e^{(j)} \in A^{(j)}, \qquad 0 \leqslant j \leqslant m, \end{aligned}$$

and take the relation

$$W(f, X; A^{(0)}, \ldots, A^{(m)}) \subseteq W(\tilde{f}, X, \ldots, X; A^{(0)}, \ldots, A^{(m)})$$

into account, then it follows that

$$\begin{aligned} &|f(u; a^{(0)}, \ldots, a^{(m)}) - \tilde{f}(x^{(1)}, \ldots, x^{(n)}; c^{(0)}, \ldots, c^{(m)})| \\ &\quad = f(u; a^{(0)}, \ldots, a^{(m)}) - \tilde{f}(x^{(1)}, \ldots, x^{(n)}; c^{(0)}, \ldots, c^{(m)}) \\ &\quad \leqslant f(u; c^{(0)}, \ldots, c^{(m)}) - \tilde{f}(x^{(1)}, \ldots, x^{(n)}; c^{(0)}, \ldots, c^{(m)}) \\ &\quad = \tilde{f}(u, \ldots, u; c^{(0)}, \ldots, c^{(m)}) - \tilde{f}(x^{(1)}, \ldots, x^{(n)}; c^{(0)}, \ldots, c^{(m)}) \\ &\quad \leqslant \gamma \max_{1 \leqslant k \leqslant n} |u - x^{(k)}| \leqslant \gamma\, d(X). \end{aligned}$$

The difference of the upper bounds of the ranges of values may be estimated in a similar way. These two estimates together prove the assertion. ■

The statements of Theorem 3 may, as the proof shows, immediately be generalized to functions of several variables $x^{(1)}, \ldots, x^{(n)}$. Instead of $\gamma\, d(X)$ we have the quantity

$$\sum_{k=1}^{n} \gamma^{(k)} d(X^{(k)}) \qquad \left(\leqslant \gamma \max_{1 \leqslant k \leqslant n} d(X^{(k)}) \right).$$

The following example demonstrates that the approximation of the range of a function f by an interval evaluation is dependent on the choice of expression $f(x; a^{(0)}, \ldots, a^{(m)})$ through which the values of f are estimated.

Example: Let $f(x) = x - x^2$ and $X = [0, 1]$; then we get

$$W(f, [0, 1]) = \{x - x^2 \mid 0 \leqslant x \leqslant 1\} = [0, \tfrac{1}{4}].$$

The following equivalent expressions result in

$$f^{(0)}(x) = x - x^2 \Rightarrow f^{(0)}([0, 1]) = [0, 1] - [0, 1] = [-1, 1],$$

$$f^{(1)}(x) = x(1 - x) \Rightarrow f^{(1)}([0, 1]) = [0, 1](1 - [0, 1]) = [0, 1],$$

$$f^{(2)}(x) = \tfrac{1}{4} - (x - \tfrac{1}{2})(x - \tfrac{1}{2}) \Rightarrow$$

$$f^{(2)}([0, 1]) = \tfrac{1}{4} - ([0, 1] - \tfrac{1}{2})([0, 1] - \tfrac{1}{2}) = [0, \tfrac{1}{2}],$$

$$f^{(3)}(x) = \tfrac{1}{4} - (x - \tfrac{1}{2})^2 \Rightarrow$$

$$f^{(3)}([0, 1]) = \tfrac{1}{4} - ([0, 1] - \tfrac{1}{2})^2 = [0, \tfrac{1}{4}] = W(f, [0, 1]).$$

For a particular expression for a function f, one may prove a sharper result than that given in Theorem 3. This expression is the so-called centered form of a function (compare with Hansen [7], Chuba and Miller [5], and Miller [16]). The centered form of a function f is a particular expression for a function that is to be evaluated over an interval X. We restrict ourselves to the case of one real variable, and we chose an arbitrary point $z \in X$. The function $f(x)$ may then be represented as

$$f(x) = f(z) + (x - z)h(x - z), \tag{3}$$

where the term $h(x - z)$ depends on the shifted variable $\tilde{z} = x - z$. We call (3) the centered form for $f(x)$ around z. For polynomials, the centered form (3) is simply the Taylor expansion of $f(x)$ around the point z written with the term $x - z$ bracketed out of the nonconstant terms.

Rational functions $f(x) = p(x)/q(x)$ may, according to Ratschek [21], be developed in a centered form as follows: Let n be the maximum degree of the polynomials $p(x)$ and $q(x)$. Then for $z \in X$ define

$$\gamma_\nu := p^{(\nu)}(z) - f(z)q^{(\nu)}(z), \qquad 1 \leqslant \nu \leqslant n.$$

The function

$$h(y) = \sum_{\nu=1}^{n} \gamma_\nu \frac{y^{\nu-1}}{\nu!} \Bigg/ \sum_{\nu=0}^{s} q^{(\nu)}(z) \frac{y^\nu}{\nu!}$$

then satisfies the functional equation

$$f(x) = f(z) + (x - z)h(x - z).$$

Theorem 4: Let f be a real function of the real variable x, and let

$$f(x) = f(z) + (x - z)h(x - z)$$

be an expression for f in the centered form. The expression $\tilde{h}(x^{(1)} - z, \ldots, x^{(n)} - z)$ is furthermore formed in an analogous manner to the expression in Theorem 3. Let the interval evaluation $f(Y)$ exist for some $Y \in I(\mathbb{R})$, and let $\tilde{h}(x^{(1)} - z, \ldots, x^{(n)} - z)$ satisfy a Lipschitz condition for each variable as in Theorem 3. It then follows that the relation

$$(4) \qquad q(W(f, X), f(X)) \leqslant c(d(X))^2, \qquad c \geqslant 0,$$

is valid for each $X \subseteq Y$.

Proof: Because of

$$\tilde{h}(x - z, \ldots, x - z) = h(x - z)$$

and

$$\tilde{f}(x^{(0)}, \ldots, x^{(n)}) = f(z) + (x^{(0)} - z)\tilde{h}(x^{(1)} - z, \ldots, x^{(n)} - z),$$

it follows that

$$\begin{aligned}\tilde{f}(x, \ldots, x) &= f(z) + (x - z)\tilde{h}(x - z, \ldots, x - z) \\ &= f(z) + (x - z)h(x - z) = f(x).\end{aligned}$$

The interval evaluation of the centered form for f may now be written

$$f(X) = W(\tilde{f}, X, \ldots, X),$$

and the assertion therefore may be stated as

$$q(W(f, X), W(\tilde{f}, X, \ldots, X)) \leqslant c\,(d(X))^2, \qquad c \geqslant 0.$$

Let

$$\begin{aligned}W(\tilde{f}, X, \ldots, X) = [&f(z) + (x^{(0)} - z)\tilde{h}(x^{(1)} - z, \ldots, x^{(n)} - z), \\ &f(z) + (y^{(0)} - z)\tilde{h}(y^{(1)} - z, \ldots, y^{(n)} - z)], \\ &x^{(k)}, y^{(k)} \in X, \qquad 0 \leqslant k \leqslant n,\end{aligned}$$

and note that

$$W(f, X) \subseteq W(\tilde{f}, X, \ldots, X).$$

We may then estimate

$$q(W(f,X), W(\tilde{f}, X, \ldots, X)) \leqslant d(W(\tilde{f}, X, \ldots, X)) - d(W(f,X))$$

using (2.21).

We now let

$$\min_{x \in X} |h(x-z)| = |h(w-z)|.$$

The relation

$$f(z) + (X-z)h(w-z) \subseteq f(z) + \{(x-z)h(x-z) \mid x \in X\} = W(f,X)$$

is easily verified considering the two cases arising from the sign of $h(w-z)$. Therefore, using both (2.9) as well as (2.14), we get

$$d(W(f,X)) \geqslant d((X-z)h(w-z)) = d(X)|h(w-z)|, \qquad w \in X.$$

We may now estimate further as follows

$$\begin{aligned}
&q(W(f,X), W(\tilde{f}, X, \ldots, X)) \\
&\quad \leqslant (y^{(0)}-z)\tilde{h}(y^{(1)}-z, \ldots, y^{(n)}-z) - (x^{(0)}-z)\tilde{h}(x^{(1)}-z, \ldots, x^{(n)}-z) \\
&\quad\quad - d(X)|h(w-z)| \\
&\quad = (y^{(0)}-z)\tilde{h}(y^{(1)}-z, \ldots, y^{(n)}-z) - (y^{(0)}-z)\tilde{h}(x^{(1)}-z, \ldots, x^{(n)}-z) \\
&\quad\quad + (y^{(0)}-z)\tilde{h}(x^{(1)}-z, \ldots, x^{(n)}-z) - (x^{(0)}-z)\tilde{h}(x^{(1)}-z, \ldots, x^{(n)}-z) \\
&\quad\quad - d(X)|h(w-z)| \\
&\quad = (y^{(0)}-z)(\tilde{h}(y^{(1)}-z, \ldots, y^{(n)}-z) - \tilde{h}(x^{(1)}-z, \ldots, x^{(n)}-z)) \\
&\quad\quad + (y^{(0)}-x^{(0)})\tilde{h}(x^{(1)}-z, \ldots, x^{(n)}-z) - d(X)|\tilde{h}(w-z, \ldots, w-z)| \\
&\quad \leqslant |y^{(0)}-z|\,|\tilde{h}(y^{(1)}-z, \ldots, y^{(n)}-z) - \tilde{h}(x^{(1)}-z, \ldots, x^{(n)}-z)| \\
&\quad\quad + |y^{(0)}-x^{(0)}|\,|\tilde{h}(x^{(1)}-z, \ldots, x^{(n)}-z)| - d(X)|\tilde{h}(w-z, \ldots, w-z)| \\
&\quad \leqslant d(X)(|\tilde{h}(y^{(1)}-z, \ldots, y^{(n)}-z) - \tilde{h}(x^{(1)}-z, \ldots, x^{(n)}-z)| \\
&\quad\quad + ||\tilde{h}(x^{(1)}-z, \ldots, x^{(n)}-z)| - |\tilde{h}(w-z, \ldots, w-z)||) \\
&\quad \leqslant d(X)\left(c^{(1)} \max_{1 \leqslant k \leqslant n} |y^{(k)}-x^{(k)}| + c^{(2)} \max_{1 \leqslant k \leqslant n} |x^{(k)}-w|\right) \\
&\quad \leqslant d(X)(c^{(1)}+c^{(2)})\,d(X) = c\,(d(X))^2.
\end{aligned}$$

Here we used the Lipschitz condition for $\tilde{h}$ as well as the implied Lipschitz condition for $|\tilde{h}|$. ■

The results of Theorem 4 may also be carried over to functions of several variables. Proofs for (4) in this case have been given by Hansen [7] and using different techniques by Chuba and Miller [5].

The following estimate for the width of an interval evaluation will be shown to be an inference from Theorem 3.

Theorem 5: Let f be a real function of the real variable x, and let $f(x)$ be an expression for f. All the assumptions of Theorem 3 are assumed to hold. It then follows that the estimate

$$d(f(X)) \leqslant c \cdot d(X), \qquad c \geqslant 0, \tag{5}$$

is valid for $X \subseteq Y$.

Proof: From Theorem 3 and (2.21) we obtain

$$\begin{aligned} d(f(X)) &\leqslant 2q(f(X), W(f, X)) + d(W(f, X)) \\ &\leqslant 2c^{(1)} d(X) + d(W(f, X)), \qquad c^{(1)} \geqslant 0. \end{aligned}$$

Because of the Lipschitz boundedness of the function f, we get

$$d(W(f, X)) = f(x) - f(y) \leqslant c^{(2)}|x - y| \qquad \text{for} \quad x, y \in X, \quad c^{(2)} \geqslant 0,$$

from which the assertion

$$d(f(X)) \leqslant 2c^{(1)} d(X) + c^{(2)} d(X) = c \cdot d(X)$$

follows. ■

The corresponding result in the case of several variables is given as

$$\begin{aligned} d(f(X^{(1)}, X^{(2)}, \ldots, X^{(n)})) &\leqslant \sum_{k=1}^{n} c^{(k)} d(X^{(k)}) \\ &\leqslant c \max_{1 \leqslant k \leqslant n} d(X^{(k)}). \end{aligned} \tag{5'}$$

We now wish to prove a result on the inclusion of the range $W(f, X)$ of a real function by an interval expression derived from the mean-value theorem.

Theorem 6: Let f be a real function of the real variable x, and let f be differentiable in the interval $X = [x_1, x_2]$. Furthermore let $f'(x)$ be an expression for f' that may be evaluated as an interval expression over X. Then, if the assumptions of Theorem 5 hold for f', it follows for $y \in X$ that

(a) $W(f, X) \subseteq f(y) + f'(X)(X - y)$,

(b) $q(W(f, X), f(y) + f'(X)(X - y)) \leqslant \tilde{c}\,(d(X))^2$

with a constant $\tilde{c} \geqslant 0$.

Proof: Of (a): From the mean-value theorem we have for $x, y \in X$ that

$$f(x) = f(y) + f'(y + \theta(x - y))(x - y), \qquad 0 < \theta < 1.$$

From

$$y + \theta(x - y) \in y + [0, 1](X - y) = X,$$

it follows using inclusion monotonicity that

$$f(x)\in f(y)+f'(X)(X-y).$$

This proves (a).

Of (b): We let

$$W(f,X)=[f(u),f(v)],\qquad u,v\in X$$

Then from the mean-value theorem it follows that

$$\begin{aligned} d(W(f,X)) &= f(v)-f(u)=|f(v)-f(u)| \\ &\geqslant |f(x_1)-f(x_2)|=|f'(\xi)|\,d(X), \qquad \xi\in X. \end{aligned}$$

From (2.12), (2.3), and (2.20) it follows that

$$\begin{aligned} d(f'(X)(X-y)) &\leqslant |f'(X)|\,d(X)+d(f'(X))|X-y| \\ &\leqslant |f'(X)|\,d(X)+d(f'(X))\,d(X). \end{aligned}$$

Since $f'(\xi)\in f'(X)$ and using (2.21) we get

$$q(f'(X)), f'(\xi))\leqslant d(f'(X)).$$

We now use the inequality

$$|f'(X)|-|f'(\xi)|\leqslant q(f'(X),f'(\xi))$$

that follows from (2.4), (2.5), and Definition 2.6. Using (a), from (2.21) as well as Theorem 5 for $f'(X)$ we get the result

$$\begin{aligned} &q(W(f,X),f(y)+f'(X)(X-y)) \\ &\quad\leqslant d(f(y)+f'(X)(X-y))-d(W(f,X)) \\ &\quad\leqslant d(f'(X))\,d(X)+(|f'(X)|-|f'(\xi)|)\,d(X) \\ &\quad\leqslant d(f'(X))\,d(X)+q(f'(X),f'(\xi))\,d(X) \\ &\quad\leqslant 2c(d(X))^2=\tilde{c}\,(d(X))^2. \quad ■ \end{aligned}$$

The content of Theorem 6 shows that the qualitative result of the centered form in Theorem 4 is already obtained using the expression

$$f(y)+f'(X)(X-y),\qquad y\in X.$$

This is important since already for polynomials one needs the complete Horner scheme in order to develop the centered form. Theorem 6 may also be generalized to functions of several variables. We omit the details.

We now consider rational functions $f(x)=p(x)/q(x)$. Under certain conditions on $p(x)=\sum_{\nu=0}^{r}a_\nu x^\nu$ and $q(x)=\sum_{\nu=0}^{s}b_\nu x^\nu$ there exist expressions that are simpler than the centered form or the mean-value form of Theorem 6 which still have the property that

(6) $$q(W(f, X), f(X)) \leqslant c\, d(X)^2, \qquad c \geqslant 0.$$

Let $c = m(x)$, the midpoint of X, and let the Taylor expansions $p(x) = \sum_{\nu=0}^{r} a'_\nu (x-c)^\nu$, $q(x) = \sum_{\nu=0}^{s} b'_\nu (x-c)^\nu$ be given. Let $b'_0 = 1$ without loss of generality and assume $0 \notin 1 + \sum_{\nu=1}^{s} b'_\nu (X-c)^\nu =: q(X)$. If now

(7) $$\operatorname{sign}(a'_1)\operatorname{sign}(b'_1 \cdot a'_0) \leqslant 0,$$

then the interval expression

$$f(X) = \sum_{\nu=0}^{r} a'_\nu (X-c)^\nu \bigg/ \left(1 + \sum_{\nu=1}^{s} b'_\nu (X-c)^\nu\right)$$

has the property (6) if it is assumed that the expressions for $p(X)$ and $q(X)$ satisfy $d(p(X)) \leqslant c_1\, d(X)$ and $d(q(X)) \leqslant c_2\, d(X)$. This is the case for both the above expressions whether one computes the value of the polynomial using the powers of $X - c$ or one uses the Horner scheme.

If now

$$0 \notin 1 + (X-c)q'(X),$$

then we again have under the assumption (7) that the expression

$$f(X) = \frac{a'_0 + (X-c)p'(X)}{1 + (X-c)q'(X)}$$

also satisfies (6). Here $p'(X)$ is an interval evaluation of the first derivative of $p(x)$ satisfying $d(p'(X)) \leqslant \alpha\, d(X)$. Similarly $q'(X)$ is an interval evaluation of the first derivative of $q(x)$ satisfying $d(q'(X)) \leqslant \beta\, d(X)$.

The proof of property (6) for the above two cases is found in Alefeld and Rokne [2], and it will not be repeated here.

In Chapter 7 we consider inclusion methods for zeros of functions that use inclusions for the slopes of the function. In the following we give a finite number of possible inclusions of the difference quotient. Furthermore these will be partially ordered. It turns out that the optimal inclusion can be described in a simple and systematic manner and that the calculation with the respective iterations can be performed with the same computational cost as the interval evaluation of the derivative. The derivation of these inclusions is done using considerations quite different from those of Hansen [8] where the same problem was treated.

The inclusions for the examples in Hansen [8] correspond exactly to the optimal inclusions for the same examples using the results that follow. The development here corresponds to that of Alefeld [1]. We are given the polynomial

$$p(x) = \sum_{\nu=0}^{n} a_\nu x^\nu.$$

The following two equations are clearly valid

(8)
$$p(x) - p(y) = \sum_{i=0}^{n} a_i(x^i - y^i)$$
$$= \left(\sum_{i=1}^{n} a_i \sum_{j=1}^{i} x^{i-j}y^{j-1} \right)(x - y)$$
$$= \left(\sum_{i=1}^{n} \left(\sum_{j=i}^{n} a_j y^{j-i} \right) x^{i-1} \right)(x - y),$$

(9)
$$p(x) - p(y) = \sum_{i=0}^{n} a_i(x^i - y^i)$$
$$= \left(\sum_{i=1}^{n} a_i \sum_{j=1}^{i} y^{i-j}x^{j-1} \right)(x - y)$$
$$= \left(\sum_{i=1}^{n} \left(\sum_{j=i}^{n} a_j x^{j-i} \right) y^{i-1} \right)(x - y).$$

For a fixed y and arbitrary $x \in X$ we get from (8) and inclusion monotonicity that

$$\frac{p(x) - p(y)}{x - y} \in \left(\sum_{i=1}^{n} c_{i-1} X^{i-1} \right)_{\mathrm{H}} =: J_1$$
$$\subseteq J_2 := \sum_{i=1}^{n} c_{i-1} X^{i-1}$$

with

$$c_{i-1} = \sum_{j=i}^{n} a_j y^{j-i}, \qquad 1 \leqslant i \leqslant n.$$

The letter H denotes here and in the sequel that the expression is evaluated according to the Horner scheme. In J_2 the powers X^r are defined by $X^0 = 1$ and $X^r = X^{r-1}X$ for $r \geqslant 1$. The inclusion $J_1 \subseteq J_2$ holds because of the subdistributive law. We always have that

$$\sum_{i=1}^{n} A_{i-1} y^{i-1} = \left(\sum_{i=1}^{n} A_{i-1} y^{i-1} \right)_{\mathrm{H}}$$

for a real number y and intervals A_j, $0 \leqslant j \leqslant n - 1$. Using subdistributivity and this equation we get for a fixed y and arbitrary $x \in X, x \neq y$, from (9) that

$$\frac{p(x) - p(y)}{x - y} \in \sum_{i=1}^{n} (C_{i-1})_{\mathrm{H}} y^{i-1} = \left(\sum_{i=1}^{n} (C_{i-1})_{\mathrm{H}} y^{i-1} \right)_{\mathrm{H}} =: J_3$$
$$\subseteq J_4 := \sum_{i=1}^{n} C_{i-1} y^{i-1} = \left(\sum_{i=1}^{n} C_{i-1} y^{i-1} \right)_{\mathrm{H}}$$

with

$$(C_{i-1})_{\mathrm{H}} = \left(\sum_{j=i}^{n} a_j X^{j-i} \right)_{\mathrm{H}}, \qquad 1 \leqslant i \leqslant n,$$

and

$$C_{i-1} = \sum_{j=i}^{n} a_j X^{j-i}, \qquad 1 \leqslant i \leqslant n.$$

We now prove the following theorem.

Theorem 7: The above expressions satisfy

(a) $J_1 \subseteq J_2 \subseteq J_4$,
(b) $J_1 \subseteq J_3 \subseteq J_4$,
(c) $J_4 \subseteq p'(X) = \sum_{\nu=1}^{n} \nu a_\nu X^{\nu-1}$.

Proof: We restrict ourselves to the proof for the case of a polynomial of degree 4 for clarity; that is, $n = 4$. The general case is handled in a completely analogous manner.

Of (a) and (c): We only have to show $J_2 \subseteq J_4 \subseteq p'(X)$. From the inclusion monotonicity and from (1.8) we obtain

$$\begin{aligned} J_2 &= \sum_{i=1}^{n} c_{i-1} X^{i-1} \\ &= (a_1 + a_2 y + a_3 y^2 + a_4 y^3)X^0 + (a_2 + a_3 y + a_4 y^2)X \\ &\quad + (a_3 + a_4 y)X^2 + a_4 X^3 \\ &\subseteq a_1 + a_2 X + a_3 X^2 + a_4 X^3 + a_2 y + a_3 y X + a_4 y X^2 \\ &\quad + a_3 y^2 + a_4 y^2 X + a_4 y^3 \\ &= a_1 + a_2 X + a_3 X^2 + a_4 X^3 + (a_2 + a_3 X + a_4 X^2) y \\ &\quad + (a_3 + a_4 X) y^2 + a_4 y^3 = J_4 \\ &\subseteq a_1 + a_2 X + a_3 X^2 + a_4 X^3 + a_2 X + a_3 X^2 + a_4 X^3 \\ &\quad + a_3 X^2 + a_4 X^3 + a_4 X^3 = p'(X). \end{aligned}$$

Of (b): We only have to show that $J_1 \subseteq J_3$.

$$\begin{aligned} J_1 &= ((c_3 X + c_2)X + c_1)X + c_0 \\ &= ((a_4 X + (a_3 + a_4 y))X + a_2 + a_3 y + a_4 y^2)X + a_1 + a_2 y + a_3 y^2 + a_4 y^3 \\ &\subseteq ((a_4 X + a_3)X + a_4 y X + a_2 + a_3 y + a_4 y^2)X + a_1 + a_2 y + a_3 y^2 + a_4 y^3 \\ &= (((a_4 X + a_3)X + a_2) + a_4 y X + a_3 y + a_4 y^2)X + a_1 + a_2 y + a_3 y^2 + a_4 y^3 \\ &= (((a_4 X + a_3)X + a_2) + (a_4 X + a_3) y + a_4 y^2)X + a_1 + a_2 y + a_3 y^2 + a_4 y^3 \\ &\subseteq ((a_4 X + a_3)X + a_2)X + (a_4 X + a_3) y X + a_4 y^2 X + a_1 + a_2 y + a_3 y^2 + a_4 y^3 \\ &= (((a_4 X + a_3)X + a_2)X + a_1) y^0 + ((a_4 X + a_3)X + a_2) y \\ &\quad + (a_4 X + a_3) y^2 + a_4 y^3 = J_3. \end{aligned}$$

This proves the theorem. ■

There is no general statement as to which of J_2 and J_3 gives the best inclusion. One may have $J_2 \subseteq J_3$ or $J_3 \subseteq J_2$. As an example let

$$p(x) = x^3 - x^2, \qquad X = [-1, 2], \qquad y = 1.$$

We then get

$$J_2 = (a_1 + a_2 y + a_3 y^2)X^0 + (a_2 + a_3 y)X + a_3 X^2 = X^2 = [-2, 4]$$

as well as

$$\begin{aligned} J_3 &= ((a_3 X + a_2)X + a_1)y^0 + (a_3 X + a_2)y + a_3 y^2 \\ &= (X - 1)X + (X - 1) + 1 = [-5, 4], \end{aligned}$$

which implies $J_2 \subset J_3$.

If, on the other hand, $y = 0$ and therefore $c_{i-1} = a_i, 1 \leqslant i \leqslant n$, then we get

$$J_2 = \sum_{i=1}^{n} a_i X^{i-1} \qquad \text{and} \qquad J_3 = \left(\sum_{i=1}^{n} a_i X^{i-1} \right)_{\mathrm{H}},$$

with $J_3 \subseteq J_2$.

We consider the above example $p(x) = x^3 - x^2$ with $y = 0$ and $X = [0, 2]$. Then we get

$$J_2 = X^2 - X = [-2, 4] \qquad \text{and} \qquad J_3 = (X - 1)X = [-2, 2],$$

and therefore $J_3 \subset J_2$.

The calculation of the intervals J_1 and J_2 from Theorem 7 requires the calculation of $c_{i-1} = \sum_{j=i}^{n} a_j y^{j-i}$, $1 \leqslant i \leqslant n$, initially. If the value of the polynomial $p(x)$ at the point y is also calculated, as for example in the iteration methods of Chapter 7, then the calculation of the c_{i-1} does not require any further arithmetic operations. The c_{i-1} are calculated during the calculation of $p(y)$. Let

$$p(x) = \sum_{i=0}^{n} a_i x^i$$

be given as above. Then calculate the Horner scheme

$$p_n := a_n$$

for $i = n(-1)1$:

$$p_{i-1} := p_i y + a_{i-1}$$

and we have $p_0 = p(y)$. From the definition

$$\begin{aligned} c_{n-1} &= a_n & (&= p_n), \\ c_{n-2} &= a_n y + a_{n-1} & (&= p_{n-1}), \\ &\vdots & &\vdots \\ c_0 &= c_1 y + a_1 & (&= p_1). \end{aligned}$$

Therefore $c_{i-1} = p_i$, $1 \leqslant i \leqslant n$.

Examples: The examples (a)–(e) are taken from Alefeld [1].

(a) $p(x) = x^4 - 1, \qquad x = [0.5, 3.5], \qquad y = 2.$

We get

$$J_1 = J_2 = J_3 = J_4 = [10.625, 89.375],$$
$$p'(X) = (p'(X))_{\mathrm{H}} = [0.5, 171.5].$$

J_1 coincides with the estimate given in Hansen [8] for this example.

(b) $p(x) = x^3 + 4x - 16, \qquad X = [-1, 3], \qquad y = 1.$

One gets

$$J_1 = J_2 = J_3 = J_4 = [1, 17],$$
$$p'(X) = (p'(X))_{\mathrm{H}} = [-5, 31],$$

which again coincides with the calculation in Hansen [8].

(c) $p(x) = \sum_{i=0}^{n} a_i x^i, \qquad 0 \in X, \qquad y = 0.$

Then we have

$$c_0 = a_1, \qquad c_1 = a_2, \ldots, c_{n-1} = a_n$$

and

$$J_1 = \left(\sum_{i=1}^{n} c_{i-1} X^{i-1}\right)_{\mathrm{H}} = \left(\sum_{i=1}^{n} a_i X^{i-1}\right)_{\mathrm{H}}.$$

(d) $p(x) = x^3 - x^2, \qquad X = [1, 3], \qquad y = 2.$

$$J_1 = J_2 = J_3 = [4, 14] \subset [2, 16] = J_4 \subset (p'(X))_{\mathrm{H}} = [1, 21] \subset [-3, 25] = p'(X),$$

which coincides with the value calculated in Hansen [8].

For $X = [-1, 2]$, $y = 1$, one, however, obtains

$$J_1 = J_2 = [-2, 4], \qquad J_3 = [-5, 4], \qquad J_4 = [-5, 7],$$
$$(p'(X))_{\mathrm{H}} = [-10, 8], \qquad p'(X) = [-10, 14].$$

(e) Let $x_0 \in X$ and $f \in C^{n+1}(X)$. Using a Taylor expansion, one obtains

$$f(x) = p(x) + \phi(x)$$

with

$$\phi(x) = \int_{x_0}^{x} \frac{(x-t)^n}{n!} f^{(n+1)}(t)\, dt$$

and

$$p(x) = \sum_{k=0}^{n} \frac{(x-x_0)^k}{k!} f^{(k)}(x_0).$$

ϕ is differentiable and

$$\phi'(x) = \int_{x_0}^{x} \frac{(x-t)^{n-1}}{(n-1)!} f^{(n+1)}(t)\, dt.$$

The mean-value theorem for integrals gives

$$\phi'(x) = f^{(n+1)}(\eta) \int_{x_0}^{x} \frac{(x-t)^{n-1}}{(n-1)!}\, dt = \frac{(x-x_0)^n}{n!} f^{(n+1)}(\eta)$$

for η between x and x_0. Applying the mean-value theorem to ϕ we get

$$\begin{aligned} f(x) - f(y) &= p(x) - p(y) + \phi(x) - \phi(y) \\ &= \left\{ \sum_{k=1}^{n} c_{k-1}(x-x_0)^{k-1} + \phi'(\xi) \right\} (x-y) \end{aligned}$$

with

$$c_{k-1} = \sum_{j=k}^{n} (y-x_0)^{j-k} \frac{f^{(k)}(x_0)}{k!}, \qquad 1 \leqslant k \leqslant n,$$

and

$$\phi'(\xi) = \frac{(\xi - x_0)^n}{n!} f^{(n+1)}(\eta),$$

where ξ is between x and y and η is between x_0 and ξ. The particular choice $y = x_0$ results in

$$c_0 = f'(x_0)/1!, \qquad \ldots, \qquad c_{n-1} = f^{(n)}(x_0)/n!.$$

If the $(n+1)$th derivative has an interval expression that can be evaluated then for $y = x_0$, we have

$$\frac{f(x) - f(y)}{x - y} \in \sum_{k=1}^{n} \frac{f^{(k)}(x_0)}{k!} (X - x_0)^{k-1} + f^{(n+1)}(X) \frac{(X - x_0)^n}{n!},$$

since $\eta, \xi \in X$. This estimate coincides with the estimate given in Hansen [8].

(f) $p(x) = x^7 + 3x^6 - 4x^5 - 12x^4 - x^3 - 3x^2 + 4x + 12.$

$$X = [1.8, 3], \qquad y = 2.$$

We get

$$J_1 = [173.2362, 2400], \qquad J_2 = [161.4762, 2411.76]$$

$$J_3 = [24.72, 2400], \qquad J_4 = [-870.2933, 3443.5296]$$

$$(p'(X))_H = [71.799808, 6520], \qquad p'(X) = [-2378.791292, 8970.592].$$

These considerations may also be carried over to the multidimensional case (compare with Alefeld [1]).

Remarks: In this section interval evaluations of real functions were considered. General maps from $I(\mathbb{R})$ to $I(\mathbb{R})$ were deliberately avoided. The applications in the later chapters require many of the properties that we are able to prove only for interval evaluations. If one allows more general maps on $I(\mathbb{R})$, then one has to specify a set of conditions for each application. The following example indicates how large the class of maps from $I(\mathbb{R})$ to $I(\mathbb{R})$ is. The only restriction here is that if the domain is restricted to $\mathbb{R}$, then the range is in $\mathbb{R}$. For this, let f be a real function with an expression $f(x)$ and an interval evaluation $f(X)$ of this expression. Then

$$\Psi(X) = f(X) + \phi(d(X))[-1, 1]$$

defines a map from $I(\mathbb{R})$ to $I(\mathbb{R})$ for every $\phi(x)$ with $\phi(0) = 0$. Clearly $\Psi([x, x]) \in \mathbb{R}$. If $\phi(x) \geqslant 0$ for $x \geqslant 0$, then we have that $W(f, X) \subseteq \Psi(X)$, and if $\phi(x)$ is monotonically nondecreasing for $x \geqslant 0$, then $\Psi(X)$ has the inclusion monotonicity property (1′). The example shows that depending on the choice of $\phi(x)$ one may generate maps Ψ with different properties. If all the properties of interval evaluations were required by maps from $I(\mathbb{R})$ to $I(\mathbb{R})$, then there were no other practical realizations other than exactly these evaluations.

We also wish to show how the proof of Theorem 4 can be shortened using Theorem 5 in a manner similar to the proof of Theorem 6. The interval evaluation of the centered form

$$f(X) = f(z) + (X - z)h(X - z)$$

satisfies

$$W(f, X) \subseteq f(X).$$

According to (2.21) we may then estimate

$$q(W(f, X), f(X)) \leqslant d(f(X)) - d(W(f, X)).$$

Let now

$$\min_{x \in X} |h(x - z)| = |h(w - z)|.$$

Then we get

$$f(z) + (X - z)h(w - z) \subseteq f(z) + \{(x - z)h(x - z) \mid x \in X\} = W(f, X).$$

Using (2.9) and (2.14), we get from the above inclusion that

$$d(W(f, X)) \geqslant d((X - z)h(w - z)) = d(X)|h(w - z)|, \qquad w \in X.$$

Considering (2.10), (2.12), (2.3), and (2.20) we get

$$\begin{aligned} d(f(X)) &= d(f(z) + (X - z)h(X - z)) \\ &= d((X - z)h(X - z)) \\ &\leqslant |X - z|\, d(h(X - z)) + d(X)|h(X - z)| \\ &\leqslant d(X)\, d(h(X - z)) + d(X)|h(X - z)|. \end{aligned}$$

Since $h(w - z) \in h(X - z)$, we get from (2.21) that

$$q(h(X - z), h(w - z)) \leqslant d(h(X - z)).$$

Definition 2.6 and (2.4), (2.5) give the inequality

$$|h(X - z)| - |h(w - z)| \leqslant q(h(X - z), h(w - z)).$$

From the above inequalities we obtain

$$\begin{aligned} q(W(f, X), f(X)) &\leqslant d(f(z) + (X - z)h(X - z)) - d(W(f, X)) \\ &\leqslant d(X)\, d(h(X - z)) + d(X)|h(X - z)| - d(X)|h(w - z)| \\ &= d(X)\, d(h(X - z)) + (|h(X - z)| - |h(w - z)|)\, d(X) \\ &\leqslant d(X)\, d(h(X - z)) + q(h(X - z), h(w - z))\, d(X) \\ &\leqslant d(X) \cdot 2 \cdot d(h(X - z)). \end{aligned}$$

Finally, by applying Theorem 5 to $h(X - z)$, we arrive at the result

$$q(W(f, X), f(X)) \leqslant d(X) \cdot 2 \cdot \tilde{c} \cdot d(X) = c(d(X))^2.$$

Under certain differentiability conditions for the function expressions one may determine general conditions under which (4) holds (see Herzberger [11]). As special cases one gets the forms given in Theorems 4 and 6. In Herzberger [12] a simple interpolation formula is given which yields the range of a set of polynomials with given coefficients. In this case one uses the fact mentioned in Theorem 2 that the interval evaluation of a function gives the exact range if the variables and the parameters occur only once in the expression.

The forthcoming book by Ratschek and Rokne [24] contains a thorough treatment of the problem of approximating the range of a function by interval evaluations. Cornelius and Lohner [23] introduced an expression $f(X)$ for which $q(W(f, X),\ f(X)) \leqslant c\, d(X)^{s+1}$, $s \geqslant 1$. For $s = 1, 2, 3$ a very simple algorithm computes $f(X)$. For details see [24].

REFERENCES

[1] G. Alefeld, Bounding the slope of polynomial and some applications. *Computing* **26**, 227–237 (1981).

[2] G. Alefeld and J. Rokne, On the evaluation of rational functions in interval arithmetic. *SIAM J. Numer. Anal.* **18**, 862–870 (1981).

[3] N. Apostolatos and U. Kulisch, Grundlagen einer Maschinenintervallarithmetik. *Computing* **2**, 89–104 (1967).

[4] N. Apostolatos and U. Kulisch, Approximation der erweiterten Intervallarithmetik durch einfache Maschinenintervallarithmetik. *Computing* **2**, 181–194 (1967).

[5] W. Chuba and W. Miller, Quadratic convergence in interval arithmetic, Part. I. *BIT* **12**, 284–290 (1972).

[6] A. J. Goldstein and P. L. Richmann: A midpoint phenomenon. *J. Assoc. Comput. Mach.* **20**, 301–304 (1973).

[7] E. Hansen, The centred form. *In* "Topics in Interval Analysis" (E. Hansen, ed.). Oxford Univ. Press (Clarendon), London and New York, 1969.

[8] E. Hansen, Interval forms of Newton's method. *Computing* **20**, 153–163 (1978).

[9] J. Herzberger, Metrische Eigenschaften von Mengensystemen und einige Anwendungen. Ph.D. Thesis, Univ. Karlsruhe (1969).

[10] J. Herzberger, Definition und Eigenschaften allgemeiner Intervallräume. *Z. Angew. Math. Mech.* **50**, T50–T51 (1970).

[11] J. Herzberger, Zur Approximation des Wertebereiches reeller Funktionen durch Intervallausdrücke. *Computing*, *Suppl.* **1**, 57–64 (1977).

[12] J. Herzberger, Note on a bounding technique for polynomial functions. *SIAM J. Appl. Math.* **34**, 685–686 (1978).

[13] U. Kulisch, Grundzüge der Intervallrechnung. *In* "Überblicke Mathematik 2." Bibliograph. Inst., Mannheim, 1969.

[14] O. Mayer, Über die in der Intervallrechnung auftretenden Räume und einige Anwendungen. Ph.D. Thesis, Univ. Karlsruhe (1968).

[15] O. Mayer, Algebraische und metrische Strukturen in der Intervallrechnung und einige Anwendungen. *Computing* **5**, 144–162 (1970).

[16] W. Miller, Quadratic convergence in interval arithmetic, Part II. *BIT* **12**, 291–298 (1972).

[17] W. Miller, More on quadratic convergence in interval arithmetic. *BIT* **13**, 76–83 (1973).

[18] S. B. Nadler, Multi-valued contraction mappings. *Pacific J. Math.* **30**, 475–488 (1969).

[19] H. Ratschek, Gleichheit von Produkt und Formalprodukt bei Intervallpolynomen. *Computing* **10**, 245–254 (1972).

[20] H. Ratschek, Ergebnisse einer Untersuchung über die Struktur von Intervallpolynomen. *Ber. Ges. Math. Dat. Nr.* 52 (1972).

[21] H. Ratschek, Centered forms. *SIAM J. Numer. Anal.* **17**, 656–662 (1980).

[22] H. Ratschek and G. Schröder, Über die Ableitung von intervallwertigen Funktionen. *Computing* **7**, 172–187 (1971).

Added in proof:

[23] H. Cornelius and R. Lohner, Die Restgliedform zur Einschließung des Wertebereichs reeller Funktionen (to be published).

[24] H. Ratschek and J. Rokne, Centered forms (unpublished).

Chapter 4 / MACHINE INTERVAL ARITHMETIC

We now turn to the realization of interval operations on a digital computer. As is well known, computers have only a finite set of numbers that are often represented in a semilogarithmic manner as fixed length floating point numbers

$$x = m \cdot b^e.$$

Here m is the mantissa, b the base, and e the exponent. The numbers are normally represented internally with a base $b = 2$ and a normalized mantissa, that is $\frac{1}{2} \leqslant |m| < 1$. The integer exponent e is bounded by $e_{\min} \leqslant e \leqslant e_{\max}$.

The set of machine numbers of the above type is denoted by $\mathbb{R}_M$ and we assume for all further considerations that $\mathbb{R}_M$ is symmetric around 0; that is,

$$\mathbb{R}_M = -\mathbb{R}_M.$$

The real numbers lying in the interval $[\min_{y \in \mathbb{R}_M} y;\ \max_{y \in \mathbb{R}_M} y]$ may be approximated effectively by machine numbers $\tilde{x} \in \mathbb{R}_M$. This approximation is achieved by a map

(1) $$fl: \mathbb{R} \ni x \to \tilde{x} = fl(x) \in \mathbb{R}_M.$$

This map is called rounding if

(2) $$x \leqslant y \Rightarrow fl(x) \leqslant fl(y) \qquad \text{(monotonicity)}$$

is satisfied. Roundings that map $\mathbb{R}_M$ into $\mathbb{R}_M$ in the sense that

(3) $$x \in \mathbb{R}_M \Rightarrow fl(x) = x$$

is satisfied, are called optimal roundings. We are here particularly interested in directed roundings that always "round up," respectively, "round down." If a rounding $\downarrow$ satisfies

(4) $$x \in \mathbb{R} \Rightarrow \downarrow x \leqslant x,$$

then one talks about a downward-directed rounding. Using the definition

(5) $$\uparrow x := -(\downarrow(-x)), \qquad x \in \mathbb{R},$$

one also gets an upward-directed rounding. The internal generation of such roundings for the various codings of numbers is described in the literature (see, for example, Knuth [11], Kupka [17], Wilkinson [29], Miranker and Kulisch [20], and Moore [22]). Algebraic and ordering properties of roundings are for example described in Kulisch [14–16]. In the same manner that one represents real numbers by machine numbers one can represent real intervals by machine intervals. One then has to represent an interval $X \in I(\mathbb{R})$ for which it is assumed that $X \subseteq [\min_{y \in \mathbb{R}_M} y, \max_{y \in \mathbb{R}_M} y]$ by a suitable machine interval from the set

$$I(\mathbb{R}_M) = \{[x_1, x_2] \mid x_1, x_2 \in \mathbb{R}_M, x_1 \leqslant x_2\} \subset I(\mathbb{R})$$

of machine intervals. The interval rounding

$$\updownarrow : I(\mathbb{R}) \ni X \to \updownarrow X \in I(\mathbb{R}_M)$$

used must satisfy

(6) $$X \in I(\mathbb{R}) \Rightarrow X \subseteq \updownarrow X,$$

as well as

(7) $$X, Y \in I(\mathbb{R}), \qquad X \subseteq Y \Rightarrow \updownarrow X \subseteq \updownarrow Y,$$

in order that the basic properties of the interval operations are carried over to the machine interval operations. If we consider the transition from an interval $X = [x_1, x_2] \in I(\mathbb{R})$ to its representation $\tilde{X} = [\tilde{x}_1, \tilde{x}_2]$, then (7) says that this must be effectuated by the rounding of the individual bounds. From (6) it follows that the bounds must be rounded in the directed sense. This implies that every interval rounding may be represented by

(8) $$\updownarrow X = \updownarrow [x_1, x_2] = [\downarrow x_1, \uparrow x_2].$$

From the previous discussion it follows that it is sufficient to have a downward-directed rounding $\downarrow$ in order to generate an interval rounding. On the other hand $\uparrow$ and $\downarrow$ must not necessarily be connected by (5).

If one operates with two machine numbers $x, y \in \mathbb{R}_M$ on a computer using an operation $* \in \{+, -, \cdot, :\}$, then the result is a new number $z \in \mathbb{R}_M$. If we ignore transgressions of the range of numbers in $\mathbb{R}_M$ (underflow and overflow), then a result of such an operation may be represented by

(9) $$z = fl(x * y)$$

using a suitable rounding *fl*. In this manner we may define the result of machine operations by the

Definition 1: Let $A, B \in I(\mathbb{R}_M)$, $* \in \{+, -, \cdot, :\}$, and let $\updownarrow$ be a given interval rounding. Then the result of operating on A and B by $*$ using $\updownarrow$ is given by

$$C = \updownarrow(A * B) \in I(\mathbb{R}_M). \quad \blacksquare \tag{10}$$

We now show that the basic properties of interval arithmetic are retained when this definition is used.

Theorem 2: For the machine interval operations from Definition 1 it holds that

$$\begin{aligned} &A^{(k)}, B^{(k)} \in I(\mathbb{R}_M),\ * \in \{+, -, \cdot, :\},\ A^{(k)} \subseteq B^{(k)},\ k = 1, 2 \\ &\Rightarrow C^{(1)} = \updownarrow(A^{(1)} * A^{(2)}) \subseteq C^{(2)} = \updownarrow(B^{(1)} * B^{(2)}). \quad \blacksquare \end{aligned} \tag{11}$$

The proof of Theorem 2 is immediate from property (7) of interval roundings.

Property (11) is nothing more than the inclusion monotonicity (1.9) for machine interval operations. The following properties are of interest for the estimation of the rounding errors.

Theorem 3: Let $\updownarrow$ be an interval rounding generated by directed roundings $\downarrow$ and $\uparrow$ according to (8) and let $* \in \{+, -, \cdot, :\}$. Then we have

$$\begin{aligned} &A, B \in I(\mathbb{R}_M) \Rightarrow A * B \subseteq C = \updownarrow(A * B) \in I(\mathbb{R}_M), \\ &a \in A, b \in B \Rightarrow a * b \in C = \updownarrow(A * B) \in I(\mathbb{R}_M). \end{aligned} \tag{12}$$

If a rounding *fl* always satisfies

$$\downarrow x \leqslant fl(x) \leqslant \uparrow x, \qquad x \in \mathbb{R}, \tag{13}$$

then for $x, y, z \in \mathbb{R}_M$ it follows that

$$z = fl(x * y) \in Z = \updownarrow([x, x] * [y, y]) \in I(\mathbb{R}_M). \quad \blacksquare$$

The elementary proofs of (12) and (13) follow immediately from the respective definitions and they are therefore omitted here.

A summary of the above results follows.

The interval evaluation of an expression for a function using the operations defined in Definition 1 gives including intervals for the interval evaluation. These including intervals are also including estimates of the range of the functions. These numerically calculated interval results furthermore satisfy the property of inclusion monotonicity.

The practical realization of the machine interval operations is done using the corresponding machine operators. These operators may either be a part of a higher programming language (see, for example, Apostolatos *et al.*

[4] and Signum Newsletter [27]) or they may be realized by subroutines which may be written for example in ALGOL. We shall consider the last case briefly. Such a set of subroutines contains in most cases an operator generating a downward-directed rounding $\downarrow$ as in Christ [6]. This is realized, for example, by the function procedure LOW. Using this procedure the interval operations ADD, SUB, MUL, and DIV are defined representing the standard interval arithmetic operations. The unary operations of Definition 1.3, the so-called elementary functions, are defined in a similar manner (see Herzberger [10] and Dewar [7]). The details of the realization of such subprograms are dealt with in Appendix B.

We now consider algorithms that are derived in the field of real numbers. Examples of such algorithms are the Horner scheme or the Gaussian algorithm. If such an algorithm is carried out on a computer using a machine arithmetic, then already the input data cannot be exactly represented in general. This may now be remedied by the use of machine interval arithmetic. The input data is simply enclosed in an interval having machine numbers as bounds. If the algorithm is now executed ignoring rounding errors, then the interval result or approximation set will still in general suffer from an increase in width not attributable to the original data as shown in Chapter 3. This property is magnified when roundoff errors are taken into consideration. We therefore discuss what increase one may expect in accuracy when one first executes an algorithm using a machine interval arithmetic with t_1 digits in the mantissa, then executes the algorithm with the number of digits in the mantissa increased to t_2, $t_2 > t_1$. It is assumed that the range of the exponent does not change when t_1 is changed to t_2. Under this assumption all numbers representable with t_1 digits are also exactly representable with t_2 digits.

Let $x \in \mathbb{R}$, $x \neq 0$, and

$$x = \left(\sum_{\nu=-1}^{-\infty} a_\nu b^\nu \right) b^e, \qquad 1 \leqslant a_{-1} \leqslant b-1, \qquad 0 \leqslant a_\nu \leqslant b-1, \qquad \nu \leqslant -2.$$

In order to guarantee uniqueness of this representation we assume that we do not have $a_\nu = b-1$, $\nu \leqslant \nu_0$, for a certain fixed ν_0. It is also assumed that x is not exactly representable using the floating point system with t_1 digits in the mantissa. (If this was the case, then the following consideration would indeed be superfluous.) It is furthermore assumed that the interval rounding (8) is carried out using an optimal rounding of the bounds. For the case $x > 0$ we get according to (8) that

$$\updownarrow x = \updownarrow [x, x] = [\downarrow x, \uparrow x]$$

with

$$\downarrow x = \left(\sum_{\nu=-1}^{-t_1} a_\nu b^\nu \right) b^e, \qquad \uparrow x = \left(\sum_{\nu=-1}^{-t_1} a_\nu b^\nu \right) b^e + b^{-t_1+e}.$$

Clearly the width of $\updownarrow x$ is

$$d(\updownarrow x) = b^{-t_1+e}.$$

The same result is obtained for the width of $\updownarrow x$ if $x < 0$.

In order that one may note the dependency of the results on the length of the mantissa, we write $fl_1(x)$ (resp. $fl_2(x)$) in the sequel. By fl we therefore mean the interval rounding of a real number (or later real interval). The above relation may now be written

$$d(fl_1(x)) = b^{-t_1+e}.$$

Analogously we get

$$d(fl_2(x)) \leqslant b^{-t_1+e-l}$$

for mantissa length $t_2 = t_1 + l$. The $<$ sign is valid for the case that x is exactly representable using mantissa length t_2. In particular we have

(14) $$d(fl_2(x)) \leqslant b^{-l} d(fl_1(x)).$$

From the assumptions made for the interval roundings it follows that

$$\updownarrow(A * B) = fl_1(A * B) = [(1 - \varepsilon_1)(A * B)_1, (1 + \varepsilon_2)(A * B)_2]$$

for two machine intervals A and B (see Definition 1). Here $(A * B)_1$ and $(A * B)_2$ calculates the bounds for the exact result and we have

$$-\varepsilon_1(A * B)_1 \leqslant 0, \qquad \varepsilon_2(A * B)_2 \geqslant 0,$$

as well as

$$|\varepsilon_1|, |\varepsilon_2| \leqslant b^{1-t_1}.$$

One may therefore also write

(15a) $$fl_1(A * B) = A * B + [-\varepsilon_1(A * B)_1, \varepsilon_2(A * B)_2].$$

For the width of the result it therefore holds that

(15b) $$d(fl_1(A * B)) \leqslant d(A * B) + 2b^{1-t_1}|A * B|.$$

This estimate shows that the absolute value of the exact interval result is responsible for the increase in width of $d(A * B)$ when a fixed mantissa length is used.

Let the real number $\tilde{x}$ have the property that $\tilde{x} \in X \in I(\mathbb{R})$. This suggests that one should choose $x \in X$ as an approximation for $\tilde{x}$. The absolute error is now given by

(16) $$|x - \tilde{x}| \leqslant d(X) =: \Delta(X),$$

and if $0 \notin X$, $\tilde{x} \neq 0$, the relative error by

(17) $$\left|\frac{x - \tilde{x}}{\tilde{x}}\right| \leqslant \frac{d(X)}{\min\{|x| \mid x \in X\}} =: \rho(X).$$

We now prove

Theorem 4: Let A, B, C, and D be real machine intervals for which

$$A \subseteq C, \qquad B \subseteq D, \tag{18}$$

as well as

$$\begin{aligned} d(C) &\leqslant s_1, & d(D) &\leqslant s_2 \\ d(A) &\leqslant b^{-l}s_1, & d(B) &\leqslant b^{-l}s_2. \end{aligned} \tag{19}$$

Let $*$ denote one of the arithmetic operations on real intervals. Then one obtains bounds for $\Delta(fl_2(A * B))$, respectively, for $\rho(fl_2(A * B))$, if $0 \notin fl_1(C * D)$, that are smaller than the bounds for $\Delta(fl_1(C * D))$, respectively, $\rho(fl_1(C * D))$, by a factor of b^{-l}.

Proof: Using (15b), (2.10), (2.12), as well as

$$d(1/X) \leqslant |1/X|^2\, d(X) \qquad (0 \notin X)$$

and the first line of (19), one gets the inequality

$$d(fl_1(C * D)) \leqslant d(C * D) + 2b^{1-t_1}|C * D|$$

$$\leqslant \left\{\begin{matrix} s_1 + s_2, & * = +, - \\ |C|s_2 + s_1|D|, & * = \cdot \\ |C|\,|1/D|^2 s_2 + |1/D|s_1, & * = : \end{matrix}\right\} + 2b^{1-t_1}|C * D|.$$

Using (18) and (19) one proves analogously that

$$d(A * B) \leqslant b^{-l}\left\{\begin{matrix} s_1 + s_2, & * = +, - \\ |C|s_2 + s_1|D|, & * = \cdot \\ |C|\,|1/D|^2 s_2 + |1/D|s_1, & * = : \end{matrix}\right\}. \tag{20}$$

Because of (18), we have from Theorem 2 the inclusion

$$fl_2(A * B) \subseteq fl_2(C * D) \subseteq fl_1(C * D),$$

since we made the assumption that the interval rounding was introduced using optimal roundings for the bounds. Therefore we get

$$\min\{|x| \mid x \in fl_2(A * B)\} \geqslant \min\{|x| \mid x \in fl_1(C * D)\}. \tag{21}$$

From (15b), (20), and the fact that $|A * B| \leqslant |C * D|$ it follows that

$$d(fl_2(A * B)) \leqslant d(A * B) + 2b^{1-t_1-l}|C * D|$$

$$\leqslant b^{-l}\left\{\begin{matrix} s_1 + s_2, & * = +, - \\ |C|s_2 + s_1|D|, & * = \cdot \\ |C|\,|1/D|^2 s_2 + |1/D|s_1, & * = : \end{matrix}\right\} + 2b^{1-t_1-l}|C * D|.$$

This proves the assertion for the upper bounds of the absolute error. From (21) we immediately get the result for the upper bounds of the relative error. ■

An elementary but important inference from this theorem is given by

Theorem 5: All the above assumptions for a machine interval arithmetic are assumed to hold. One now executes an algorithm defined in the field of real numbers on a computer using machine interval arithmetic with mantissa length t_1. If the algorithm is then executed using machine interval arithmetic with mantissa length t_2, $t_2 = t_1 + l, l > 0$, then the bounds for both the absolute and the relative error are reduced by the factor b^{-l}. (An algorithm is here a uniquely defined sequence of arithmetic operations with given input data.)

Proof: From (14) it follows that the interval rounding of the input satisfies the important assumption (19) from Theorem 4. The properties of the interval arithmetic validates (18). The result then follows from Theorem 4 using complete induction. ■

We get from Theorem 5 an indication of how to calculate the output within an a priori given absolute or relative accuracy. Let, for example, the largest width of the resulting intervals calculated using t_1 digits in the mantissa be d_1, and let the required absolute accuracy be ε. If $d_1 \leqslant \varepsilon$, then one is finished. Otherwise one increases the number of digits by l, where l satisfies

$$b^{-l} d_1 \leqslant \varepsilon.$$

(With this choice one is not guaranteed that the absolute error is reduced by b^{-l}. According to Theorem 5 this holds only for the upper bound of the absolute error.)

The facts discussed and proved in Theorem 5 were investigated theoretically by Rump [26]. He also gives illustrative numerical examples. As a concrete example we chose the set of equations formed by the 7×7 Hilbert matrix with right-hand side $(1, 1, \ldots, 1)^T$ (see also Rump [26]). The Gaussian algorithm was executed using machine interval arithmetic with 15, 20, 25, 30,

Table 1

The Upper Bound $\rho(X_i)$ for the Relative Error in the Gaussian Algorithm

No. digits in mantissa, i	15	20	25	30	35
1	$> 1^a$	0.11×10^{-3}	0.11×10^{-8}	0.11×10^{-13}	0.11×10^{-18}
2	0.34×10^{0}	0.29×10^{-5}	0.29×10^{-10}	0.29×10^{-15}	0.29×10^{-20}
3	0.18×10^{-1}	0.17×10^{-6}	0.17×10^{-11}	0.17×10^{-16}	0.17×10^{-21}
4	0.16×10^{-2}	0.16×10^{-7}	0.16×10^{-12}	0.16×10^{-17}	0.16×10^{-22}
5	0.26×10^{-3}	0.25×10^{-8}	0.25×10^{-13}	0.25×10^{-18}	0.25×10^{-23}
6	0.64×10^{-4}	0.64×10^{-9}	0.64×10^{-14}	0.64×10^{-19}	0.64×10^{-24}
7	0.58×10^{-4}	0.58×10^{-9}	0.58×10^{-14}	0.58×10^{-19}	0.58×10^{-24}

[a] The entry $\rho(X_1) > 1$ for 15 digits means that the interval X_1 contains 0.

and 35 decimal digits in the mantissa. The results are repeated in Table 1, where only the upper bound $\rho(X_i)$ for the relative error is given for each component of the solution vector. The details are found in Rump [26]. A simple example illustrating the contents of Theorem 5 is also found in Moore [23, page 60].

We now consider the following problem: Let there be given machine intervals (this means real intervals having machine numbers as endpoints), say,

$$C_0, A_0, B_0, D_0, A_1, B_1, D_1, \dots, A_{n-1}, B_{n-1}, D_{n-1}$$

and a machine number a_n.

The expression

$$R_n = (1/a_n)\{C_0 - A_0(B_0 - D_0) - A_1(B_1 - D_1) - \cdots - A_{n-1}(B_{n-1} - D_{n-1})\}$$

now has to be computed.

Theoretically we can use the following algorithm:

$$(S) \qquad \begin{aligned} S_0 &:= C_0, \\ S_i &:= S_{i-1} - A_{i-1}(B_{i-1} - D_{i-1}), \qquad 1 \leqslant i \leqslant n, \\ R_n &:= S_n/a_n. \end{aligned}$$

In practice, however, we are actually performing the following operations:

$$(\bar{S}) \qquad \begin{aligned} \bar{S}_0 &:= S_0 := C_0 \\ \bar{S}_i &:= fl(\bar{S}_{i-1} - fl(A_{i-1} fl(B_{i-1} - D_{i-1}))), \qquad 1 \leqslant i \leqslant n, \\ \bar{R}_n &:= fl(\bar{S}_n/a_n). \end{aligned}$$

We start with (15a) setting eps $:= \frac{1}{2}b^{1-t}$ and obtain for general intervals A and B that

$$(22) \qquad fl(A * B) \subseteq A * B + [-\varepsilon, \varepsilon] A * B$$

with $\max\{|\varepsilon_1|, |\varepsilon_2|\} \leqslant \varepsilon = 2\text{eps}$ holds.

Assume for the moment that

$$\bar{S}_0 = S_0 = C_0, \bar{S}_1, \dots, \bar{S}_{n-1}$$

have already been computed. Then we have from (22) that

$$\begin{aligned} & fl(B_{n-1} - D_{n-1}) \subseteq B_{n-1} - D_{n-1} + |B_{n-1} - D_{n-1}|[-\varepsilon, \varepsilon], \\ & fl(A_{n-1} fl(B_{n-1} - D_{n-1})) \\ & \quad \subseteq A_{n-1}(B_{n-1} - D_{n-1} + |B_{n-1} - D_{n-1}|[-\varepsilon, \varepsilon]) \\ & \qquad + |A_{n-1}(B_{n-1} - D_{n-1} + |B_{n-1} - D_{n-1}|[-\varepsilon, \varepsilon])|[-\varepsilon, \varepsilon] \\ & \quad \subseteq A_{n-1}(B_{n-1} - D_{n-1}) + |A_{n-1}||B_{n-1} - D_{n-1}|[-2\varepsilon - \varepsilon^2, 2\varepsilon + \varepsilon^2], \end{aligned}$$

and therefore

$$(23)\qquad \begin{aligned}\bar{S}_n \subseteq \bar{S}_{n-1} &- A_{n-1}(B_{n-1} - D_{n-1})\\ &- |A_{n-1}|\,|B_{n-1} - D_{n-1}|\,[-2\varepsilon - \varepsilon^2, 2\varepsilon + \varepsilon^2]\\ &+ |\bar{S}_{n-1} - A_{n-1}(B_{n-1} - D_{n-1})\\ &- |A_{n-1}|\,|B_{n-1} - D_{n-1}|\,[-2\varepsilon - \varepsilon^2, 2\varepsilon + \varepsilon^2]|\,[-\varepsilon, \varepsilon]\\ \subseteq \bar{S}_{n-1} &- A_{n-1}(B_{n-1} - D_{n-1}) + |\bar{S}_{n-1}|\,[-\varepsilon, \varepsilon]\\ &+ |A_{n-1}|\,|B_{n-1} - D_{n-1}|\,[-3\varepsilon - 3\varepsilon^2 - \varepsilon^3, 3\varepsilon + 3\varepsilon^2 + \varepsilon^3].\end{aligned}$$

Using mathematical induction we now show that

$$(24)\qquad \begin{aligned}\bar{S}_n \subseteq S_n + [-\varepsilon, \varepsilon] \sum_{i=0}^{n-1} |\bar{S}_i| &+ [-3\varepsilon - 3\varepsilon^2 - \varepsilon^3, 3\varepsilon + 3\varepsilon^2 + \varepsilon^3]\\ &\times \sum_{i=0}^{n-1} |A_i|\,|B_i - D_i|\end{aligned}$$

holds. For $n = 1$ we have from (23) using $\bar{S}_0 = S_0 = C_0$

$$\begin{aligned}\bar{S}_1 &\subseteq \bar{S}_0 - A_0(B_0 - D_0) + |\bar{S}_0|\,[-\varepsilon, \varepsilon]\\ &\quad + |A_0|\,|B_0 - D_0|\,[-3\varepsilon - 3\varepsilon^2 - \varepsilon^3, 3\varepsilon + 3\varepsilon^2 + \varepsilon^3]\\ &= S_1 + [-\varepsilon, \varepsilon]\,|\bar{S}_0| + [-3\varepsilon - 3\varepsilon^2 - \varepsilon^3, 3\varepsilon + 3\varepsilon^2 + \varepsilon^3]\,|A_0|\,|B_0 - D_0|,\end{aligned}$$

and therefore the assertion holds for $n = 1$. If (24) holds for some $n \geqslant 1$, then replacing n by $n + 1$ in (23) and using (S) we have

$$\begin{aligned}\bar{S}_{n+1} &\subseteq \bar{S}_n - A_n(B_n - D_n) + [-\varepsilon, \varepsilon]\,|\bar{S}_n|\\ &\qquad + [-3\varepsilon - 3\varepsilon^2 - \varepsilon^3, 3\varepsilon + 3\varepsilon^2 + \varepsilon^3]\,|A_n|\,|B_n - D_n|\\ &\subseteq S_{n+1} + [-\varepsilon, \varepsilon] \sum_{i=0}^{n} |\bar{S}_i| + [-3\varepsilon - 3\varepsilon^2 - \varepsilon^3, 3\varepsilon + 3\varepsilon^2 + \varepsilon^3]\\ &\qquad \times \sum_{i=0}^{n} |A_i|\,|B_i - D_i|,\end{aligned}$$

which is (24) with n replaced by $n + 1$. Employing (22) once more we have the final result

$$(25)\qquad \bar{R}_n \subseteq \bar{S}_n/a_n + (|\bar{S}_n|/|a_n|)\,[-\varepsilon, \varepsilon].$$

The inequalities (24) and (25) are used in Chapter 18.

Remarks: The concept of rounding employed here is discussed in detail in Miranker and Kulisch [20]. The inequality (15b), which is the starting point of the discussion on the influence of rounding errors, is found in Wallisch and

Grützmann [28]. Estimate (24) was proved in Alefeld and Rokne [1]. We shall refer to this estimate in Chapter 18. From a remark in Moore [22, p. 70] we conclude that the content of Theorem 5 was already proved in Moore [21, p. 102].

REFERENCES

[1] G. Alefeld and J. G. Rokne, On the improvement of approximate triangular factorizations. *Beitr. Numer. Math.* (to be published).

[2] N. Apostolatos and U. Kulisch, Grundlagen einer Maschinenintervallarithmetik. *Computing* **2**, 89–104 (1967).

[3] N. Apostolatos and U. Kulisch, Approximation der erweiterten Intervallarithmetik durch die einfache Maschinenintervallarithmetik. *Computing* **2**, 181–194 (1967).

[4] N. Apostolatos *et al.*, The algorithmic language Triplex-Algol 60. *Numer. Math.* **11**, 175–180 (1968).

[5] R. Boche, Some observations on the economics of interval arithmetic. *Comm. ACM* **8**, 649 (1965).

[6] H. Christ, Realisierung einer Maschinenintervallarithmetik mit beliebigen ALGOL-60 Compilern. *Elektron. Rech.* **10**, 217–222 (1968).

[7] J. K. S. Dewar, Procedures for interval arithmetic. *Comput. J.* **14**, 447–450 (1970).

[8] D. I. Good and R. L. London, Computer interval arithmetic: Definition and proof of correct implementation. *J. Assoc. Comput. Mach.* **17**, 603–612 (1970).

[9] J. Herzberger, Definition und Eigenschaften allgemeiner Intervallräume. *Z. Angew. Math. Mech.* **50**, T50–T51 (1970).

[10] J. Herzberger, Intervallmäßige Auswertung von Standardfunktionen in ALGOL-60. *Computing* **5**, 377–384 (1970).

[11] D. E. Knuth, "The Art of Computer Programming," Vol. 2. Addison-Wesley, Reading, Massachusetts, 1969.

[12] F. Krückeberg, "Numerische Intervallrechnung und deren Anwendung." Bonn, 1966.

[13] U. Kulisch, Grundzüge der Intervallrechnung. *In* "Überblicke Mathematik," Vol. 2. Bibliograph. Inst., Mannheim, 1969.

[14] U. Kulisch, An axiomatic approach to rounded computation. *Numer. Math.* **18**, 1–17 (1971).

[15] U. Kulisch, On the concept of a screen. *Z. Angew. Math. Mech.* **53**, 115–119 (1973).

[16] U. Kulisch, Implementation and formalization of floating-point arithmetics. *Caratheodory Symp., Athen* (1973).

[17] I. Kupka, Simulation reeller Arithmetik und reeller Funktionen in endlichen Mengen. *Numer. Math.* **17**, 143–152 (1971).

[18] D. P. Laurie, INTFORT-Interval Arithmetic Interpreter and Subroutine Package for the Use with FORTRAN. *Comput. Progr. Sci. Technol.* (*CPST*) **1**, (2), 41 (1971).

[19] O. Mayer, Über die in der Intervallrechnung auftretenden Räume und einige Anwendungen. Ph.D. Thesis, Univ. Karlsruhe (1968).

[20] W. L. Miranker and U. Kulisch, Computer Arithmetic in Theory and Practice. Research Rep. RC 7776 (33658), July 24, Mathematics, IBM Thomas J. Watson Research Center, Yorktown Heights (1979).

[21] R. E. Moore, Interval Arithmetic and Automatic Error Analysis in Digital Computing. Ph. Thesis, Mathematics Dept., Stanford Univ., October (1962).

[22] R. E. Moore, "Elements of Scientific Computing." Holt, New York, 1975.

[23] R. E. Moore, "Methods and Applications of Interval Analysis," SIAM Studies in Applied Mathematics. SIAM, Philadelphia, Pennsylvania, 1979.

[24] P. Rechenberg, "Grundzüge digitaler Rechenautomaten." Oldenbourg Verlag, München, 1964.

[25] P. L. Richman, Automatic error analysis for determining precision. *Comm. ACM* **15**, 813–817 (1972).

[26] S. M. Rump, Kleine Fehlerschranken bei Matrixproblemen. Ph.D. Thesis, Fakultät für Mathematik, Univ. Karlsruhe (1980).

[27] Signum Newsletter, Special Issue, October (1979).

[28] W. Wallisch and J. Grützmann, Intervallanalytische Fehleranalyse. *Beitr. Numer. Math.* **3**, 163–171 (1975).

[29] J. H. Wilkinson, "Rounding Errors in Algebraic Processes," H. M. Stationery Office, London, 1968.

[30] H. W. Wippermann, Realisierung einer Intervallarithmetik in einem ALGOL-60 System. *Elektron. Rech.* **9**, 224–233 (1967).

[31] H. W. Wippermann, Definition von Schrankenzahlen in TRIPLEX-ALGOL 60. *Computing* **3**, 99–109 (1968).

Chapter 5 / COMPLEX INTERVAL ARITHMETIC

We now wish to define and use a so-called complex interval arithmetic. It will be shown that many of the properties and results for real interval arithmetic can be carried over to a complex interval arithmetic. In order to do this we have to define the sets of complex numbers that will constitute our complex intervals. There are two reasonable choices which will now be considered here:

A. RECTANGLES AS COMPLEX INTERVALS

Definition 1: Let $A_1, A_2 \in I(\mathbb{R})$. Then the set

$$A = \{a = a_1 + ia_2 \mid a_1 \in A_1, a_2 \in A_2\} \qquad (i = \sqrt{-1})$$

of complex numbers is called a complex interval. ■

Sets of complex numbers as per Definition 1 constitute rectangles in the complex plane with sides parallel to the coordinate axes. The set of such complex intervals is denoted by $R(\mathbb{C})$ and the members of this set by capital letters $A, B, C, \ldots, X, Y, Z$. If $A \in R(\mathbb{C})$, then we may write A as $A = A_1 + iA_2$ where $A_1, A_2 \in I(\mathbb{R})$. A complex number $a = a_1 + ia_2$ may be considered to be a complex point interval

$$A = [a_1, a_1] + i[a_2, a_2] \in R(\mathbb{C}).$$

Every member $A_1 \in I(\mathbb{R})$ may be considered to be an element $A = A_1 + i[0,0] \in R(\mathbb{C})$ from which clearly $I(\mathbb{R}) \subset R(\mathbb{C})$.

Definition 2: Let $A = A_1 + iA_2$ and $B = B_1 + iB_2$ be two members of $R(\mathbb{C})$. Then A and B are equal, written $A = B$, iff

$$A_1 = B_1 \qquad \text{and} \qquad A_2 = B_2$$

(see also Definition 1.1). ■

The relation $=$ between elements from $R(\mathbb{C})$ as defined above is reflexive, symmetric, and transitive.

We now generalize the arithmetic on the complex numbers to an arithmetic on $R(\mathbb{C})$.

Definition 3: Let $* \in \{+, -, \cdot, :\}$ be a binary operation on elements from $I(\mathbb{R})$ (as in Definition 1.2). Then if

$$A = A_1 + iA_2, \qquad B = B_1 + iB_2 \in R(\mathbb{C}),$$

we define

$$A \pm B = A_1 \pm B_1 + i(A_2 \pm B_2),$$

$$A \cdot B = A_1 B_1 - A_2 B_2 + i(A_1 B_2 + A_2 B_1),$$

$$A : B = (A_1 B_1 + A_2 B_2) : (B_1^2 + B_2^2) + i(A_2 B_1 - A_1 B_2) : (B_1^2 + B_2^2). \quad ■$$

It is assumed that $0 \notin B_1^2 + B_2^2$ in the case of division. This may not be the case even if $0 \notin B_1 + iB_2$ if the powers are computed as $B_1^2 = B_1 B_1$ and $B_2^2 = B_2 B_2$ according to Definition 1.2. If this results in $0 \in B_1^2 + B_2^2$, then the division is not defined.

We consider the following example in order to illustrate this point.

Example: Let

$$B = [-1, 1] + i[1, 3].$$

Then we get

$$0 \in [0, 10] = [-1, 1] + [1, 9] = B_1 B_1 + B_2 B_2.$$

We therefore stipulate that in Definition 3 when a division is to be performed on two elements from $R(\mathbb{C})$ the expression $B_1^2 + B_2^2$ is to be calculated as

$$B_1^2 + B_2^2 = \{b_1^2 \mid b_1 \in B_1\} + \{b_2^2 \mid b_2 \in B_2\}$$

(see also Definition 1.3).

In the above example we then get

$$B_1^2 + B_2^2 = [0, 1] + [1, 9] = [1, 10].$$

We shall now take a closer look at the properties of the complex interval arithmetic introduced above.

It is immediately obvious that if $A, B \in R(\mathbb{C})$, then

$$A \pm B = \{a \pm b \mid a \in A, b \in B\}$$

holds for the addition (resp. subtraction) on $R(\mathbb{C})$. This is in general not the case for multiplication and division as shown in the following simple

Example: Let

$$A = [2,4] + i[0,0], \qquad B = [1,1] + i[1,1].$$

From Definition 3 we get

$$AB = [2,4] + i[2,4].$$

On the other hand we obtain

$$\{ab \mid a \in A, b \in B\} = \{s(1+i) \mid s \in \mathbb{R}, 2 \leqslant s \leqslant 4\} \subset AB.$$

The following theorem is, however, valid.

Theorem 4: The operations introduced in Definition 3 satisfy

$$\{a * b \mid a \in A, b \in B\} \subseteq A * B.$$

For addition and subtraction, the inclusion may be replaced by equality. For multiplication it holds that

$$AB = \inf\{X \in R(\mathbb{C}) \mid \{a \cdot b \mid a \in A, b \in B\} \subseteq X\},$$

where the infimum is taken with respect to the partial order on $R(\mathbb{C})$ defined by set theoretic inclusion.

Proof: The assertion with respect to addition and subtraction was already mentioned above. Let now $a \in A$ and $b \in B$. Using inclusion monotonicity for real intervals we have with $a = a_1 + ia_2$ and $b = b_1 + ib_2$ that

$$\begin{aligned} ab &= a_1 b_1 - a_2 b_2 + i(a_1 b_2 + a_2 b_1) \\ &\in A_1 B_1 - A_2 B_2 + i(A_1 B_2 + A_2 B_1) = AB. \end{aligned}$$

Since each variable occurs only once in the expression $a_1 b_1 - a_2 b_2$ we get that

$$\{a_1 b_1 - a_2 b_2 \mid a_k \in A_k, b_k \in B_k, k = 1,2\} = A_1 B_1 - A_2 B_2.$$

For the same reason, it follows that

$$\{a_1 b_2 + a_2 b_1 \mid a_k \in A_k, b_k \in B_k, k = 1,2\} = A_1 B_2 + A_2 B_1.$$

The last two relations show that for every real number

$$c_1 = a_1 b_1 - a_2 b_2 \in A_1 B_1 - A_2 B_2, \qquad a_k \in A_k, \qquad b_k \in B_k, \qquad k = 1,2,$$

one can find a real number

$$c_2 = a_2 b_1 + a_1 b_2 \in A_2 B_1 + A_1 B_2, \qquad a_k \in A_k, \qquad b_k \in B_k, \qquad k = 1, 2,$$

such that $c = c_1 + ic_2 \in AB$, which was to be shown. The relation

$$\{a : b \mid a \in A, b \in B\} \subseteq A : B$$

follows from inclusion monotonicity. ■

The result of Theorem 4 for multiplication does not, in general, hold for division. One may, however, obtain an "improvement" in the sense of inclusion if one defines

$$A : B = A \cdot \frac{1}{B}$$

and then calculates $1/B$ as

$$1/B = \inf\{X \in R(\mathbb{C}) \mid \{1/b \mid b \in B\} \subseteq X\}.$$

This possibility is given in Rokne and Lancaster [6] as a set of formulas requiring considerable computational effort. FORTRAN subroutines are given in Rokne and Lancaster [7].

B. CIRCULAR REGIONS AS COMPLEX INTERVALS

Definition 5: Let $a \in \mathbb{C}$ be a complex number and let $r \geqslant 0$. We call

$$Z = \{z \in \mathbb{C} \mid |z - a| \leqslant r\}$$

a circular disk, circular interval, or simply a complex interval when there is no confusion with rectangular intervals. ■

The set of circular disks is denoted by $K(\mathbb{C})$ and elements from $K(\mathbb{C})$ by capital letters $A, B, C, \ldots, X, Y, Z$. Circular disks Z with center a and radius r are also written

$$Z = \langle a, r \rangle.$$

Complex numbers may be considered to be special elements from $K(\mathbb{C})$ of the form $\langle a, 0 \rangle$. Clearly $\mathbb{C} \subset K(\mathbb{C})$.

Definition 6: Two circular disks $A = \langle a, r_1 \rangle$ and $B = \langle b, r_2 \rangle$ are called equal, that is, $A = B$, iff there is set theoretic equality between them. In this case $a = b$ and $r_1 = r_2$. ■

This equality relation is again reflexive, symmetric, and transitive.

The operations on $K(\mathbb{C})$ are introduced as generalizations of operations on real numbers in the following manner.

Definition 7: Let $* \in \{+, -, \cdot, :\}$ be a binary operation on the complex numbers. Then if $A = \langle a, r_1 \rangle$ and $B = \langle b, r_2 \rangle$ we define

$$A \pm B = \langle a \pm b, r_1 + r_2 \rangle,$$

$$A \cdot B = \langle ab, |a|r_2 + |b|r_1 + r_1 r_2 \rangle,$$

$$\frac{1}{B} = \left\langle \frac{\bar{b}}{b\bar{b} - r_2^2}, \frac{r_2}{b\bar{b} - r_2^2} \right\rangle \quad \text{for} \quad 0 \notin B,$$

$$A : B = A \cdot \frac{1}{B} \quad \text{for} \quad 0 \notin B.$$

Here $|a| = \sqrt{a_1^2 + a_2^2}$ denotes the euclidean norm of a complex number $a = a_1 + ia_2$ and $\bar{b} = b_1 - ib_2$ is the complex conjugate of $b = b_1 + ib_2$. ■

For addition and subtraction of circular disks it clearly follows that

$$A \pm B = \{a \pm b \mid a \in A, b \in B\}.$$

The same holds for the inverse of a circular disk: If we apply the theory of conformal mappings to the mapping of a circular disk not containing zero using the mapping $w = 1/z$ then we get another circular disk. That is,

$$1/B = \{1/b \mid b \in B\}.$$

Elementary calculations verify the formulas of Definition 7 for the center and radius of $\{1/b \mid b \in B\}$.

For multiplication (and therefore also for division) of two elements from $K(\mathbb{C})$ according to Definition 7, it is in general only true that

$$\{z_1 z_2 \mid z_1 \in A, z_2 \in B\} \subseteq AB.$$

This follows from the following inequalities

$$\begin{aligned} |z_1 z_2 - ab| &= |a(z_2 - b) + b(z_1 - a) + (z_1 - a)(z_2 - b)| \\ &\leqslant |a|\,|z_2 - b| + |b|\,|z_1 - a| + |z_1 - a|\,|z_2 - b| \\ &\leqslant |a|r_2 + |b|r_1 + r_1 r_2. \end{aligned}$$

An "improvement" of the multiplication of elements from $K(\mathbb{C})$ in the sense of inclusion is discussed in Krier [5]. A great deal of calculation is necessary for this improvement and it requires among other things the calculation of the roots of a third degree polynomial. The property of having the product of the centers being the center of the product is also lost.

Corresponding to Theorem 1.4 we now collect together the most important properties for the operations on $R(\mathbb{C})$, respectively, $K(\mathbb{C})$. Unless otherwise mentioned the notation $I(\mathbb{C})$ may be considered to be the set $R(\mathbb{C})$ with the operations from Definition 3 or $K(\mathbb{C})$ using Definition 7 for the operations.

Theorem 8: Let $A, B, C \in I(\mathbb{C})$. Then we have

(1) $A + B = B + A, \qquad AB = BA$ (commutativity),

(2) $(A + B) + C = A + (B + C),$

$(AB)C = A(BC)$ for $A, B, C \in K(\mathbb{C})$ (associativity),

(3) $X = [0, 0] + i[0, 0] \in R(\mathbb{C})$, resp., $X = \langle 0, 0\rangle \in K(\mathbb{C})$,

and

$X = [1, 1] + i[0, 0] \in R(\mathbb{C})$, resp., $X = \langle 1, 0\rangle \in K(\mathbb{C})$,

are the uniquely determined neutral elements with respect to addition and multiplication.

(4) $I(\mathbb{C})$ has no zero divisors.

(5) An element Z from $I(\mathbb{C})$ has an inverse element with respect to addition, respectively, multiplication, iff $Z \in \mathbb{C}$ and in case of multiplication $Z \neq 0$. It is, however, the case that $0 \in A - A$ and $1 \in A : A$.

(6) $A(B + C) \subseteq AB + AC$ (subdistributivity),

$a(B + C) = aB + aC$ for $a \in \mathbb{C}$.

Proof: The proof of these statements follows from the definitions of the operations in Definitions 3 and 7. As an example we prove (6) for $K(\mathbb{C})$. If $A = \langle a, r_1\rangle$, $B = \langle b, r_2\rangle$, $C = \langle c, r_3\rangle \in K(\mathbb{C})$, then one obtains

$$\begin{aligned} A(B + C) &= \langle a, r_1\rangle\langle b + c, r_2 + r_3\rangle \\ &= \langle a(b + c), |a|(r_2 + r_3) + |b + c|r_1 + r_1(r_2 + r_3)\rangle \\ &\subseteq \langle ab + ac, |a|r_2 + |a|r_3 + |b|r_1 + |c|r_1 + r_1r_2 + r_1r_3\rangle \\ &= \langle ab, |a|r_2 + |b|r_1 + r_1r_2\rangle + \langle ac, |a|r_3 + |c|r_1 + r_1r_3\rangle \\ &= AB + AC. \end{aligned}$$

For the case $A = \langle a, 0\rangle$, that is, $r_1 = 0$, the proof shows that

$$a(B + C) = aB + aC. \quad \blacksquare$$

We strongly point out that the associative law (2) is in general not valid for multiplication of elements from $R(\mathbb{C})$. This is shown by the following example.

Example:

$$A = [2, 4] + i[0, 0], \qquad B = [1, 1] + i[1, 1], \qquad C = [1, 1] + i[1, 1],$$

$$(AB)C = ([2, 4] + i[2, 4])([1, 1] + i[1, 1]) = [-2, 2] + i[4, 8],$$

$$A(BC) = ([2, 4] + i[0, 0])([0, 0] + i[2, 2]) = [0, 0] + i[4, 8].$$

Inclusion monotonicity is also valid in $I(\mathbb{C})$.

Theorem 9: Let $A^{(k)}, B^{(k)} \in I(\mathbb{C})$, $k = 1, 2$, be such that

$$A^{(k)} \subseteq B^{(k)}, \qquad k = 1, 2.$$

Then

$$A^{(1)} * A^{(2)} \subseteq B^{(1)} * B^{(2)}$$

holds for the operations $* \in \{+, -, \cdot, :\}$.

Proof: This is true for $R(\mathbb{C})$ since the inclusion monotonicity holds for elements from $I(\mathbb{R})$ (see Theorem 1.5).

In the case of addition and subtraction in $K(\mathbb{C})$ we have

$$\begin{aligned} A^{(1)} \pm A^{(2)} &= \{z = x \pm y \mid x \in A^{(1)}, y \in A^{(2)}\} \\ &\subseteq \{w = u \pm v \mid u \in B^{(1)}, v \in B^{(2)}\} = B^{(1)} \pm B^{(2)}. \end{aligned}$$

Consider now the multiplication in $K(\mathbb{C})$, and let

$$A^{(k)} = \langle a^{(k)}, r^{(k)} \rangle, \qquad B^{(k)} = \langle b^{(k)}, s^{(k)} \rangle, \qquad k = 1, 2.$$

The assumption $A^{(k)} \subseteq B^{(k)}$, $k = 1, 2$, is equivalent to

$$|a^{(k)} - b^{(k)}| \leqslant s^{(k)} - r^{(k)}, \qquad k = 1, 2.$$

Furthermore

$$\begin{aligned} A^{(1)}A^{(2)} &= \langle a^{(1)}a^{(2)}, |a^{(1)}|r^{(2)} + |a^{(2)}|r^{(1)} + r^{(1)}r^{(2)} \rangle, \\ B^{(1)}B^{(2)} &= \langle b^{(1)}b^{(2)}, |b^{(1)}|s^{(2)} + |b^{(2)}|s^{(1)} + s^{(1)}s^{(2)} \rangle. \end{aligned}$$

It is to be shown that

$$\begin{aligned} |a^{(1)}a^{(2)} - b^{(1)}b^{(2)}| \leqslant{} & |b^{(1)}|s^{(2)} + |b^{(2)}|s^{(1)} + s^{(1)}s^{(2)} \\ & - (|a^{(1)}|r^{(2)} + |a^{(2)}|r^{(1)} + r^{(1)}r^{(2)}). \end{aligned}$$

From the triangle inequality, we get

$$\begin{aligned} -|b^{(2)}| &\leqslant -|a^{(2)}| + |a^{(2)} - b^{(2)}|, \\ -|b^{(1)}| &\leqslant -|a^{(1)}| + |a^{(1)} - b^{(1)}|, \end{aligned}$$

and because of

$$|a^{(k)} - b^{(k)}| \leqslant s^{(k)} - r^{(k)}, \qquad k = 1, 2,$$

we get

$$\begin{aligned} -|b^{(2)}|r^{(1)} &\leqslant -|a^{(2)}|r^{(1)} + r^{(1)}(s^{(2)} - r^{(2)}) \\ &= -|a^{(2)}|r^{(1)} + r^{(1)}s^{(2)} - r^{(1)}r^{(2)}, \\ -|b^{(1)}|r^{(2)} &\leqslant -|a^{(1)}|r^{(2)} + r^{(2)}(s^{(1)} - r^{(1)}) \\ &= -|a^{(1)}|r^{(2)} + r^{(2)}s^{(1)} - r^{(1)}r^{(2)}. \end{aligned}$$

From this, one gets

$$\begin{aligned}|a^{(1)}a^{(2)} - b^{(1)}b^{(2)}| &\leqslant |b^{(2)}|\,|a^{(1)} - b^{(1)}| + |b^{(1)}|\,|a^{(2)} - b^{(2)}| \\ &\quad + |a^{(1)} - b^{(1)}|\,|a^{(2)} - b^{(2)}| \\ &\leqslant |b^{(2)}|(s^{(1)} - r^{(1)}) + |b^{(1)}|(s^{(2)} - r^{(2)}) \\ &\quad + (s^{(1)} - r^{(1)})(s^{(2)} - r^{(2)}) \\ &\leqslant |b^{(2)}|s^{(1)} + |b^{(1)}|s^{(2)} + s^{(1)}s^{(2)} \\ &\quad - (|a^{(2)}|r^{(1)} + |a^{(1)}|r^{(2)} + r^{(1)}r^{(2)}),\end{aligned}$$

which proves the assertion for the multiplication.

From

$$1/A^{(2)} = \{z = 1/x \mid x \in A^{(2)}\} \subseteq \{w = 1/u \mid u \in B^{(2)}\} = 1/B^{(2)},$$

it follows that

$$A^{(1)} : A^{(2)} = A^{(1)} \cdot \frac{1}{A^{(2)}} \subseteq B^{(1)} \cdot \frac{1}{B^{(2)}} = B^{(1)} : B^{(2)}.$$

This proves the theorem. ■

As a special case of Theorem 9 one has

Corollary 10: Let $A, B \in I(\mathbb{C})$ and let $a \in A, b \in B$. Then

$$a * b \in A * B$$

for $* \in \{+, -, \cdot, :\}$. ■

Remarks: The systematic use of circular disks for complex intervals was first considered by Gargantini and Henrici [3, 4]. The circular arithmetic considered in this section was introduced in Gargantini and Henrici [3], where it was also applied to the simultaneous inclusion of polynomial roots (see Chapter 9). Further applications of circular arithmetic are found in Chapters 12, 15, and 16. The important property of inclusion monotonicity in the form given in Theorem 9 was first proven here. As already mentioned, the paper [5] considers the problem of defining a multiplication of circular disks that results in smaller sets. The arithmetic in $R(\mathbb{C})$ that reduces to real arithmetic under certain conditions was introduced in Alefeld [1]. The properties of this arithmetic were also investigated in Boche [2] and Rokne and Lancaster [6].

As already mentioned several times, inclusion monotonicity (Theorem 9) forms the basis for almost all applications of interval arithmetic. The multiplication introduced in Krier [5] does not satisfy this property as shown in the counterexample of Krier [5, p. 76].

The approximation of the arithmetic in $R(\mathbb{C})$ on a digital computer causes no problems since the operations in $R(\mathbb{C})$ are reduced to operations in $I(\mathbb{R})$. The operations in $I(\mathbb{R})$ were shown to be approximable on a digital computer in Chapter 4 without relinquishing the most important properties of the arithmetic, which means they are also approximable in $R(\mathbb{C})$. In [B 305] there is a discussion of the possibility of approximating the arithmetic in $K(\mathbb{C})$ on a computer.

REFERENCES

[1] G. Alefeld, Intervallrechnung über den komplexen Zahlen und einige Anwendungen. Ph.D. Thesis, Univ. Karlsruhe (1968).

[2] R. Boche, Complex interval arithmetic with some applications. Lockheed Missiles and Space Company, 4-22-66-1, Sunnyvale, California (1966).

[3] J. Gargantini and P. Henrici, Circular arithmetic and the determination of polynomial zeros. *Numer. Math.* **18**, 305–320 (1972).

[4] P. Henrici, Circular arithmetic and the determination of polynomial zeros. Springer Lecture Notes 228, pp. 86–92 (1971).

[5] R. Krier, Komplexe Kreisarithmetik. Ph.D. Thesis, Univ. Karlsruhe (1973).

[6] J. Rokne and P. Lancaster, Complex interval arithmetic. *Comm. ACM* **14**, 111–112 (1971).

[7] J. Rokne and P. Lancaster, Complex interval arithmetic. Algorithm 86, *Comput. J.* **18**, 83–86 (1975).

Chapter 6 / METRIC, ABSOLUTE VALUE, AND WIDTH IN $I(\mathbb{C})$

In this chapter, q denotes the metric on $I(\mathbb{R})$ introduced in Definition 2.1. A metric is introduced in $R(\mathbb{C})$ by

Definition 1: Let $A = A_1 + iA_2, B = B_1 + iB_2 \in R(\mathbb{C})$. Then

$$p(A, B) = q(A_1, B_1) + q(A_2, B_2)$$

defines the distance between the elements A and B. ■

The restriction of p to $I(\mathbb{R})$ results in p having the same values as q from Definition 2.1. In the following we therefore denote the distance in $R(\mathbb{C})$ by q, and it follows that

$$q(A, B) = q(A_1, B_1) + q(A_2, B_2).$$

It is easy to prove that q is a metric in $R(\mathbb{C})$ using the fact that q is a metric in $I(\mathbb{R})$. The introduction of a metric q in $R(\mathbb{C})$ makes $R(\mathbb{C})$ into a topological space. If we now introduce the concept of convergence in the usual manner for a metric space, then we have that a sequence $\{A^{(k)}\}_{k=0}^{\infty}$ with $A^{(k)} = A_1^{(k)} + iA_2^{(k)}$ from $R(\mathbb{C})$ converges to an element $A = A_1 + iA_2$ from $R(\mathbb{C})$ iff

$$(1) \qquad \lim_{k\to\infty} A_1^{(k)} = A_1 \qquad \text{and} \qquad \lim_{k\to\infty} A_2^{(k)} = A_2.$$

Using the fact that the metric space $(I(\mathbb{R}), q)$ is complete, one shows that $R(\mathbb{C})$ with the metric q is also a complete metric space.

Definition 2: Let $A = A_1 + iA_2 \in R(\mathbb{C})$. Then

$$|A| = q(A,0) = |A_1| + |A_2| = q(A_1,0) + q(A_2,0)$$

is called the absolute value of A. ■

If $A = [a_1, a_1] + i[a_2, a_2] = a_1 + ia_2 = a$, in particular, then we have

(2) $$|A| = |a| = |a_1| + |a_2|.$$

The absolute value of a member $A \in R(\mathbb{C})$ therefore does not reduce to the euclidean absolute value of a complex number. In the following it will be clear from the context whether we use the euclidean absolute value or the absolute value from Definition 2. Finally we mention that using (2) the relation

$$|A| = \max_{a \in A} |a|$$

holds.

Let d denote the width of a real interval as introduced in Definition 2.8. We then have

Definition 3: Let $A = A_1 + iA_2 \in R(\mathbb{C})$. Then

$$d(A) = d(A_1) + d(A_2)$$

is called the width of A. ■

We now introduce the corresponding concepts in $K(\mathbb{C})$ by

Definition 4: Let $A = \langle a, r_1 \rangle$, $B = \langle b, r_2 \rangle \in K(\mathbb{C})$. Then

(a) $q(A,B) = |a - b| + |r_1 - r_2|$ is called the distance between A and B;
(b) $|A| = |a| + r_1$ is called the absolute value of A, and
(c) $d(A) = 2r_1$ is called the width of A. ■

The euclidean metric is used in Definition 4 to define the distance between two circular intervals in the complex plane. The absolute value of a circular interval reduces to the euclidean absolute value when restricted to complex numbers. We also note that the relation

$$|A| = \max_{a \in A} |a|$$

holds.

The completeness of $K(\mathbb{C})$ with the metric q is easily verified when the convergence of sequences from $K(\mathbb{C})$ is, as usual, defined using this metric. With this definition we get that

(3) $$\lim_{k\to\infty} A^{(k)} = A \qquad \text{iff} \qquad \lim_{k\to\infty} a^{(k)} = a, \ \lim_{k\to\infty} r^{(k)} = r,$$

holds for $\{A^{(k)}\}_{k=0}^{\infty} = \{\langle a^{(k)}, r^{(k)} \rangle\}_{k=0}^{\infty}$, $A = \langle a, r \rangle$. We now collect together the most important properties of the metric, the absolute value, and the width for the sets $R(\mathbb{C})$ and $K(\mathbb{C})$.

Theorem 5: Let $A, B, C, D \in I(\mathbb{C})$. Then the following are true:

(4) $$q(A + B, A + C) = q(B, C),$$

(5) $$q(A + B, C + D) \leqslant q(A, C) + q(B, D),$$

(6) $$q(aB, aC) \leqslant |a| q(B, C), \qquad a \in \mathbb{C}.$$

Equality always holds in (6) when $B, C \in K(\mathbb{C})$.

(7) $$q(AB, AC) \leqslant |A| q(B, C),$$

(8) $$|A| \geqslant 0, \qquad |A| = 0 \Leftrightarrow A = 0,$$

(9) $$|A + B| \leqslant |A| + |B|,$$

(10) $$|aB| \leqslant |a|\,|B|, \qquad a \in \mathbb{C}.$$

Equality always holds in (10) when $B \in K(\mathbb{C})$.

(11) $$|AB| \leqslant |A||B|,$$

(12) $$d(aB) = |a|\, d(B), \qquad a \in \mathbb{C},$$

(13) $$d(AB) \leqslant |A|\, d(B) + |B|\, d(A),$$

(14) $$d(A) = |A - A|,$$

(15) $$d(AB) \geqslant |A|\, d(B),$$

(16) $$d(A \pm B) = d(A) + d(B),$$

(17) $$A \subseteq B \Rightarrow \tfrac{1}{2}(d(B) - d(A)) \leqslant q(A, B) \leqslant d(B) - d(A).$$

Proof: We first prove these properties for $R(\mathbb{C})$. The verification of properties (4)–(7) follows in a simple manner from the corresponding properties (2.4)–(2.7) in Theorem 2.7 for real intervals. Let therefore

$$A = A_1 + iA_2, \qquad B = B_1 + iB_2,$$
$$C = C_1 + iC_2, \qquad D = D_1 + iD_2 \in R(\mathbb{C}).$$

Of (4):

$$\begin{aligned} q(A + B, A + C) &= q(A_1 + B_1 + i(A_2 + B_2), A_1 + C_1 + i(A_2 + C_2)) \\ &= q(A_1 + B_1, A_1 + C_1) + q(A_2 + B_2, A_2 + C_2) \\ &= q(B_1, C_1) + q(B_2, C_2) = q(B, C). \end{aligned}$$

Of (5):

$$q(A+B,C+D) = q(A_1+B_1,C_1+D_1) + q(A_2+B_2,C_2+D_2)$$
$$\leqslant q(A_1,C_1) + q(B_1,D_1) + q(A_2,C_2) + q(B_2,D_2)$$
$$= q(A,C) + q(B,D).$$

Of (6), (7):

$$q(AB,AC) = q(A_1B_1 - A_2B_2, A_1C_1 - A_2C_2)$$
$$+ q(A_1B_2 + A_2B_1, A_1C_2 + A_2C_1)$$
$$\leqslant |A_1|q(B_1,C_1) + |A_2|q(B_2,C_2) + |A_1|q(B_2,C_2) + |A_2|q(B_1,C_1)$$
$$= (|A_1| + |A_2|)q(B,C) = |A|q(B,C).$$

The results (8)–(11) are proved using the definition of $|A|$.

Of (8):

$$|A| = q(A,0) = q(A_1,0) + q(A_2,0) = |A_1| + |A_2| \geqslant 0,$$
$$|A| = 0 \Leftrightarrow |A_1| = |A_2| = 0 \Leftrightarrow A = 0.$$

Of (9):

$$|A+B| = q(A+B,0) \leqslant q(A,0) + q(B,0) = |A| + |B|$$

[according to (5)].

Of (10), (11):

$$|AB| = q(AB,0) = q(AB, A\cdot 0) \leqslant |A|q(B,0) = |A|\,|B|$$

[from (6), (7)].

Of (12): Let $a = a_1 + ia_2 \in \mathbb{C}$. From Definition 5.3 we get

$$aB = a_1B_1 - a_2B_2 + i(a_1B_2 + a_2B_1)$$

and using (2) we get

$$d(aB) = d(a_1B_1 - a_2B_2) + d(a_1B_2 + a_2B_1)$$
$$= d(a_1B_1) + d(a_2B_2) + d(a_1B_2) + d(a_2B_1)$$
$$= |a_1|\,d(B_1) + |a_2|\,d(B_2) + |a_1|\,d(B_2) + |a_2|\,d(B_1)$$
$$= (|a_1| + |a_2|)(d(B_1) + d(B_2)) = |a|\,d(B).$$

Of (13):

$$d(AB) = d(A_1B_1 - A_2B_2) + d(A_1B_2 + A_2B_1)$$
$$= d(A_1B_1) + d(A_2B_2) + d(A_1B_2) + d(A_2B_1)$$
$$\leqslant |A_1|\,d(B_1) + |B_1|\,d(A_1) + |A_2|\,d(B_2) + |B_2|\,d(A_2)$$
$$+ |A_1|\,d(B_2) + |B_2|\,d(A_1) + |A_2|\,d(B_1) + |B_1|\,d(A_2)$$
$$= (|A_1| + |A_2|)(d(B_1) + d(B_2)) + (|B_1| + |B_2|)(d(A_1) + d(A_2))$$
$$= |A|\,d(B) + |B|\,d(A).$$

Of (14):

$$d(A) = d(A_1) + d(A_2) = |A_1 - A_1| + |A_2 - A_2| = |A - A|.$$

Of (15):

$$\begin{aligned} d(AB) &= d(A_1B_1 - A_2B_2) + d(A_1B_2 + A_2B_1) \\ &\geq |A_1|\,d(B_1) + |A_2|\,d(B_2) + |A_1|\,d(B_2) + |A_2|\,d(B_1) \\ &= (|A_1| + |A_2|)(d(B_1) + d(B_2)) = |A|\,d(B). \end{aligned}$$

Of (16):

$$\begin{aligned} d(A \pm B) &= d(A_1 \pm B_1) + d(A_2 \pm B_2) \\ &= d(A_1) + d(A_2) + d(B_1) + d(B_2) \\ &= d(A) + d(B). \end{aligned}$$

Of (17): This result is a direct consequence of (2.21).
Let now

$$A = \langle a, r_1 \rangle, \qquad B = \langle b, r_2 \rangle,$$
$$C = \langle c, r_3 \rangle, \qquad D = \langle d, r_4 \rangle \in K(\mathbb{C}).$$

Of (4):

$$\begin{aligned} q(A + B, A + C) &= |a + b - (a + c)| + |r_1 + r_2 - (r_1 + r_3)| \\ &= |b - c| + |r_2 - r_3| = q(B, C). \end{aligned}$$

Of (5):

$$\begin{aligned} q(A + B, C + D) &= |a + b - (c + d)| + |r_1 + r_2 - (r_3 + r_4)| \\ &\leq |a - c| + |r_1 - r_3| + |b - d| + |r_2 - r_4| \\ &= q(A, C) + q(B, D). \end{aligned}$$

Of (6):

$$\begin{aligned} q(aB, aC) &= |ab - ac| + \big| |a|r_2 - |a|r_3 \big| \\ &= |a|\{|b - c| + |r_2 - r_3|\} = |a|q(B, C). \end{aligned}$$

Of (7):

$$\begin{aligned} q(AB, AC) &= |ab - ac| + \big| |a|r_2 + |b|r_1 + r_1r_2 - (|a|r_3 + |c|r_1 + r_1r_3) \big| \\ &\leq |a|\,|b - c| + |a|\,|r_2 - r_3| + r_1\big| |b| - |c| \big| + r_1|r_2 - r_3| \\ &\leq (|a| + r_1)(|b - c| + |r_2 - r_3|) = |A|q(B, C). \end{aligned}$$

Of (8):

$$|A| = |a| + r_1 \geq 0, \qquad |A| = 0 \Leftrightarrow (a = 0,\ r_1 = 0).$$

Of (9):

$$|A + B| = |a + b| + |r_1 + r_2| \leqslant |a| + r_1 + |b| + r_2 = |A| + |B|.$$

Of (10):

$$|aB| = |ab| + |a|r_2 = |a|\,|B|.$$

Of (11):

$$|AB| = q(AB, 0) = q(AB, A \cdot 0) \leqslant |A|q(B, 0) = |A|\,|B|$$

[from (7)].

Of (12):

$$d(aB) = 2|a|r_2 = |a|\,d(B).$$

Of (13):

$$\begin{aligned} d(AB) &= 2\{|a|r_2 + |b|r_1 + r_1 r_2\} \\ &= 2\{(|a| + r_1)r_2 + |b|r_1\} \\ &\leqslant 2\{(|a| + r_1)r_2 + (|b| + r_2)r_1\} \\ &= |A|\,d(B) + |B|\,d(A). \end{aligned}$$

Of (14):

$$d(A) = 2r_1 = |\langle 0, 2r_1\rangle| = |A - A|.$$

Of (15):

$$\begin{aligned} d(AB) &= 2\{|a|r_2 + |b|r_1 + r_1 r_2\} \\ &= 2\{(|a| + r_1)r_2 + |b|r_1\} \\ &\geqslant 2(|a| + r_1)r_2 = |A|\,d(B). \end{aligned}$$

Of (16):

$$d(A \pm B) = d(\langle a \pm b, r_1 + r_2\rangle) = 2(r_1 + r_2) = d(A) + d(B).$$

Of (17): $A \subseteq B$ iff $|a - b| \leqslant r_2 - r_1$. Therefore

$$\begin{aligned} \tfrac{1}{2}(d(B) - d(A)) &= |r_2| - |r_1| \leqslant |r_2 - r_1| \leqslant |a - b| + |r_1 - r_2| \\ &= q(A, B) \leqslant r_2 - r_1 + |r_2 - r_1| = d(B) - d(A). \quad \blacksquare \end{aligned}$$

Corresponding to Theorem 2.4 we have

Theorem 6: The operations $\{+, -, \cdot, :\}$ on $R(\mathbb{C})$ according to Definition 5.3 and on $K(\mathbb{C})$ according to Definition 5.7 are continuous mappings.

Proof: Let $\{A^{(k)}\}_{k=0}^{\infty}$, $\{B^{(k)}\}_{k=0}^{\infty}$ be sequences for which

$$A^{(k)} = A_1^{(k)} + iA_2^{(k)}, \qquad B^{(k)} = B_1^{(k)} + iB_2^{(k)} \in R(\mathbb{C})$$

and

$$\lim_{k\to\infty} A^{(k)} = A = A_1 + iA_2, \qquad \lim_{k\to\infty} B^{(k)} = B = B_1 + iB_2.$$

We prove that multiplication is a continuous operation. We therefore calculate that

$$\begin{aligned}\lim_{k\to\infty} A^{(k)}B^{(k)} &= \lim_{k\to\infty}\{A_1^{(k)}B_1^{(k)} - A_2^{(k)}B_2^{(k)} + i(A_1^{(k)}B_2^{(k)} + A_2^{(k)}B_1^{(k)})\}\\ &= \lim_{k\to\infty}(A_1^{(k)}B_1^{(k)} - A_2^{(k)}B_2^{(k)}) + i\lim_{k\to\infty}(A_1^{(k)}B_2^{(k)} + A_2^{(k)}B_1^{(k)})\\ &= A_1B_1 - A_2B_2 + i(A_1B_2 + A_2B_1) = AB,\end{aligned}$$

since the operations of taking real and imaginary parts of a complex number are continuous operations in $I(\mathbb{R})$.

Similar proofs may be supplied for the other operations in $R(\mathbb{C})$ as well as for the operations in $K(\mathbb{C})$. ∎

Analogous to (2.22) we now introduce another binary operation in $R(\mathbb{C})$. For two intervals $A, B \in R(\mathbb{C})$ we call the set theoretic intersection

(18) $$A \cap B = \{c \mid c \in A, c \in B\}$$

the intersection of A and B. The intersection of two elements A and B is in $R(\mathbb{C})$ if the set theoretic intersection is not empty. If $A = A_1 + iA_2$ and $B = B_1 + iB_2$, then we have that

(19) $$A \cap B = A_1 \cap B_1 + i(A_2 \cap B_2),$$

where $A_i \cap B_i$ must be formed according to (2.23).

Corresponding to Corollary 2.12 we get

Corollary 7: Let $A, B, C, D \in R(\mathbb{C})$. Then we have

(20) $A \subseteq C,\ B \subseteq D \Rightarrow A \cap B \subseteq C \cap D$ (inclusion monotonicity),

and the intersection operation is a continuous operation if the result is in $R(\mathbb{C})$. ∎

This corollary may be proved using Corollary 2.12 applied to the real and the imaginary parts.

Remarks: Most of the properties given in this section for $R(\mathbb{C})$ are found in Alefeld [1]. The corresponding properties for the metric in $K(\mathbb{C})$ are collected here for the first time even though they were partly used implicitly in Gargantini and Henrici [2]. These properties will be used in particular in Chapters 11–17.

REFERENCES

[1] G. Alefeld, Intervallrechnung über den komplexen Zahlen und einige Anwendungen. Ph.D. Thesis, Univ. Karlsruhe (1968).

[2] I. Gargantini and P. Henrici, Circular arithmetic and the determination of polynomial zeros. *Numer. Math.* **18**, 305–320 (1972).

Chapter 7 / INCLUSION OF ZEROS OF A FUNCTION OF ONE REAL VARIABLE

In this chapter we shall consider methods for including zeros of a real function f of one real variable x. The methods will allow us to find a set of intervals of smallest possible width such that each interval includes one or more zeros of f from a given interval $X^{(0)} \in I(\mathbb{R})$. The case of one isolated zero in $X^{(0)}$ is of particular interest. In developing such methods we consider two particular points. The methods should on the one hand be applicable to large classes of functions under easily verifiable conditions. On the other hand, the inclusions for the zeros should also be guaranteed when the methods are implemented on a computer where instead of the regular interval arithmetic we have a machine interval arithmetic as given in Chapter 4. These methods therefore differ fundamentally from the methods given for particular classes of functions (Schmidt [42]) and from other general procedures given in Brent [11] and Chien [13].

Simple realizations of such methods are given by the so-called subdivision methods. These are interval versions of the binary search method or of other search methods. We shall give a short explanation of such a procedure. It is required only that there exists an interval evaluation for the function f (see Chapter 3) in the interval $X^{(0)}$. In order to improve the inclusion of all the zeros in $X^{(0)}$ we subdivide $X^{(0)}$ at the point

$$m(X^{(0)}) = \tfrac{1}{2}(x_1^{(0)} + x_2^{(0)})$$

into $U^{(0)}$ and $V^{(0)}$ such that

$$X^{(0)} = U^{(0)} \cup V^{(0)} = [x_1^{(0)}, m(X^{(0)})] \cup [m(X^{(0)}), x_2^{(0)}].$$

If $0 \in f(U^{(0)})$, then it is possible that $U^{(0)}$ contains a zero of f, and we therefore repeat the procedure on $U^{(0)}$. If $0 \in f(V^{(0)})$, we similarly repeat the procedure on $V^{(0)}$. On the other hand if we have $0 \notin f(U^{(0)})$ or $0 \notin f(V^{(0)})$, then we disregard the respective subinterval since it cannot contain a zero of f because of (3.1). Such a subinterval is therefore omitted from the further calculations. This iteration therefore generates a sequence of subintervals of $X^{(0)}$ that are "suspected of having a zero of f." The widths of these intervals tend to zero since the widths are halved at each step. Because of (3.5) we have that these stepwise calculated intervals will necessarily converge to the zeros of f in $X^{(0)}$. See Apostolatos *et al.* [7] and Moore [36].

In order to prevent the number of "suspect" intervals from increasing too much one can introduce the following modification. At each step one investigates either only the right or the left half of the interval. If at any step we have $0 \notin f(Y)$ for this half interval Y, then the procedure is restarted with the interval $[x_1^{(0)}, y_1] \subset X^{(0)}$ (resp., $[y_2, x_2^{(0)}] \subset X^{(0)}$). In this manner we calculate the individual zeros of f in the order right to left (resp., left to right) in a sequence. In this manner one avoids the problem of storing large numbers of "suspect" intervals [4, 20].

A. NEWTON-LIKE METHODS

In this and the following sections we investigate interval modifications of Newton's method. For this purpose we consider a continuous function f that has a zero in a given interval $X^{(0)} = [x_1^{(0)}, x_2^{(0)}]$; that is,

$$f(\xi) = 0$$

for some $\xi \in X^{(0)}$. Let

$$(1) \qquad f(x_1^{(0)}) < 0 \qquad \text{and} \qquad f(x_2^{(0)}) > 0$$

at the boundary points of $X^{(0)}$. Further let m_1, m_2 be bounds for the divided differences

$$(2) \qquad 0 < m_1 \leqslant \frac{f(x) - f(\xi)}{x - \xi} = \frac{f(x)}{x - \xi} \leqslant m_2 < \infty, \qquad \xi \neq x \in X^{(0)}.$$

These bounds are combined into the interval $M = [m_1, m_2] \in I(\mathbb{R})$. (A similar representation is also possible if we assume $f(x_1^{(0)}) > 0$ and $f(x_2^{(0)}) < 0$ as well as $m_2 < 0$.) Under the above assumptions it is clear that f has no other zeros in $X^{(0)}$.

Starting from the initial including interval $X^{(0)} \ni \xi$ we now calculate new intervals $X^{(k)}$, $k \geqslant 1$, iteratively according to the following procedure:

$$(3) \quad \begin{cases} X^{(k+1)} = \{m(X^{(k)}) - f(m(X^{(k)}))/M\} \cap X^{(k)}, & k \geqslant 0, \\ & \text{with} \quad m(X^{(k)}) \in X^{(k)}. \end{cases}$$

Figure 1 clarifies the first step.

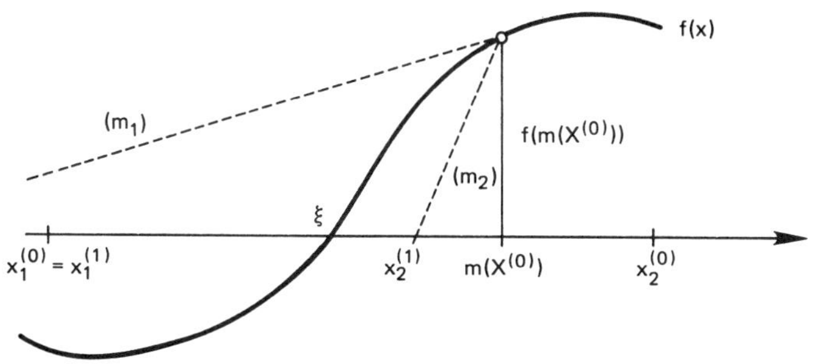

Fig. 1

The iteration (3) may be written without using interval operations as

$$(3') \quad \begin{cases} x_1^{(k+1)} = \begin{cases} \max\{x_1^{(k)}, m(X^{(k)}) - f(m(X^{(k)}))/m_1\} & \text{if } f(m(X^{(k)})) \geqslant 0 \\ m(X^{(k)}) - f(m(X^{(k)}))/m_2 & \text{if } f(m(X^{(k)})) \leqslant 0 \end{cases} \\ x_2^{(k+1)} = \begin{cases} m(X^{(k)}) - f(m(X^{(k)}))/m_2 & \text{if } f(m(X^{(k)})) \geqslant 0 \\ \min\{x_2^{(k)}, m(X^{(k)}) - f(m(X^{(k)}))/m_1\} & \text{if } f(m(X^{(k)})) \leqslant 0. \end{cases} \end{cases}$$

In both the formulation (3) as well as in (3′) the notation

$$m\colon I(\mathbb{R}) \ni X \to m(X) \in \mathbb{R}$$

implies a selection procedure for a real number m from an interval. The midpoint

$$(4) \qquad m(X) = \tfrac{1}{2}(x_1 + x_2)$$

is often used.

We now wish to list the most important properties of the sequence $\{X^{(k)}\}_{k=0}^{\infty}$ of iterates.

Theorem 1: Let f be a continuous function and ξ a zero of f in the interval $X^{(0)}$. (1) holds, and (2) is valid for the interval $M = [m_1, m_2]$, $m_1 > 0$. The sequence $\{X^{(k)}\}_{k=0}^{\infty}$ calculated according to (3) then has the following properties:

$$(5) \qquad \xi \in X^{(k)}, \qquad k \geqslant 0,$$

$$(6) \qquad X^{(0)} \supset X^{(1)} \supset X^{(2)} \supset \cdots \qquad \text{with} \quad \lim_{k \to \infty} X^{(k)} = \xi,$$

or the sequence comes to rest after a finite number of steps at the point $[\xi, \xi]$,

$$(7) \qquad d(X^{(k+1)}) \leqslant (1 - m_1/m_2)\, d(X^{(k)}).$$

Proof: Of (5): From (2) and Corollary 1.6 we get

$$\xi = m(X^{(0)}) - \frac{f(m(X^{(0)}))}{f(m(X^{(0)}))/(m(x^{(0)}) - \xi)}$$

$$\in \{m(X^{(0)}) - f(m(X^{(0)}))/M\} \cap X^{(0)} = X^{(1)}.$$

For $k > 1$ the proof follows using complete induction.

Of (6), (7): We assume that $f(m(X^{(k)})) > 0$. If now $f(m(X^{(k)})) \geqslant (m(X^{(k)}) - x_1^{(k)})m_1$ is valid, then using (3′) we get

$$\begin{aligned} d(X^{(k+1)}) &= x_2^{(k+1)} - x_1^{(k+1)} = m(X^{(k)}) - f(m(X^{(k)}))/m_2 - x_1^{(k)} \\ &\leqslant (m(X^{(k)}) - x_1^{(k)}) - (m(X^{(k)}) - x_1^{(k)})m_1/m_2 \\ &= (m(X^{(k)}) - x_1^{(k)})(1 - m_1/m_2) \leqslant d(X^{(k)})(1 - m_1/m_2). \end{aligned}$$

If now $f(m(X^{(k)})) \leqslant (m(X^{(k)}) - x_1^{(k)})m_1$, then using (3′) we get

$$\begin{aligned} d(X^{(k+1)}) &= x_2^{(k+1)} - x_1^{(k+1)} \\ &= m(X^{(k)}) - f(m(X^{(k)}))/m_2 - m(X^{(k)}) + f(m(X^{(k)}))/m_1 \\ &= \frac{f(m(X^{(k)}))}{m_1}\left(1 - \frac{m_1}{m_2}\right) \leqslant (m(X^{(k)}) - x_1^{(k)})\left(1 - \frac{m_1}{m_2}\right) \\ &\leqslant d(X^{(k)})(1 - m_1/m_2). \end{aligned}$$

The case $f(m(X^{(k)})) < 0$ is proved in a similar manner. If $f(m(X^{(k)})) = 0$, however, then $m(X^{(k)}) = \xi$ and therefore $d(X^{(k+1)}) = 0$ as well as $X^{(k+i)} = \xi$, $i \geqslant 1$. This proves (7). Because of $m_1 \leqslant m_2$ we get

$$d(X^{(k+1)}) \leqslant \gamma^{k+1}\, d(X^{(0)}), \qquad 0 \leqslant \gamma = (1 - m_1/m_2) < 1,$$

which implies

$$\lim_{k\to\infty} d(X^{(k+1)}) = 0.$$

Since (5) implies $\xi \in X^{(k)}$, $k \geqslant 0$, it follows that $\lim_{k\to\infty} X^{(k)} = \xi$, unless $X^{(k_0+i)} = \xi$, $i \geqslant 1$, already for some k_0. The first part of (6) is an immediate consequence of the procedure (3). ■

Therefore Theorem 1 guarantees that under the given assumptions the iterates $X^{(k)}$, $k \geqslant 0$, converge to the zero ξ of f. Each of the iterated intervals then includes the desired zero. If, on the other hand, one uses (3) on an interval $X^{(0)}$ where $\xi \notin X^{(0)}$, then there exists an index k_0 for which the intersection (3) is empty. This may be shown through a simple proof by contradiction using (7) and the assumption that the intersection is nonempty. An iteration of the type (3) is found in Sunaga [45] and in particular in Moore [36]. This iteration method was investigated in depth by Chernous'ko [12] in a general setting. It is also connected to fixed point iterations by monotone functions (see, for example, Haaser and Sullivan [23, Chapter 5, Section 2]). Similar procedures for polynomials were used by Bauer [10] already in 1917.

We now consider two modifications of (3) that arise from particular choices of the point m. We first fix the choice of m and get

Corollary 2: The notations and assumptions are as in Theorem 1. In addition the choice

$$m(X^{(k)}) = \tfrac{1}{2}(x_1^{(k)} + x_2^{(k)}), \qquad k \geqslant 0,$$

is made. Then the inequality

$$d(X^{(k+1)}) \leqslant \tfrac{1}{2}(1 - m_1/m_2)\, d(X^{(k)}), \tag{8}$$

valid for the sequence $\{X^{(k)}\}_{k=0}^{\infty}$ of iterates, is an improvement on (7).

Proof: In the proof of (7) in Theorem 1 we have from the particular choice of $m(X^{(k)})$ that

$$m(X^{(k)}) - x_1^{(k)} = \tfrac{1}{2}\, d(X^{(k)}).$$

From this we get (8). ■

If the midpoint is chosen for $m(X^{(k)})$, then it is guaranteed that in every iteration step the width of the inclusion is at least halved. The method considered by Moore [36] uses this choice of m.

Other possibilities for the choice of $m(X^{(k)})$ are also considered. Nickel [37] for example uses

$$m(X^{(k)}) = m(X^{(k-1)}) - f(m(X^{(k-1)}))/m_0 \qquad \text{with} \quad m_0 \in M,$$

respectively,

$$m(X^{(k)}) \in \{x_1^{(k)}, x_2^{(k)}\} \qquad \text{if} \quad m(X^{(k)}) \notin X^{(k)}, \quad k \geqslant 1.$$

The interval M bounding the divided difference (3) is required both in Theorem 1 as well as Corollary 2. It may be chosen as

$$M = \left[\inf_{y \in X^{(0)}} f'(y),\ \sup_{y \in X^{(0)}} f'(y)\right]$$

using the mean-value theorem if the function f is continuously differentiable (if $f'(x) \neq 0$ for $x \in X^{(0)}$). In general one is only able to estimate a superset of this interval, for example, through the interval evaluation of f'; that is, one evaluates

$$M = f'(X^{(0)}).$$

The condition $m_1 > 0$ can then be guaranteed if necessary using an a priori estimation of $\inf_{y \in X^{(0)}} f'(y)$.

B. DETERMINATION OF AN OPTIMAL METHOD

The iteration (3) considered in Section A has a certain degree of freedom in choosing $m(X^{(k)}) \in X^{(k)}$. Depending on the choice of $m(X^{(k)})$ from $X^{(k)}$ we get different sequences of including intervals $\{X^{(k)}\}_{k=0}^{\infty}$. These sequences can in general not be compared element by element with respect to inclusion. Clearly one must therefore aim for methods for choosing $m(X^{(k)}) \in X^{(k)}$ that generate sequences $\{X^{(k)}\}_{k=0}^{\infty}$ where the width of the individual elements is as small as possible. We wish to define this more clearly and we therefore denote by $\phi[X]$ the class of all functions f with the properties:

(α) $f(x_1) < 0$ and $f(x_2) > 0$ for $X = [x_1, x_2]$,

(β) for the interval $M = [m_1, m_2]$ with $m_1 > 0$ it holds that $m_1 \leqslant (f(x) - f(y))/(x - y) \leqslant m_2$ for $x \neq y, x, y \in X$.

Clearly every function $f \in \phi[X]$ has one and only one zero ξ in the interval X. All assumptions required for the iteration (3) are furthermore satisfied and all the statements of Theorem 1 are valid.

In order to determine a suitable procedure $m(X^{(k)}) \in X^{(k)}$ we proceed in a stepwise manner as Chernous'ko [12]. The sequence of iterates according to (3) is denoted by $\{X^{(k)}\}_{k=0}^{\infty}$. We need the quantities $m(X^{(k)})$ and $f(m(X^{(k)}))$ in order to calculate the new iterate $X^{(k+1)}$. If we now fix $m(X^{(k)}) = x \in X^{(k)}$ in $X^{(k)}$, then $X^{(k+1)}$ is only dependent on $f(m(X^{(k)}))$. This function value can vary, however, only between $y_1^{(k)}$ and $y_2^{(k)}$ since $f \in \phi[X]$ and since $f(m(X^{(i)}))$, $0 \leqslant i < k$, are fixed. This allows us to determine the largest possible width

$$\max\{d(X^{(k+1)}) \mid m(X^{(k)}) = x, y_1^{(k)} \leqslant f(m(X^{(k)})) \leqslant y_2^{(k)}\}.$$

This is the "worst" case that may happen for a function $f \in \phi[X]$.

We shall now determine $\tilde{x} = m(X^{(k)}) \in X^{(k)}$ in such a manner that this largest width is minimized. That is, the value

$$\min_{x \in X^{(k)}} \{\max d(X^{(k+1)}) \mid m(X^{(k)}) = x, y_1^{(k)} \leqslant f(m(X^{(k)})) \leqslant y_2^{(k)}\}$$

is calculated and the corresponding value $\tilde{x}$ is chosen as the value of $m(X^{(k)})$. The determination of $m(X^{(k)})$ is therefore made by minimizing the "worst" case.

A detailed description of the above procedure is now given. Without restricting the generality of the development we consider the case when $f(m(X^{(k-1)})) > 0$. The cross-hatched area in Fig. 2 shows the range of the function values $f(m(X^{(k)}))$ if it is assumed that $f \in \phi[X]$ with $f(m(X^{(k-1)})) > 0$.

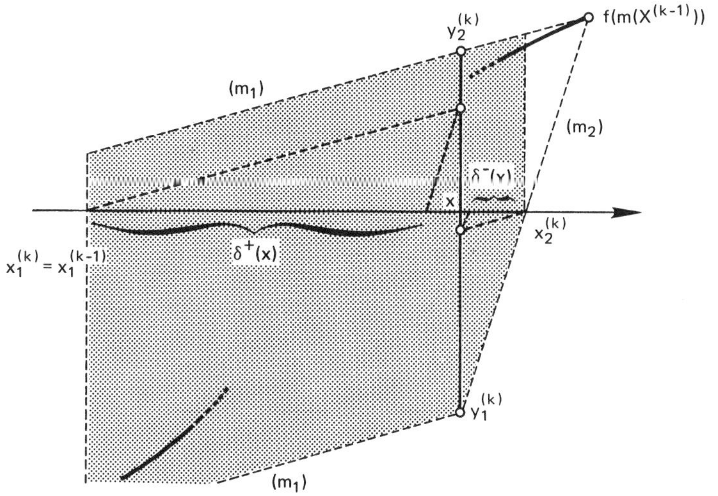

Fig. 2

The lower limiting line (m_1) may disappear in Fig. 2 if, for example, $f(m(X^{(i)})) > 0$, $0 \leqslant i < k - 1$. The other values $f(m(X^{(i)}))$, $0 \leqslant i < k - 1$, do not place any additional restrictions on this range.

The possible values of $d(X^{(k+1)})$ will now be evaluated when $m(X^{(k)}) = x \in X^{(k)}$ is determined. First, let $f(m(X^{(k)})) \geqslant 0$. For all the values

$$0 \leqslant f(x) \leqslant (x - x_1^{(k)})m_1,$$

we get from (3′) that

$$d(X^{(k+1)}) = x - \frac{f(x)}{m_2} - x + \frac{f(x)}{m_1} = f(x)\left(\frac{1}{m_1} - \frac{1}{m_2}\right).$$

Similarly for all the values

$$(x - x_1^{(k)})m_1 \leqslant f(x) \leqslant y_2^{(k)},$$

(3′) implies that

$$d(X^{(k+1)}) = x - f(x)/m_2 - x_1^{(k)}.$$

(Note that since $x_1^{(k)} = \max\{x^{(k-1)}, m(X^{(k-1)}) - f(m(X^{(k-1)}))/m_1\}$, one always has $y_2^{(k)} \geqslant (x - x_1^{(k)})m_1$.) In the first case $d(X^{(k+1)})$ is a monotonically increasing and in the second case a monotonically decreasing function of $f(x)$. For $f(x) = (x - x_1^{(k)})m_1$ we have the maximum

$$\delta^+(x) = (x - x_1^{(k)})(1 - m_1/m_2).$$

The remaining cases $f(m(X^{(k)})) \leqslant 0$ are handled analogously giving a maximum value of $d(X^{(k+1)})$ as

$$\delta^-(x) = (x_2^{(k)} - x)(1 - m_1/m_2).$$

Figure 2 shows the two possibilities for the evaluation of $X^{(k+1)}$, which lead to the maximal width $\delta^+(x)$ (resp., $\delta^-(x)$).

We may now determine the minimum

$$\min_{x \in X^{(k)}} \max\{\delta^+(x), \delta^-(x)\}.$$

Both expressions $\delta^+(x)$ and $\delta^-(x)$ satisfy the requirement

$$\delta^+(\tfrac{1}{2}(x_1^{(k)} + x_2^{(k)}) - t) = \delta^-(\tfrac{1}{2}(x_1^{(k)} + x_2^{(k)}) + t)$$

for $0 \leqslant |t| \leqslant \frac{1}{2}(x_2^{(k)} - x_1^{(k)})$. The minimum therefore is

$$d(X^{(k+1)}) = \tfrac{1}{2}d(X^{(k)})(1 - m_1/m_2)$$

at the point

$$\tilde{x} = \tfrac{1}{2}(x_1^{(k)} + x_2^{(k)}).$$

(Compare with Corollary 2.)

We now wish to extend our optimization principle for the calculation of $X^{(k+1)}$ to the determination of the values $m(X^{(i)})$, $0 \leqslant i < k$. We aim at determining the values $m(X^{(0)}) = x^{(0)}, \ldots, m(X^{(k)}) = x^{(k)}$ in such a manner that the values

$$\min_{x^{(0)} \in X^{(0)}} \quad \max_{y_1^{(0)} \leqslant f(x^{(0)}) \leqslant y_2^{(0)}} \cdots \min_{x^{(k)} \in X^{(k)}} \quad \max_{y_1^{(k)} \leqslant f(x^{(k)}) \leqslant y_2^{(k)}} \quad d(X^{(k+1)})$$

are found. This turns out to be simple since the optimal value of $d(X^{(k+1)})$ is proportional to $d(X^{(k)})$ for fixed $m(X^{(k-1)})$. The allowable range of function values of $f(m(X^{(k-1)}))$ is determined solely from the value $f(m(X^{(k-2)}))$. One may therefore carry through the same arguments for $m(X^{(k-1)})$ as for $m(X^{(k)})$ obtaining the optimal value

$$m(X^{(k-1)}) = \tfrac{1}{2}(x_1^{(k-1)} + x_2^{(k-1)}).$$

Correspondingly one gets the results for

$$m(X^{(i)}), \qquad i = k-2, k-3, \ldots, 0$$

in that order.

Theorem 3: The iteration (3) is applied to functions $f \in \phi[X]$. If one uses the rule

$$m(X^{(k)}) = \tfrac{1}{2}(x_1^{(k)} + x_2^{(k)}), \qquad 0 \leqslant k \leqslant i, \quad i \geqslant 0,$$

then the maximal width $d(X^{(i+1)})$ for the functions $f \in \phi[X]$ is smaller than for all other choices of $m(X^{(k)})$. If $f \in \phi[X]$, then we have

$$d(X^{(i+1)}) \leqslant \frac{1}{2^{i+1}}(1 - m_1/m_2)^{i+1}\, d(X^{(0)}).$$

Furthermore, there exists a $g \in \phi[X]$ for which equality holds in the last relation. ■

This theorem was proved in the above discussion. It should be pointed out that the function $g \in \phi[X]$ may be chosen as a piecewise linear function through the points $(m(X^{(k)}), f(m(X^{(k)})))$, $0 \leqslant k \leqslant i$.

C. QUADRATICALLY CONVERGENT METHODS

In order to use the method (3) we needed a fixed pair m_1, m_2 of bounds for the divided differences of f. This procedure corresponds to an interval version of the simplified Newton iteration. If we assume that f is continuously differentiable and that the derivative f' has an interval evaluation $f'(X)$ (see Chapter 3), then we may also define an interval version of the usual Newton iteration. The new procedure is arrived at by modifying the iteration (3) by evaluating the interval M in each iteration step as

$$M^{(k)} = f'(X^{(k)}). \tag{9}$$

If one knows a priori bounds

$$0 < l_1 \leqslant f'(x) \leqslant l_2, \qquad x \in X^{(0)},$$

then one may guarantee that $m_1 > 0$ and one may use the expression

$$M^{(k)} = [m_1^{(k)}, m_2^{(k)}] = f'(X^{(k)}) \cap L, \qquad L = [l_1, l_2]. \tag{10}$$

We therefore get

$$\left\{ \begin{aligned} X^{(k+1)} &= \{m(X^{(k)}) - f(m(X^{(k)}))/M^{(k)}\} \cap X^{(k)}, \qquad k \geqslant 0, \\ &\qquad\qquad\qquad\qquad \text{with} \quad m(X^{(k)}) \in X^{(k)}. \end{aligned} \right. \tag{11}$$

When (11) is used a sequence of intervals $\{X^{(k)}\}_{k=0}^{\infty}$ is generated for which we may prove similar statements as in Theorem 1.

Theorem 4: Let f be a continuously differentiable function, and let f' satisfy the assumptions of Theorem 3.5 in the interval $X^{(0)}$. Furthermore, let the relation (1) hold in $X^{(0)}$. The zero of f in $X^{(0)}$ is denoted by ξ, and the intervals $M^{(k)}$ are defined by expression (9) or (10). The sequence $\{X^{(k)}\}_{k=0}^{\infty}$ then satisfies

$$\xi \in X^{(k)}, \qquad k \geqslant 0, \tag{5}$$

$$X^{(0)} \supset X^{(1)} \supset X^{(2)} \supset \cdots \qquad \text{and} \qquad \lim_{k \to \infty} X^{(k)} = \xi, \tag{6}$$

or the sequence comes to rest at $[\xi, \xi]$ after a finite number of steps.

(12) $$d(X^{(k+1)}) \leqslant (1 - m_1^{(k)}/m_2^{(k)})\, d(X^{(k)}) \leqslant \beta\,(d(X^{(k)}))^2, \qquad \beta \geqslant 0,$$

that is the R order of the iteration (11) (see Appendix A) satisfies

$$O_R((11), \xi) \geqslant 2.$$

Proof: For $x \in X^{(k)}$ it holds that

$$\frac{f(x)}{x-\xi} = \frac{f(x) - f(\xi)}{x - \xi} = f'(\eta) \in M^{(k)}, \qquad \eta = x + \theta(\xi - x), \qquad 0 < \theta < 1.$$

The analogous conclusions for $M^{(k)}$ may therefore be proved as in Theorem 1.

It remains to show the validity of (12). As in the proof of Theorem 1 one obtains

$$d(X^{(k+1)}) \leqslant \left(1 - \frac{m_1^{(k)}}{m_2^{(k)}}\right) d(X^{(k)}) = \frac{m_2^{(k)} - m_1^{(k)}}{m_2^{(k)}}\, d(X^{(k)})$$

and therefore using (2.9) and Theorem 3.5

$$d(X^{(k+1)}) \leqslant \frac{d(M^{(k)})}{m_1^{(0)}}\, d(X^{(k)}) \leqslant \frac{d(f'(X^{(k)}))}{m_1^{(0)}}\, d(X^{(k)})$$

$$\leqslant (c/m_1^{(0)})(d(X^{(k)}))^2, \qquad c/m_1^{(0)} \geqslant 0. \quad \blacksquare$$

The iteration (11) and Theorem 4 correspond to the statements in Moore [36]. Here we modify this iteration somewhat more. For this we note that depending on whether $f(m(X^{(k)})) > 0$ or $f(m(X^{(k)})) < 0$ the wanted zero ξ must lie in the interval $[x_1^{(k)}, m(X^{(k)})]$, respectively, $[m(X^{(k)}), x_2^{(k)}]$. If $f(m(X^{(k)})) = 0$ then $m(X^{(k)}) = \xi$ and the iteration comes to a rest. It is therefore sufficient to set

$$M^{(k)} = f'(Y^{(k)}) \cap L, \qquad L \text{ as in (10)},$$

with

(13) $$Y^{(k)} = \begin{cases} [x_1^{(k)}, m(X^{(k)})] & \text{if } f(m(X^{(k)})) > 0 \\ [m(X^{(k)}), x_2^{(k)}] & \text{if } f(m(X^{(k)})) < 0 \\ X^{(k)} & \text{otherwise,} \end{cases}$$

in (11). We then have $f'(Y^{(k)}) \subseteq f'(X^{(k)})$ and $d(Y^{(k)}) \leqslant d(X^{(k)})$ and the condition $m_1^{(k)} > 0$ may also in this manner be more easily satisfied. Theorem 4 is also valid for the choice of (13).

With respect to the choice of $m(X^{(k)}) \in X^{(k)}$ for the method (11) we have a statement similar to the statement of Corollary 2 and we may carry through a similar discussion as in Section B (see also Herzberger [28]). We shall not discuss this further.

We now clarify these interval Newton iterations by some numerical examples.

Example: (α) The function

$$f(x) = x^2(\tfrac{1}{3}x^2 + \sqrt{2}\sin x) - \sqrt{3}/19$$

has a zero ξ in the interval $X^{(0)} = [0.1, 1]$. The derivative

$$f'(x) = x(\tfrac{4}{3}x^2 + \sqrt{2}(2\sin x + x\cos x))$$

may be estimated in $X^{(0)}$ by

$$l_1 = 0.0013 \leqslant f'(x) \leqslant l_2, \qquad x \in X^{(0)}.$$

Including iterated intervals

$$X^{(k)}, \quad k \geqslant 0 \quad \text{using} \quad (13),$$
$$Y^{(k)}, \quad k \geqslant 0 \quad \text{using} \quad (10),$$

were calculated on a computer according to the procedure (11) until no improvements were shown. The results are given in Table 1.

Table 1

$X^{(k)}$	
[0.09999999999990, 0.4384388546433]	
[0.3382030708107, 0.4384388546433]	
[0.3915056049954, 0.3924484948316]	
[0.3923789206719, 0.3923799504692]	
[0.3923795071350, 0.3923795071378]	
$Y^{(k)}$	$d(X^{(k)})/d(Y^{(k)})$
[0.09999999999990, 0.5181776715881]	0.809
[0.3455588336928, 0.5181776715881]	0.581
[0.3739864679691, 0.4075613703040]	0.028
[0.3922481030413, 0.3925441306206]	0.004
[0.3923794945039, 0.3923795211850]	0.001
[0.3923795071350, 0.3923795071378]	–

(β) The polynomial

$$p(x) = x(x^9 - 1) - 1$$

has a unique zero ξ in the interval $X^{(0)} = [1, 1.5]$. The interval evaluation $p'(X)$ of the derivative

$$p'(x) = 10x^9 - 1$$

satisfies $0 \notin p'(X)$ for $X \subseteq X^{(0)}$. Iterated including intervals

$$X^{(k)}, \quad k \geqslant 0, \quad \text{using (13)} \quad \text{with} \quad L = p'(X^{(0)}),$$
$$Y^{(k)}, \quad k \geqslant 0, \quad \text{using (9)},$$

were calculated according to the procedure (11) using (4). The values in Table 2 were obtained.

Table 2

$X^{(k)}$
[1.000000000000, 1.153909281002]
[1.074525733152, 1.075772270022]
[1.075764355129, 1.075767749943]
[1.075766066086, 1.075766066088]

$Y^{(k)}$	$d(X^{(k)})/d(Y^{(k)})$
[1.000000000000, 1.231579011696]	0.665
[1.018539065305, 1.102153489956]	0.015
[1.071809768336, 1.084762444669]	3×10^{-4}
[1.075647094319, 1.075931180877]	6×10^{-9}
[1.075766039501, 1.075766097327]	...
[1.075766066085, 1.075766066090]	...
[1.075766066085, 1.075766066088]	...

The determining condition for the iteration (11) is in practice that $m_1 > 0$ must hold. We have shown that we may achieve this with (10) using a known lower bound l_1 for the derivative $f'(x)$ in $X^{(0)}$. If such a bound l_1 is not known and if $0 \in f'(X^{(0)})$, then the procedure (11) cannot be started. In order to start the iteration one may therefore first execute a few steps of the method of subdivision of intervals as described in the introduction to this section. In this manner one may find an interval $Y^{(0)} \subset X^{(0)}$ for which $0 \notin f'(Y^{(0)})$ holds.

There is yet another modification of the interval Newton method that is applicable even in the above case where $0 \in f'(X^{(0)})$. This method, also applicable to the case where several zeros of f exist in $X^{(0)}$, is now outlined here. If $0 \notin f'(X^{(0)})$, then this method is identical to the iteration (11). Assume therefore $0 \in f(X^{(0)})$. Subdivide $X^{(0)}$ into subintervals

$$U^{(1)} = [x_1^{(0)}, m(X^{(0)}) - |f(m(X^{(0)}))|/m_2^{(0)}],$$

$$V^{(1)} = [m(X^{(0)}) + |f(m(X^{(0)}))|/m_2^{(0)}, x_2^{(0)}],$$

assuming $f(m(X^{(0)})) \neq 0$. All zeros of f in $X^{(0)}$ must also lie in $U^{(1)} \cup V^{(1)}$. A zero $\xi \in X^{(0)}$ must therefore satisfy

$$\left| \frac{f(m(X^{(0)}))}{\xi - m(X^{(0)})} \right| \leqslant m_2^{(0)},$$

from which

$$|f(m(X^{(0)}))|/m_2^{(0)} \leqslant |\xi - m(X^{(0)})|$$

and

$$\xi \geqslant m(X^{(0)}) + \frac{|f(m(X^{(0)}))|}{m_2^{(0)}} \quad \text{or} \quad \xi \leqslant m(X^{(0)}) - \frac{|f(m(X^{(0)}))|}{m_2^{(0)}}$$

follows. The last inequality implies, however, that $\xi \in U^{(1)} \cup V^{(1)}$. Furthermore, it holds that

$$d(U^{(1)} \cup V^{(1)}) = d(X^{(0)}) - 2|f(m(X^{(0)}))|/m_2^{(0)} < d(X^{(0)}),$$

provided that $f(m(X^{(0)})) \neq 0$.

This procedure may now be repeated for the subintervals $U^{(1)}$ and $V^{(1)}$, etc. The total width of these intervals tends to zero. If f has only simple zeros in $X^{(0)}$ then after a certain step in the iteration these will lie in disjoint subintervals. Furthermore, for a certain index k the procedure reverts to iteration (11). According to this iteration either the subintervals tend to an interval that contains a zero or there is an empty intersection at some point.

Instead of using $M^{(k)} := f'(X^{(k)})$ in (11) according to (3) one may for polynomials use the intervals J_1, J_2, J_3, and J_4 from Theorem 3.7 with $y := m(X^{(k)})$ and $X := X^{(k)}$ for the inclusion of the derivative. All the statements of Theorem 4 are still valid. Since it was shown in Theorem 3.7 that J_1 was the optimal inclusion it is reasonable to take this interval for the inclusion of the derivative in order to get the best inclusions for the zeros at each step.

In this connection we consider the following

Example: Let

$$p(x) = x^7 + 3x^6 - 4x^5 - 12x^4 - x^3 - 3x^2 + 4x + 12$$

be a polynomial from [30] having a zero ξ in $X^{(0)} = [1.8, 2.4]$. Using iteration (11) we calculate including intervals for the zero calculating $M^{(k)} := p'(X^{(k)})$ according to the Horner scheme. Table 3 contains the calculated intervals.

If $p'(X^{(k)})$ is replaced by J_1, we get in a similar manner Table 4. We also display in Table 5 the quotient $d_1^{(k)}/d_2^{(k)}$ of the widths of the first iterated interval

Table 3

k	$X^{(k)}$
0	[1.8, 2.4]
1	[1.8, 2.0727618077482]
2	[1.9742900052812, 2.0727618077842]
3	[1.9948757147483, 2.0059215482353]
4	[1.9999888234200, 2.0000115390070]
5	[1.9999999999894, 2.0000000000107]
6	[2.0, 2.0]

divided by the width of the second iterated interval at each step of the iteration. This example was calculated on the computer CDC 6500 (mantissa length 48 bits) of the Computing Center of the Technical University of Berlin.

Table 4

k	$X^{(k)}$
0	$[1.8, 2.4]$
1	$[1.9419538108826, 2.0566964050488]$
2	$[1.9999999975872, 2.0001112993369]$
3	$[1.9999999975872, 2.0000000029595]$
4	$[2.0, 2.0]$

Table 5

k	0	1	2	3
$d_1^{(k)}/d_2^{(k)}$	1	2.37	492.35	3.7×10^6

D. HIGHER ORDER METHODS

We now develop higher order methods for finding a zero ξ in an interval $X^{(0)} = [x_1^{(0)}, x_2^{(0)}]$ for a strictly monotonically increasing or decreasing real function having continuous derivatives of sufficiently high order. These methods are always convergent. The principle used for the construction is due to Ehrmann [21]. Using interval analytic tools as well as this principle, methods may be developed for which the convergence is always forced. As in parts A, B, and C we assume without loss of generality that

$$(1)\quad f(x_1^{(0)}) < 0 \qquad \text{and} \qquad f(x_2^{(0)}) > 0.$$

Furthermore, let m_1 and m_2 again be bounds for the divided difference

$$(2)\quad 0 < m_1 \leqslant \frac{f(x) - f(\xi)}{x - \xi} = \frac{f(x)}{x - \xi} \leqslant m_2 < \infty, \qquad \xi \neq x \in X^{(0)}.$$

The bounds m_1 and m_2 are combined into the interval $M = [m_1, m_2]$. The function f is $(p + 1)$-times continuously differentiable and for $F_i \in I(\mathbb{R})$, $2 \leqslant i \leqslant p + 1$, it holds that

$$(14)\qquad f^{(i)}(x) \in F_i, \qquad x \in X^{(0)}.$$

The intervals F_i may be calculated from the interval evaluation of the derivatives of f over $X^{(0)}$ for example. If the interval expressions for the derivatives are not defined (for example, because of dividing by X when $0 \in X$) then one may, for example, subdivide the interval $X^{(0)}$ and then construct intervals F_i by taking the union of the evaluations over the subdivisions.

We now consider the following iteration

$$
(15)\quad \begin{cases}
x^{(k)} = m(X^{(k)}) \in X^{(k)}, \\
X^{(k+1,0)} = \{x^{(k)} - f(x^{(k)})/M\} \cap X^{(k)}, \\
X^{(k+1,i)} = \left\{x^{(k)} - \dfrac{1}{f'(x^{(k)})}\left[f(x^{(k)}) + \displaystyle\sum_{v=2}^{i} \dfrac{f^{(v)}(x^{(k)})}{v!}(X^{(k+1,i-1)} - x^{(k)})^{v}\right.\right. \\
\qquad \left.\left. + \dfrac{1}{(i+1)!}F_{i+1}(X^{(k+1,i-1)} - x^{(k)})^{i+1}\right]\right\} \cap X^{(k+1,i-1)}, \\
\qquad\qquad 1 \leqslant i \leqslant p, \\
X^{(k+1)} = X^{(k+1,p)}, \qquad k \geqslant 0.
\end{cases}
$$

As in Section A of this chapter, $m(X)$ means an arbitrary choice of a real number from the interval X. The iteration given above requires the calculation of the values of $f(x^{(k)}), f'(x^{(k)}), \ldots, f^{(p)}(x^{(k)})$ at each step and it has the following properties.

Theorem 5: Let f be a $(p + 1)$-times continuously differentiable function, $p \geqslant 1$, and let the relation (1) be valid in $X^{(0)}$. Let ξ be the zero of f in $X^{(0)}$, and let the interval $M = [m_1, m_2]$ be defined by (2). Furthermore, let (14) hold, then for the iteration (15) it holds that

$$(16)\qquad \xi \in X^{(k)}, \qquad k \geqslant 0,$$

$$(17)\qquad X^{(0)} \supset X^{(1)} \supset X^{(2)} \supset \cdots \qquad \text{and} \qquad \lim_{k\to\infty} X^{(k)} = \xi,$$

or the sequence comes to a rest at the point $[\xi, \xi]$ after a finite number of steps,

$$(18)\qquad d(X^{(k+1)}) \leqslant \gamma(d(X^{(k)}))^{p+1},$$

with $\gamma \geqslant 0$. Therefore, according to Theorem 2, Appendix A, the R order of the sequence $\{X^{(k)}\}_{k=0}^{\infty}$ is at least $p + 1$.

Proof: Of (16): Assume that $\xi \in X^{(k)}$ for some $k \geqslant 0$. This holds true for $k = 0$ by the assumption of the theorem. As in Theorem 1 one shows that

$$\xi \in X^{(k+1,0)}.$$

Suppose $\xi \in X^{(k+1,i)}$ for some $i \geqslant 0$. This holds true again for $i = 0$. Then we have

$$\xi - x^{(k)} \in X^{(k+1,i)} - x^{(k)}.$$

From Taylor's formula we get

$$0 = f(\xi) = f(x^{(k)}) + f'(x^{(k)})(\xi - x^{(k)}) + \cdots + \frac{1}{(i+1)!} f^{(i+1)}(x^{(k)})(\xi - x^{(k)})^{i+1}$$
$$+ \frac{1}{(i+2)!} f^{(i+2)}(\eta_{i+2})(\xi - x^{(k)})^{i+2},$$

for some η_{i+2} between $x^{(k)}$ and ξ. Inclusion monotonicity then gives us the relation

$$\xi = x^{(k)} - \frac{1}{f'(x^{(k)})}\left[f(x^{(k)}) + \sum_{v=2}^{i+1} \frac{f^{(v)}(x^{(k)})}{v!}(\xi - x^{(k)})^v + \frac{f^{(i+2)}(\eta_{i+2})}{(i+2)!}(\xi - x^{(k)})^{i+2}\right]$$
$$\in \left\{x^{(k)} - \frac{1}{f'(x^{(k)})}\left[f(x^{(k)}) + \sum_{v=2}^{i+1} \frac{f^{(v)}(x^{(k)})}{v!}(X^{(k+1,i)} - x^{(k)})^v + \frac{F_{i+2}}{(i+2)!}(X^{(k+1,i)} - x^{(k)})^{i+2}\right]\right\} \cap X^{(k+1,i)}$$
$$= X^{(k+1,i+1)}.$$

We therefore have $\xi \in X^{(k+1,i)}$, $0 \leqslant i \leqslant p$, and $\xi \in X^{(k+1)} = X^{(k+1,p)}$.

Of (17): In the same manner as in Theorem 1 one shows that $X^{(k)} \supset X^{(k+1,0)}$ and since we take intersections in method (15) we get $X^{(k)} \supset X^{(k+1)}$, $k \geqslant 0$. Furthermore, we show as in Theorem 1 that

$$d(X^{(k+1,0)}) \leqslant (1 - m_1/m_2)\, d(X^{(k)}).$$

Since we take intersections in method (15) we get

$$d(X^{(k+1)}) \leqslant (1 - m_1/m_2)\, d(X^{(k)}), \qquad k \geqslant 0.$$

As in Theorem 1 we get the convergence statement $\lim_{k\to\infty} X^{(k)} = \xi$. The remainder of (17) is shown as in Theorem 1.

Of (18): We have $d(X^{(k+1,0)}) \leqslant d(X^{(k)})$, and therefore

$$d(X^{(k+1,1)}) \leqslant d\left(x^{(k)} - \frac{1}{f'(x^{(k)})}(f(x^{(k)}) + \tfrac{1}{2}F_2(X^{(k+1,0)} - x^{(k)})^2)\right)$$
$$\leqslant \frac{1}{2} d\left(\frac{F_2}{f'(x^{(k)})}(X^{(k)} - X^{(k)})^2\right)$$
$$\leqslant \frac{1}{2} d\left(\frac{F_2}{M}[-(d(X^{(k)}))^2, (d(X^{(k)}))^2]\right)$$
$$= |F_2/M|(d(X^{(k)}))^2 = \gamma_1 (d(X^{(k)}))^2$$

with a constant $\gamma_1 = |F_2/M|$ independent of k.

Assume that for some $i \geqslant 1$ we have

$$d(X^{(k+1,i)}) \leqslant \gamma_i (d(X^{(k)}))^{i+1},$$

where γ_i is independent of k. This was proved above for $i = 1$. For $i > 1$ we have, from the iteration (15) using the rules for width from Chapter 2, that

$$\begin{aligned}
d(X^{(k+1,i+1)}) &\leqslant d\left(\sum_{\nu=2}^{i+1} \frac{f^{(\nu)}(x^{(k)})}{\nu! f'(x^{(k)})} (X^{(k+1,i)} - x^{(k)})^{\nu} \right. \\
&\quad \left. + \frac{1}{(i+2)!} \frac{F_{i+2}}{f'(x^{(k)})} (X^{(k+1,i)} - x^{(k)})^{i+2} \right) \\
&\leqslant \sum_{\nu=2}^{i+1} \frac{1}{\nu!} \left| \frac{f^{(\nu)}(x^{(k)})}{f'(x^{(k)})} \right| d((X^{(k+1,i)} - x^{(k)})^{\nu}) \\
&\quad + \frac{1}{(i+2)!} d\left(\frac{F_{i+2}}{f'(x^{(k)})} (X^{(k+1,i)} - x^{(k)})^{i+2} \right) \\
&\leqslant \sum_{\nu=2}^{i+1} \frac{1}{\nu!} \left| \frac{F_{\nu}}{M} \right| \nu |X^{(k+1,i)} - x^{(k)}|^{\nu-1} d(X^{(k+1,i)} - x^{(k)}) \\
&\quad + \frac{1}{(i+2)!} d\left(\frac{F_{i+2}}{f'(x^{(k)})} (X^{(k+1,i)} - x^{(k)})^{i+2} \right) \\
&\leqslant \sum_{\nu=2}^{i+1} \frac{1}{(\nu-1)!} \left| \frac{F_{\nu}}{M} \right| |X^{(k)} - X^{(k)}|^{\nu-1} d(X^{(k+1,i)}) \\
&\quad + \frac{1}{(i+2)!} d\left(\frac{F_{i+2}}{M} (X^{(k)} - X^{(k)})^{i+2} \right) \\
&\leqslant \sum_{\nu=2}^{i+1} \frac{1}{(\nu-1)!} \left| \frac{F_{\nu}}{M} \right| (d(X^{(k)}))^{\nu-1} \gamma_i (d(X^{(k)}))^{i+1} \\
&\quad + \frac{1}{(i+2)!} d\left(\frac{F_{i+2}}{M} [-(d(X^{(k)}))^{i+2}, (d(X^{(k)}))^{i+2}] \right) \\
&= (d(X^{(k)}))^{i+2} \sum_{\nu=2}^{i+1} \frac{1}{(\nu-1)!} \left| \frac{F_{\nu}}{M} \right| \gamma_i (d(X^{(k)}))^{\nu-2} \\
&\quad + \frac{2}{(i+2)!} \left| \frac{F_{i+2}}{M} \right| (d(X^{(k)}))^{i+2} \\
&\leqslant \left(\sum_{\nu=2}^{i+1} \frac{1}{(\nu-1)!} \left| \frac{F_{\nu}}{M} \right| \gamma_i (d(X^{(0)}))^{\nu-2} \right. \\
&\quad \left. + \frac{2}{(i+2)!} \left| \frac{F_{i+2}}{M} \right| \right) (d(X^{(k)}))^{i+2} \\
&= \gamma_{i+1} (d(X^{(k)}))^{i+2}
\end{aligned}$$

with a constant

$$\gamma_{i+1} = \gamma_i \sum_{\nu=2}^{i+1} \frac{1}{(\nu-1)!}\left|\frac{F_\nu}{M}\right|(d(X^{(0)}))^{\nu-2} + \frac{2}{(i+2)!}\left|\frac{F_{i+2}}{M}\right|$$

independent of k. The relation

$$d(X^{(k+1,i)}) \leqslant \gamma_i (d(X^{(k)}))^{i+1}$$

is therefore valid for $1 \leqslant i \leqslant p$. Furthermore,

$$d(X^{(k+1)}) = d(X^{(k+1,p)}) \leqslant \gamma_p(d(X^{(k)}))^{p+1}$$

with γ_p independent of k. This is the statement of (18) with $\gamma = \gamma_p$ and the theorem is proved. ■

We now wish to investigate the case $p = 1$ in some further detail. For $p = 1$ the iteration (15) may be written

$$\begin{cases} x^{(k)} = m(X^{(k)}) \in X^{(k)}, \\ X^{(k+1,0)} = \{x^{(k)} - f(x^{(k)})/M\} \cap X^{(k)}, \\ X^{(k+1,1)} = \{x^{(k)} - (1/f'(x^{(k)}))(f(x^{(k)}) + \frac{1}{2}F_2(X^{(k+1,0)} - x^{(k)})^2)\} \cap X^{(k+1,0)}, \\ X^{(k+1)} = X^{(k+1,1)}, \qquad k \geqslant 0. \end{cases}$$

The method has the same properties as Moore's method given in Section C. Apart from some additional arithmetic operations it requires less work than Moore's method since both the value and the derivative have to be calculated at the point $x^{(k)}$. In the case of Moore's method the derivative has to be evaluated using the interval $X^{(k)}$. This generally needs more computational work than the point evaluation at $x^{(k)}$. If the interval F_2 can be calculated simply, then method (15) with $p = 1$ is preferable to Moore's method. These results only hold true theoretically when one calculates exactly. If one wishes to guarantee the inclusion of the zeros when the method is implemented on a computer, then one must include the effect of rounding errors. This is achieved by implementing all operations as machine interval operations. In particular, one has to evaluate $f'(x^{(k)})$ using machine interval arithmetic. In this case, method (15) with $p = 1$ requires, apart from a few arithmetic operations, the same amount of calculation as Moore's method. Since one also has to calculate the interval F_2, one would prefer Moore's method if rounding errors are to be accounted for.

We also wish to mention at this point that in Krawzcyk [34, Section 9, Theorem 3] the following method was considered

$$\begin{cases} x^{(k)} = m(X^{(k)}) \in X^{(k)} \\ X^{(k+1)} = \{x^{(k)} - (1/f'(x^{(k)}))(f(x^{(k)}) + \frac{1}{2}f''(X^{(k)})(X^{(k)} - x^{(k)})^2)\} \cap X^{(k)}, \qquad k \geqslant 0, \end{cases}$$

under the assumption that f is twice differentiable. We have $\xi \in X^{(k)}$, $k \geqslant 0$.

Conditions for convergence $\lim_{k\to\infty} X^{(k)} = \xi$ are not given. If the method converges, then the sequence of widths of the iterates converges to zero quadratically if $f'(\xi) \neq 0$. Comparing to method (15) with $p = 1$, we have to evaluate the second derivative over $X^{(k)}$ in the interval sense in each step. This reduces the constant of convergence but does not improve the order of convergence. (The same is true for the iteration (15) with $p = 1$ if the constant interval F_2 is replaced by $f''(X^{(k)})$ in each step.) The practical execution of this method, that is, if rounding errors are to be considered in the calculation of $f(x^{(k)})$ and $f'(x^{(k)})$, requires another third as many calculations. Since convergence is not assured the method is rather less attractive.

When method (15) is used one must decide on a particular order. We also remark that under certain assumptions one has the result that method (15) is optimal for $p = 2$, that is the third order method. The details may be proved as in Ehrmann [21].

E. INTERPOLATORY METHODS

As in the previous section we shall consider convergent methods of higher order. The underlying principle is the well-known interpolation method for finding zeros of functions. This method is now modified using interval analytic tools in such a manner that we always get a monotonic inclusion of the zeros. In the same manner as in Section D we also need interval estimates of the higher derivatives of the function f.

The various methods are determined from a set of $n + 1$ nonnegative integer parameters

$$m_0, m_1, \ldots, m_n.$$

Define

$$r = \sum_{j=0}^{n} m_j,$$

and assume that

$$m_0 m_n > 0,$$

which implies that $r > 0$. The function f for which we wish to find the zero $\xi \in X^{(0)} = [x_1^{(0)}, x_2^{(0)}]$ is assumed differentiable on $X^{(0)}$ as often as necessary. Furthermore, intervals H and K are determined such that

$$f'(x) \in H, \quad x \in X^{(0)}, \qquad \text{with} \quad 0 \notin H = [h_1, h_2],$$
$$f^{(r)}(x) \in K, \qquad x \in X^{(0)},$$

hold.

In order to determine the general step of the iteration – here the $k + 1$ step – we assume that we already have $n + 1$ pairwise different approximations for ξ,

$$x^{(k)}, x^{(k-1)}, \ldots, x^{(k-n)} \quad \text{in} \quad X^{(0)}$$

and that the including interval $X^{(k)}$ last determined has the form

$$X^{(k)} = [x^{(k)} - \varepsilon^{(k)}, x^{(k)} + \varepsilon^{(k)}]$$

for an $\varepsilon^{(k)} > 0$. We assume in addition that

$$\xi \neq x_1^{(0)} \qquad \text{and} \qquad \xi \neq x_2^{(0)}$$

(see, for example, Cornelius [15]). By executing steps (S1)–(S5) below a new approximation $X^{(k+1)}$ is determined as an improved including interval. These steps are described in the following:

(S1) *The determination of the unique Hermite interpolation polynomial*

$$p_k(x) = p_{(m_0, m_1, \ldots, m_n)}(x; x^{(k)}, \ldots, x^{(k-n)})$$

satisfying the interpolation conditions

$$p_k^{(j)}(x^{(k-i)}) = f^{(j)}(x^{(k-i)}), \qquad 0 \leqslant i \leqslant n, \quad 0 \leqslant j \leqslant m_i - 1.$$

(We set $f^{(0)} = f$, and if $m_i = 0$ then the conditions at the point $x^{(k-i)}$ are empty.) An interval $Z^{(k)} \subset X^{(k)}$ is determined according to the procedure

$$Z^{(k)} = \begin{cases} [x^{(k)} - \varepsilon^{(k)}, x^{(k)}] & \text{if } f(x^{(k)})h_1 > 0, \\ [x^{(k)}, x^{(k)} + \varepsilon^{(k)}] & \text{if } f(x^{(k)})h_1 < 0. \end{cases}$$

(S2) *The determination of a real zero $y^{(k)}$ of the polynomial $p_k(x)$ in the interval*

$$[x^{(k)} - 2\varepsilon^{(k)}, x^{(k)} + 2\varepsilon^{(k)}] \cap X^{(0)}.$$

If there is no such zero one goes directly to step (S5) *with*

$$\tilde{X}^{(k+1)} := [\tilde{x}_1^{(k+1)}, \tilde{x}_1^{(k+1)}] = Z^{(k)}.$$

(S3) *Calculation of an including interval $F^{(k)}$ for the function value $f(y^{(k)})$ using the expression*

$$F^{(k)} = \frac{K}{r!} \prod_{j=0}^{n} (y^{(k)} - x^{(k-j)})^{m_j}.$$

(S4) *Calculation of the improved including interval according to*

$$\tilde{X}^{(k+1)} = \{y^{(k)} - F^{(k)}/H\} \cap Z^{(k)}.$$

(S5) *Evaluating the new approximation*

$$x^{(k+1)} = (\tilde{x}_1^{(k+1)} + \tilde{x}_2^{(k+1)})/2$$

and the new value

$$\varepsilon^{(k+1)} = (\tilde{x}_2^{(k+1)} - \tilde{x}_1^{(k+1)})/2,$$

as well as the new interval

$$X^{(k+1)} = [x^{(k+1)} - \varepsilon^{(k+1)}, x^{(k+1)} + \varepsilon^{(k+1)}] = \tilde{X}^{(k+1)}.$$

(If the points $x^{(k+1)}, x^{(k)}, \ldots, x^{(k-n+1)}$ are no longer pairwise different then one may put $x^{(k+1)} \in \tilde{X}^{(k+1)}$ with

$$x^{(k+1)} = \begin{cases} \frac{1}{2}(\tilde{x}_1^{(k+1)} + \tilde{x}_2^{(k+1)}) + \tilde{\varepsilon}^{(k)} & \text{if } f(x^{(k)})h_1 > 0, \\ \frac{1}{2}(\tilde{x}_1^{(k+1)} + \tilde{x}_2^{(k+1)}) - \tilde{\varepsilon}^{(k)} & \text{if } f(x^{(k)})h_1 < 0, \end{cases}$$

with a suitable $\tilde{\varepsilon}^{(k)}$ that guarantees the pairwise difference as well as $x^{(k+1)} \in \tilde{X}^{(k+1)}$. Such a choice is always possible as long as $x^{(k-i)} \neq \xi$, $i = 0, 1, \ldots, n$. Finally the new value of $\varepsilon^{(k+1)}$ is chosen as

$$\varepsilon^{(k+1)} = \max\{\tilde{x}_2^{(k+1)} - x^{(k+1)}, x^{(k+1)} - \tilde{x}_1^{(k+1)}\}$$

leading to

$$X^{(k+1)} = [x^{(k+1)} - \varepsilon^{(k+1)}, x^{(k+1)} + \varepsilon^{(k+1)}] \supset \tilde{X}^{(k+1)}.)$$

It is to be noted that the determination of the including intervals $\{X^{(k)}\}$ does not use a change of sign for the calculated function value. This means that several sequentially calculated function values may have the same sign, and in spite of this a new including interval is always calculated. It will also be shown that parts (S3) and (S4) will only be skipped a finite number of times. Furthermore, it is not required that the including interval $X^{(i)}$ contains any of the previous approximations except for the approximation $x^{(i)}$. The properties of the algorithm defined above are collected in the following theorem.

Theorem 6: Let f be a real function with a zero ξ. Assume that an including interval

$$X^{(0)} = \{x \mid |x^{(0)} - x| \leqslant \varepsilon^{(0)}\} \ni \xi$$

with

$$\xi \neq x_1^{(0)}, \ \xi \neq x_2^{(0)} \qquad (X^{(0)} = [x_1^{(0)}, x_2^{(0)}])$$

is given for ξ. Assume furthermore that pairwise different approximations

$$x^{(-n)}, x^{(-n+1)}, \ldots, x^{(0)} \in X^{(0)}$$

are given for ξ. The conditions on the bounds

$$f'(x) \in H \qquad \text{with} \quad 0 \notin H, \quad x \in X^{(0)},$$

and

$$f^{(r)}(x) \in K, \qquad x \in X^{(0)},$$

are valid in the interval $X^{(0)}$ for the intervals H and K where the integer, nonnegative parameters

$$m_0, m_1, \ldots, m_n \qquad \text{with} \quad r = \sum_{j=0}^{n} m_j, \quad m_0 m_n > 0,$$

are specified. The following statements are then valid for the iteration defined in parts (S1)–(S5):

(19) $$\xi \in X^{(k)}, \qquad k \geqslant 0,$$

(20) $$X^{(0)} \supset X^{(1)} \supset X^{(2)} \supset \cdots \qquad \text{and} \qquad \lim_{k \to \infty} X^{(k)} = \xi,$$

or the sequence comes to rest at the point $[\xi, \xi]$ after a finite number of steps.

(21) The R order of the iteration (S1)–(S5) (see Appendix A) is $O_R((\text{S1})\text{–}(\text{S5}), \xi) \geqslant s$, where s is the unique positive root of the polynomial

$$p(s) = s^{n+1} - \sum_{j=0}^{n} m_j s^{n-j}.$$

Proof: Of (19): According to (S1) we have $\xi \in Z^{(k)}$. The remainder term of the Hermite interpolation formula gives

$$f(x) = p_k(x) + \frac{f^{(r)}(\eta)}{r!} \prod_{j=0}^{n} (x - x^{(k-j)})^{m_j},$$

where η is a value in the interval formed from the points

$$x, x^{(k)}, \ldots, x^{(k-n)}$$

in $X^{(k-n)}$. Since $p_k(y^{(k)}) = 0$ it follows that

$$f(y^{(k)}) \in \frac{K}{r!} \prod_{j=0}^{n} (y^{(k)} - x^{(k-j)})^{m_j} = F^{(k)},$$

using the inclusion property of interval arithmetic. Since (S4) is a step of the simplified Newton method we have $\xi \in \tilde{X}^{(k+1)}$ and therefore $\xi \in X^{(k+1)}$.

Of (20): This follows immediately from step (S1) together with (19) as well from (S5) where $X^{(k+1)}$ may always be chosen in such a manner that

$$d(X^{(k+1)}) \leqslant c \cdot d(X^{(k)})$$

with $\frac{1}{2} \leqslant c < 1$.

Of (21): It may first be shown that there always exists an index k_1 such that for all steps $i \geqslant k_1$ there is a real zero in the interval $[x^{(k)} - 2\varepsilon^{(k)}, x^{(k)} + 2\varepsilon^{(k)}]$.

The proof depends on an estimate that shows that if $\varepsilon^{(k)}$ is small enough then $p_k(x)$ either changes sign between $x^{(k)}$ and $x^{(k)} + 2\varepsilon^{(k)}$ or between $x^{(k)}$ and $x^{(k)} - 2\varepsilon^{(k)}$. The details of the proof are similar to the proof in Traub [46] and will not be given here. Consequently we can therefore always assume that parts (S3) and (S4) are executed for the purpose of evaluating the order of convergence. From (S4) we get the estimate

$$d(\tilde{X}^{(k+1)}) \leqslant d(y^{(k)} - F^{(k)}/H) = d(F^{(k)}/H)$$

of the width. Using (S3) and this estimate we obtain

$$d(\tilde{X}^{(k+1)}) \leqslant \frac{1}{r!} \prod_{j=0}^{n} |y^{(k)} - x^{(k-j)}|^{m_j} d\left(\frac{K}{H}\right).$$

Since the relation

$$y^{(k)} \in [x^{(k-j)} - 2\varepsilon^{(k-j)}, x^{(k-j)} + 2\varepsilon^{(k-j)}], \qquad 0 \leqslant j \leqslant n,$$

is valid it finally follows that

$$d(\tilde{X}^{(k+1)}) \leqslant \tilde{c} \prod_{j=0}^{n} d(X^{(k-j)})^{m_j}$$

as well as

$$d(X^{(k+1)}) \leqslant c \prod_{j=0}^{n} d(X^{(k-j)})^{m_j}.$$

Using Theorem 3, Appendix A, we obtain the final result. ■

We now discuss yet another modified family of inclusion methods. This family is obtained by modifying (S2) when the required zero $y^{(k)}$ is not available. In order to be able to execute (S4) we therefore must determine $y^{(k)}$ in a different manner. This is done by

(S2′) *The determination of a real zero $y^{(k)}$ of the polynomial $p_k(x)$ in the interval $[x^{(k)} - 2\varepsilon^{(k)}, x^{(k)} + 2\varepsilon^{(k)}] \cap X^{(0)}$. If such a zero does not exist then one puts*

$$y^{(k)} = \begin{cases} x^{(k)} - \varepsilon^{(k)} & \textit{if} \quad f(x^{(k)})h_1 > 0 \\ x^{(k)} + \varepsilon^{(k)} & \textit{if} \quad f(x^{(k)})h_1 < 0. \end{cases}$$

(S3′) *The calculation of an including interval $F^{(k)}$ for the function value $f(y^{(k)})$ using the expression*

$$F^{(k)} = \begin{cases} \dfrac{K}{r!} \displaystyle\prod_{j=0}^{n} (y^{(k)} - x^{(k-j)})^{m_j}, & \textit{if} \quad p_k(y^{(k)}) = 0 \\ p_k(y^{(k)}) + \dfrac{K}{r!} \displaystyle\prod_{j=0}^{n} (y^{(k)} - x^{(k-j)})^{m_j}, & \textit{otherwise.} \end{cases}$$

It is immediately clear that Theorem 6 is valid without any modification for the iteration (S1), (S2′), (S3′), (S4), (S5).

Further modifications of this iteration are given by Cornelius in [15] relating to step (S1).

The following modification is considered there:

(S1′) *The determination of the unique Hermite interpolation polynomial*

$$p_k(x) = p_{(m_0, m_1, \ldots, m_n)}(x; x^{(k)}, \ldots, x^{(k-n)})$$

satisfying the interpolatory conditions

$$p_k^{(j)}(x^{(k-i)}) = f^{(j)}(x^{(k-i)}), \qquad 0 \leqslant i \leqslant n, \quad 0 \leqslant j \leqslant m_i - 1.$$

(We set $f^{(0)} = f$, and for the case $m_i = 0$ the conditions at the relevant point are empty.) We now compute an interval $Z^{(k)} \subset X^{(k)}$ according to

$$Z^{(k)} = \{x^{(k)} - f(x^{(k)})/H\} \cap X^{(k)}.$$

It is immediately clear that Theorem 6 holds without change for the iteration (S1′), (S2′), (S3′), (S4), (S5) as well.

The methods for computing zeros treated above contain as special cases among others an interval version of the secant method for $n = 1$ and $m_0 = m_1 = 1$. In the case of $n = 2$ and $m_0 = m_1 = m_2 = 1$ we get an interval version of Muller's method.

For the interval secant method we now consider a numerical example.

Example: Consider the function

$$f(x) = 2xe^{-5} + 1 - 2e^{-5x}$$

on the interval $X^{(0)} = [0, 1]$. The interval secant method applied to this function gives the iterates in Table 6. In addition to the interval secant method described above, the so-called interval regula falsi method is described in Cornelius [15]. This method requires similar assumptions on f as the interval secant method. It turns out that this method is also an interpolation method. In

Table 6

$X^{(k)}$
[0.0000000000000, 1.000000000000]
[0.0000000000000, 0.2703167011351]
[0.1351583505675, 0.1417860581860]
[0.1380542457667, 0.1382588014849]
[0.1382505159257, 0.1382572086567]
[0.1382571542348, 0.1382572086567]
[0.1382571550288, 0.1382571550581]
[0.1382571550553, 0.1382571550581]

contrast to the above methods it is developed using Newton's divided differences. We now give a short description of this method.

We assume that the function f is twice continuously differentiable in the interval X and that f only has a simple real zero ξ in X. Furthermore, we have intervals H, K satisfying the conditions

$$f'(x)\in H, \qquad x\in X \quad \text{with} \quad 0\notin H,$$
$$f''(x)\in K, \qquad x\in X.$$

The interval *regula falsi method*, abbreviated RF, is then described by

$$X^{(0)} = X,\ x^{(0)} = m(X^{(0)}) \qquad (\text{midpoint of } X^{(0)}),$$
$$X^{(1)} = \{x^{(0)} - f(x^{(0)})/H\} \cap X^{(0)},$$

$$\text{RF}\begin{cases} x^{(k)} = m(X^{(k)}) \qquad (\text{midpoint of } X^{(k)}),\\ Z^{(k+1)} = \{x^{(k)} - f(x^{(k)})/H\} \cap X^{(k)},\\ X^{(k+1)} = \begin{cases} \{x^{(k)} - \dfrac{x^{(k)} - x^{(k-1)}}{f(x^{(k)}) - f(x^{(k-1)})}(f(x^{(k)}) + \frac{1}{2}K(Z^{(k+1)} - x^{(k)}) \\ \times (Z^{(k+1)} - x^{(k-1)}))\} \cap Z^{(k+1)} & \text{if } f(x^{(k)}) \neq 0\\ Z^{(k+1)} & \text{otherwise,} \end{cases} \end{cases}$$

$$k \geqslant 1.$$

The properties of the algorithm RF are now collected in

Theorem 7: Let the twice differentiable function f have a simple zero ξ in the interval X. Furthermore, assume that the conditions

$$f'(x)\in H, \qquad x\in X \quad \text{with} \quad 0\notin H,$$
$$f''(x)\in K, \qquad x\in X,$$

are satisfied. The sequence $\{X^{(k)}\}$ calculated according to the procedure RF then satisfies

(22) $$\xi \in X^{(k)}, \qquad k \geqslant 0;$$

(23) $$X^{(0)} \supset X^{(1)} \supset X^{(2)} \supset \cdots \qquad \text{with} \quad \lim_{k\to\infty} X^{(k)} = \xi,$$

or the sequence comes to rest at the point $[\xi, \xi]$ after finitely many steps.

(24) The relation

$$d(X^{(k+1)}) \leqslant \gamma\, d(X^{(k)})\, d(X^{(k-1)})$$

holds for a $\gamma \geqslant 0$, that is (see Appendix A),

$$O_R((\text{RF}), \xi) \geqslant \tfrac{1}{2}(1 + \sqrt{5}).$$

Proof: Of (22): The proof is by complete induction on the index k. For $k = 1$ the assertion $\xi \in X^{(1)}$ is trivial. Suppose therefore that for a fixed k we have $x^{(k)} = m(X^{(k)}) \neq \xi$ and $\xi \in X^{(k)}$. Then we have $\xi \in Z^{(k+1)}$ and furthermore $x_k \neq x_{k-1}$. From Newton's interpolation formula it follows that

$$f(\xi) = f(x^{(k)}) + \frac{f(x^{(k)}) - f(x^{(k-1)})}{x^{(k)} - x^{(k-1)}}(\xi - x^{(k)})$$

$$+ \tfrac{1}{2}f''(\eta)(\xi - x^{(k)})(\xi - x^{(k-1)})$$

with η in the interval formed by $x^{(k)}, x^{(k-1)}$, and ξ. Since $f(\xi) = 0$ it follows that

$$\xi = x^{(k)} - \frac{x^{(k)} - x^{(k-1)}}{f(x^{(k)}) - f(x^{(k-1)})}\left(f(x^{(k)}) + \frac{f''(\eta)}{2}(\xi - x^{(k)})(\xi - x^{(k-1)})\right).$$

From the assumption that $f''(\eta) \in K$ and the fact that $\xi \in Z^{(k+1)}$ it follows from the inclusion property of interval arithmetic that

$$\xi \in x^{(k)} - \frac{x^{(k)} - x^{(k-1)}}{f(x^{(k)}) - f(x^{(k-1)})}\left(f(x^{(k)}) + \frac{K}{2}(Z^{(k+1)} - x^{(k)})(Z^{(k+1)} - x^{(k-1)})\right)$$

and therefore since $\xi \in Z^{(k+1)}$ we get $\xi \in X^{(k+1)}$.

Of (23): This property follows from the fact that $X^{(k)} \subset Z^{(k)}$ and from the construction of $Z^{(k)}$.

Of (24): Let $x^{(k)} \neq \xi$. Then we get

$$d(X^{(k+1)}) \leqslant d\left(x^{(k)} - \frac{x^{(k)} - x^{(k-1)}}{f(x^{(k)}) - f(x^{(k-1)})}(f(x^{(k)})\right.$$

$$\left. + \tfrac{1}{2}K(Z^{(k+1)} - x^{(k)})(Z^{(k+1)} - x^{(k-1)}))\right)$$

$$= d\left(-\frac{x^{(k)} - x^{(k-1)}}{f(x^{(k)}) - f(x^{(k-1)})} \cdot \frac{1}{2}K(Z^{(k+1)} - x^{(k)})(Z^{(k+1)} - x^{(k-1)})\right).$$

From

$$\frac{x^{(k)} - x^{(k-1)}}{f(x^{(k)}) - f(x^{(k-1)})} = (f'(\tau))^{-1} \in \frac{1}{H}, \qquad \tau \in X,$$

it follows that

$$d(X^{(k+1)}) \leqslant \tfrac{1}{2}d((K/H)(Z^{(k+1)} - x^{(k)})(Z^{(k+1)} - x^{(k-1)})).$$

Because of

$$X^{(k-1)} \supset X^{(k)} \supset Z^{(k+1)} \supseteq X^{(k+1)},$$

it finally follows that

$$d(X^{(k+1)}) \leqslant \tfrac{1}{2}d((K/H)(X^{(k)} - x^{(k)})(X^{(k-1)} - x^{(k-1)})).$$

An interval $A \in I(\mathbb{R})$ with center $m(A)$ satisfies the relation

$$A - m(A) = [-d(A)/2, d(A)/2],$$

which implies that the relation

$$(X^{(k)} - x^{(k)})(X^{(k-1)} - x^{(k-1)}) = [-d(X^{(k)})\, d(X^{(k-1)})/4, d(X^{(k)})\, d(X^{(k-1)})/4]$$

is valid.

If this is applied to the above inequality we get

$$d(X^{(k+1)}) \leqslant \tfrac{1}{2} d((K/H)[-d(X^{(k)})\, d(X^{(k-1)})/4, d(X^{(k)})\, d(X^{(k-1)})/4]),$$

from which

$$\begin{aligned} d(X^{(k+1)}) &\leqslant \tfrac{1}{2}|K/H| \cdot 2 \cdot \tfrac{1}{4} d(X^{(k)})\, d(X^{(k-1)}) \\ &= \gamma\, d(X^{(k)})\, d(X^{(k-1)}), \end{aligned}$$

where $\gamma = \frac{1}{4}|K/H|$. From Theorem 3, Appendix A, we then obtain the required inequality for the R order. ■

By using higher-order divided differences for the given function f we are able to construct further methods whose order of convergence also lie between 1 and 2 and that only require one new function value per iteration step. (See Cornelius [15] for a further discussion of this.)

It is also possible to construct interval versions of the methods in [41] using the interval regula falsi (RF) method. These methods then have higher-order convergence even though they only use function values for f. We shall describe such a method here representing a direct generalization of the interval regula falsi method.

The function f has again a simple zero ξ in the interval X. The intervals H, K satisfy the conditions from Theorem 7. The parameter p is given as an integer number for which $p \geqslant 1$. The *parameter regula falsi* method, p-RF, is now formulated as follows:

$$p\text{-RF} \left\{ \begin{aligned} & X^{(0)} = X, \qquad x^{(0)} = m(X^{(0)}), \\ & X^{(1)} = \{x^{(0)} - f(x^{(0)})/H\} \cap X^{(0)}. \\ & \text{For } k \geqslant 1 \text{ calculate:} \\ & x^{(k)} = m(X^{(k)}) \qquad \text{(midpoint of } X^{(k)}), \\ & X^{(k+1,0)} = \{x^{(k)} - f(x^{(k)})/H\} \cap X^{(k)}, \\ & X^{(k+1,1)} = \begin{cases} \left\{ x^{(k)} - \dfrac{x^{(k)} - x^{(k-1)}}{f(x^{(k)}) - f(x^{(k-1)})} \left(f(x^{(k)}) + \dfrac{1}{2} K (X^{(k+1,0)} - x^{(k)}) \right. \right. \\ \left. \left. \cdot (X^{(k+1,0)} - x^{(k-1)}) \right) \right\} \cap X^{(k+1,0)} & \text{if } f(x^{(k)}) \neq 0 \\ X^{(k+1,0)} & \text{otherwise.} \end{cases} \\ & \vdots \end{aligned} \right.$$

$$
p\text{-RF}\begin{cases}
\text{Then calculate for } i = 2, 3, \ldots, p \text{ (the following specifications are empty for } p = 1\text{):} \\
z^{(i)} = m(X^{(k+1,i-1)}) \\
X^{(k+1,i)} = \begin{cases} \left\{z^{(i)} - \dfrac{z^{(i)} - x^{(k)}}{f(z^{(i)}) - f(x^{(k)})}\left(f(z^{(i)}) + \dfrac{1}{2}K(X^{(k+1,i-1)} - z^{(i)}) \right.\right. \\ \left.\left. \cdot\,(X^{(k+1,i-1)} - x^{(k)})\right)\right\} \cap X^{(k+1,i-1)} & \text{if } f(x^{(k)}) \neq 0 \\ X^{(k+1,i-1)} & \text{otherwise,} \end{cases} \\
X^{(k+1)} = X^{(k+1,p)}.
\end{cases}
$$

The following properties are valid for the p-RF method.

Theorem 8: Let the function f be twice continuously differentiable in the interval X and assume that f has a zero ξ in X. Let the conditions

$$f'(x) \in H, \qquad x \in X \quad \text{with } 0 \notin H,$$

$$f''(x) \in K, \qquad x \in X,$$

be satisfied. Then the sequence $\{X^{(k)}\}$ calculated according to p-RF satisfies for $p \geqslant 1$:

(25) $$\xi \in X^{(k)}, \qquad k \geqslant 0,$$

(26) $$X^{(0)} \supset X^{(1)} \supset X^{(2)} \supset \cdots \qquad \text{with} \quad \lim_{k \to \infty} X^{(k)} = \xi,$$

or the sequence comes to rest at $[\xi, \xi]$ after a finite number of steps.

(27) The estimate

$$d(X^{(k+1)}) \leqslant \gamma\, d(X^{(k)})^p\, d(X^{(k-1)})$$

holds for some $\gamma \geqslant 0$ and furthermore

$$O_R((p\text{-RF}), \xi) \geqslant \tfrac{1}{2}(p + \sqrt{p^2 + 4}).$$

Proof: Theorem 8 reduces to Theorem 7 for $p = 1$. We therefore always assume $p \geqslant 2$.

Of (25): We only sketch the induction proof. The assertion is trivially true for $k = 0, 1$. We now assume the result for a fixed k and prove the result for $k + 1$. Assume therefore the validity of (25) for k. Then it follows that

$$\xi \in X^{(k+1,0)} \text{ and } X^{(k+1,0)} = [\xi, \xi] \qquad \text{for} \quad x^{(k)} = \xi.$$

In the case of $x^{(k)} = \xi$ we have $f(x^{(k)}) = 0$ from which it follows that

$$X^{(k+1,i)} = [\xi, \xi]$$

for $i = 2, 3, \ldots, p$. We therefore have

$$\xi \in X^{(k+1)} = X^{(k+1,p)} = [\xi, \xi].$$

If now $x^{(k)} \neq \xi$, then $f(x^{(k)}) \neq 0$ and one gets the relation

$$0 = f(\xi) = f(x^{(k)}) + \frac{f(x^{(k)}) - f(x^{(k-1)})}{x^{(k)} - x^{(k-1)}}(\xi - x^{(k)})$$
$$+ \tfrac{1}{2} f''(\eta)(\xi - x^{(k)})(\xi - x^{(k-1)})$$

with $x^{(k)} \neq x^{(k-1)}$, $x^{(k-1)} \neq \xi$.

In an analogous manner as in the proof of Theorem 7 it follows that

$$\xi \in x^{(k)} - \frac{x^{(k)} - x^{(k-1)}}{f(x^{(k)}) - f(x^{(k-1)})}(f(x^{(k)}) + \tfrac{1}{2}K(X^{(k+1,0)} - x^{(k)})(X^{(k+1,0)} - x^{(k-1)}))$$

and finally $\xi \in X^{(k+1,1)}$. It remains to show that

$$\xi \in X^{(k+1,i)}, \qquad i = 2, 3, \ldots, p.$$

For $x^{(k)} \neq \xi$ and $z^{(i)} \neq \xi$ one shows again using the remainder term of the Newton interpolation formula that from $\xi \in X^{(k+1,i-1)}$ it always follows that $\xi \in X^{(k+1,i)}$.

This is also trivially true for $x^{(k)} \neq \xi$ and $z^{(i)} = \xi$. Since

$$\xi \in X^{(k+1,1)}$$

one gets therefore

$$\xi \in X^{(k+1,i)}, \qquad i = 1, 2, \ldots, p,$$

and therefore also

$$\xi \in X^{(k+1)} = X^{(k+1,p)}.$$

Of (26): This follows immediately from the way $X^{(k+1,0)}$ is calculated. The elementary verification is omitted here.

Of (27): It can be assumed without loss of generality that $x^{(k)} \neq \xi$. It then follows immediately from the procedure that

$$d(X^{(k+1,0)}) < \tfrac{1}{2} d(X^{(k)}).$$

Analogous to the proof of statement (24) in Theorem 7 one gets

$$d(X^{(k+1,1)}) \leqslant \tfrac{1}{4}|K/H|\, d(X^{(k)})\, d(X^{(k-1)}),$$

and again in an analogous manner

$$d(X^{(k+1,i)}) \leqslant \tfrac{1}{4}|K/H|\, d(X^{(k+1,i-1)})\, d(X^{(k)}), \qquad i = 2, 3, \ldots, p.$$

The solution to this simple recursion is the relation

$$d(X^{(k+1,i)}) \leqslant \beta_i\, d(X^{(k)})^i\, d(X^{(k-1)}),$$
$$\beta_i = (\tfrac{1}{4}|K/H|)^i, \qquad i = 1, 2, \ldots, p.$$

Using $X^{(k+1)} = X^{(k+1,p)}$ we get the relation

$$d(X^{(k+1)}) \leqslant \beta_p d(X^{(k)})^p d(X^{(k-1)}),$$

from which we again get the required inequality for the R order using Theorem 3 from Appendix A. ■

Remarks: Various modifications of the methods (3), respectively, (11), were given and investigated. The version given in Krawczyk [34] was investigated in Ris [40]. A similar procedure to (3′) was given in [35] where it was used for improving the bounds for all the zeros in a prescribed interval $[x_1^{(0)}, x_2^{(0)}]$. The improvements on the lower bounds $x_1^{(k+1)}$ are calculated separately from the improvements on the upper bounds $x_2^{(k+1)}$. For this procedure we require a function value $f(x_1^{(k)})$ (resp. $f(x_2^{(k)})$) as well as a bound $m = |M|$ for $|f'(x)|$, $x \in [x_1^{(k)}, x_2^{(k)}]$. The sequence of lower bounds $x_1^{(0)}, x_1^{(1)}, \ldots$ then converges toward the smallest zero in $[x_1^{(0)}, x_2^{(0)}]$. The sequence of upper bounds correspondingly converges to the largest zero of $f(x)$ in $[x_1^{(0)}, x_2^{(0)}]$. An analogous method for polynomials was developed by Bauer [10].

An implicit representation of method (3) (resp. (11)) is used by Barth [8]. The function $f(x)$ is enclosed by the interval expression

$$I(x) = f(x^{(k)}) + (x - x^{(k)})M^{(k)}, \qquad x^{(k)} \in [x_1^{(k)}, x_2^{(k)}] = X^{(k)}$$

with $M^{(k)}$ defined similarly to (9). The intervals $[x_1^{(k+1)}, x_2^{(k+1)}]$ are then calculated using the requirement

$$[x_1^{(k+1)}, x_2^{(k+1)}] := \{x \mid x \in X^{(k)}, 0 \in I(x)\} \cap X^{(k)}.$$

Different cases arise depending on whether or not $0 \in M^{(k)}$. This method converges under certain conditions, and when there are several zeros, there are also several subsequences. A recursive procedure for this case is found in Barth [9]. This method is analogous to the method given explicitly by Alefeld [2] that was sketched at the end of Section C. It is also found in a closed form in the thesis by Kaucher [B 148]. This method was also recently considered by Hansen [25] and was applied to the determination of zeros of the derivative of a twice continuously differentiable function over an interval in order to find the global minimum of the function.

The results on methods (3) (resp. (11)) given in Theorem 1 (resp. Theorem 4) may be extended as follows. It is assumed that there exists an interval M, $0 \notin M$, for the function f in the interval $X^{(0)}$ having the property that

$$\frac{f(x) - f(y)}{x - y} \in M \qquad \text{for} \quad x, y \in X^{(0)}, \quad x \neq y. \tag{28}$$

If it holds that

$$X^{(1)} = m(X^{(0)}) - f(m(X^{(0)}))/M \subseteq X^{(0)},$$

then it follows that there is a $\xi \in X^{(1)}$ with $f(\xi) = 0$. This follows if one assumes without loss of generality that $f(m(X^{(0)})) > 0$, $m_1 > 0$, and then assumes $f(x_1^{(1)}) > 0$. One then arrives at a contradiction by considering

$$x_1^{(1)} > m(X^{(0)}) - \frac{f(m(X^{(0)}))}{(f(m(X^{(0)})) - f(x_1^{(1)}))/(m(X^{(0)}) - x_1^{(1)})} \geqslant x_1^{(1)}.$$

Therefore we must have that $f(x_1^{(1)}) \leqslant 0$, which implies that $X^{(0)}$ and as shown in Theorem 1 also $X^{(1)}$ contains a zero ξ. Existence statements of this type are found, for example, in Kahan [B 143] and Stetter [B 306]. See also Lemma 11.10.

We now wish to give a short derivation of an existence statement for a zero of the equation $f(x) = 0$ in the interval $X^{(0)} = [x^{(0)} - r, x^{(0)} + r]$ using interval arithmetic means. We assume that f is twice differentiable and that

$$|f''(x)| \leqslant \gamma, \qquad x \in X^{(0)}.$$

Furthermore, assume that $f'(x^{(0)}) \neq 0$, and set

$$\left|\frac{f(x^{(0)})}{f'(x^{(0)})}\right| = \eta, \qquad \left|\frac{1}{f'(x^{(0)})}\right| = \beta.$$

Then for $y \in X^{(0)}, x^{(0)} \neq y$, and a θ between $x^{(0)}$ and y we have

$$\frac{f(x^{(0)}) - f(y)}{x^{(0)} - y} = f'(x^{(0)}) + \tfrac{1}{2}f''(\theta)(y - x^{(0)})$$

$$\in f'(x^{(0)}) + \tfrac{1}{2}\gamma[-r, r] =: M^{(0)}.$$

If now $0 \notin M^{(0)}$ and

$$X^{(1)} = x^{(0)} - f(x^{(0)})/M^{(0)} \subseteq X^{(0)},$$

then one shows in the same manner as above that there is a zero $\xi \in X^{(1)} \subseteq X^{(0)}$. The requirement $X^{(1)} \subseteq X^{(0)}$ is satisfied iff

$$\tfrac{1}{2}\beta\gamma r^2 - r + \eta \leqslant 0.$$

This is equivalent to

(29) $$\beta\gamma\eta \leqslant \tfrac{1}{2}$$

together with

(30) $$(1 - \sqrt{1 - 2\beta\gamma\eta})/\beta\gamma \leqslant r \leqslant (1 + \sqrt{1 - 2\beta\gamma\eta})/\beta\gamma.$$

The inequalities (29) and (30) guarantee that the assumption of the Kantorovich theorem regarding the existence of a zero in the interval $X^{(0)}$ are satisfied (see [B 250, Theorem 22.4]).

These questions will be discussed again in Chapter 19. If we have $X^{(1)} \cap X^{(0)} = \varnothing$, then f has no zero in $X^{(0)}$. The condition $X^{(1)} \cap X^{(0)} = \varnothing$ is satisfied iff

$$\tfrac{1}{2}\beta\gamma r^2 + r - \eta < 0;$$

that is, if $\eta \neq 0$, iff

$$0 \leqslant r < (-1 + \sqrt{1 + 2\beta\gamma\eta})/\beta\gamma.$$

This statement may be used in a similar way to prove an exclusion theorem in Banach spaces. See Alefeld [3].

The theorem of Kantorovich is used by Rokne and Lancaster [B 274] in order to find an inclusion for a zero of a function f that also includes the effect of all the roundoff errors. Compare also with [B 273].

REFERENCES

[1] G. Alefeld, Intervallrechnung über den komplexen Zahlen und einige Anwendungen. Ph.D. Thesis, Univ. Karlsruhe (1968).

[2] G. Alefeld, Eine Modifikation des Newtonverfahrens zur Bestimmung der reellen Nullstellen einer reellen Funktion. *Z. Angew. Math. Mech.* **50**, T 32–T 33 (1970).

[3] G. Alefeld, An exclusion theorem for the solutions of operator equations. *Beitr. Numer. Math.* **6**, 7–10 (1977).

[4] G. Alefeld and N. Apostolatos, Auflösung nichtlinearer Gleichungssysteme mit zwei Unbekannten. Report, Rechenzentrum Inst. Angew. Math., Univ. Karlsruhe (1967).

[5] G. Alefeld and N. Apostolatos, Praktische Anwendung von Abschätzungsformeln bei Iterationsverfahren. *Z. Angew. Math. Mech.* **48**, T 46–T 49 (1968).

[6] N. Apostolatos, Ein Einschließungsverfahren für Nullstellen. *Z. Angew. Math. Mech.* **47**, T 80 (1967).

[7] N. Apostolatos, U. Kulisch, and K. Nickel, Ein Einschließungsverfahren für Nullstellen. *Computing* **2**, 185–201 (1967).

[8] W. Barth, Nullstellenbestimmung mit der Intervallrechnung. *Computing* **8**, 320–326 (1971).

[9] W. Barth, Ein Algorithmus zur Berechnung aller Nullstellen in einem Intervall. *Computing* **9**, 327–333 (1972).

[10] M. Bauer, Zur Bestimmung der reellen Wurzeln einer algebraischen Gleichung durch Iteration. *Jber. Deutsch. Math. Verein.* **25**, 294–301 (1917).

[11] R. P. Brent, An algorithm with guaranteed convergence for finding a zero of a function. *Comput. J.* **14**, 422–425 (1972).

[12] F. L. Chernous'ko, An optimal algorithm for finding the roots of an approximately computed function. *Zh. Vychisl. Mat. Mat. Fiz.* **8**, 705–724 (1968).

[13] H. H. Y. Chien, A multiphase algorithm for single variable equation solving. *J. Inst. Math. Appl.* **9**, 290–298 (1972).

[14] L. Collatz, "Funktionalanalysis und numerische Mathematik." Springer-Verlag, Berlin and New York, 1968.

[15] H. Cornelius, Intervallrechnung und Nullstellenbestimmung. Diplomarbeit TU Berlin (1979) (Not available).

[16] H. Cornelius, Ableitungsfreie Iterationsverfahren zur globalkonvergenten Nullstelleneinschließung. *Beitr. Numer. Math.* **11** (to appear).

[17] H. Cornelius, Inverse Interpolation zur globalkonvergenten Nullstellenberechnung. *Beitr. Numer. Math.* **10**, 23–35 (1981).

[18] R. H. Dargel, F. R. Loscalzo, and T. H. Witt, Automatic error bounds on real zeros of rational functions. *Comm. Assoc. Comput. Mach.* **9**, 806–809 (1966).

[19] M. Dowell and P. Jarratt, A modified regula falsi method for computing the root of an equation. *BIT* **11**, 168–174 (1971).

[20] H.-P. Dyck, Algorithmen zur numerischen Behandlung von Funktionen mit Fehlererfassung. Diplomarbeit, Inst. Angew. Math., Univ. Karlsruhe (1969).

[21] H. Ehrmann, Konstruktion und Durchführung von Iterationsverfahren höherer Ordnung. *Arch. Rational Mech. Anal.* **4**, 65–88 (1959).

[22] G. S. Ganshin, Extension of the convergence region of Newton's method. *Zh. Vychisl. Mat. mat. Fiz.* **11**, 1294–1296 (1971) [*English transl.*: *USSR J. Comp. Math. Math. Phys.* **11**(5), 249–252].

[23] N. B. Haaser and J. A. Sullivan, "Real Analysis," Chapter 5, Sect. 2. Van Nostrand Reinhold, New York, 1971.

[24] R. J. Hanson, Automatic error bounds for real roots of polynomials having interval coefficients. *Comput. J.* **13**, 284–288 (1970).

[25] E. R. Hansen, A globally convergent interval method for computing and bounding real roots. *BIT* **18**, 415–424 (1978).

[26] E. R. Hansen, Global optimization using interval analysis: The one-dimensional case. *J. Optim. Theory Appl.* **29**, 331–344 (1979).

[27] P. Henrici, Methods of search for solving polynomial equations. *J. Assoc. Comput. Mach.* **17**, 273–283 (1970).

[28] J. Herzberger, Über optimale Verfahren zur numerischen Bestimmung reeller Nullstellen mit Fehlerschranken. Habilitationsschrift, Univ. Karlsruhe (1972).

[29] J. Herzberger, Über die Nullstellenbestimmung bei näherungsweise berechneten Funktionen. *Computing* **10**, 23–31 (1972).

[30] J. Herzberger, Bemerkungen zu einem Verfahren von R. E. Moore. *Z. Angew. Math. Mech.* **53**, 356–358 (1973).

[31] J. Herzberger, Global konvergente Interpolationsmethoden zur Nullstellenbestimmung. *Beitr. Numer. Math.* **7**, 65–74 (1979).

[32] A. S. Householder, "The Numerical Treatment of a Single Nonlinear Equation." McGraw-Hill, New York, 1970.

[33] R. Klatte, Anwendung einer Verallgemeinerung der Modifikation des Newtonverfahrens. Diplomarbeit, Inst. Angew. Math., Univ. Karlsruhe (1969).

[34] R. Krawczyk, Newton-Algorithmen zur Bestimmung von Nullstellen mit Fehlerschranken. *Computing* **4**, 187–201 (1969).

[35] R. Krawczyk, Einschließung von Nullstellen mit Hilfe einer Intervallarithmetik. *Computing* **5**, 356–370 (1970).

[36] R. E. Moore, "Interval Analysis." Prentice-Hall, Englewood Cliffs, New Jersey, 1966.

[37] K. Nickel, Triplex-Algol and applications. *In* "Topics in Interval Analysis" (E. Hansen, ed.). Oxford Univ. Press (Clarendon), London and New York, 1969.

[38] A. Ostrowski, "Solution of Equations and Systems of Equations." Academic Press, New York, 1960.

[39] G. Peters and J. H. Wilkinson, Practical problems arising in the solution of polynomial equations. *J. Inst. Math. Appl.* **8**, 16–35 (1971).

[40] F. Ris, Interval Analysis and Applications to Linear Algebra. Ph.D. Thesis, Oxford Univ. (1972).

[41] J. W. Schmidt and H. Schwetlick, Ableitungsfreie Verfahren mit höherer Konvergenzgeschwindigkeit. *Computing* **3**, 215–226 (1968).

[42] J. W. Schmidt, Monotone Einschließung mit der Regula falsi bei konvexen Funktionen. *Z. Angew. Math. Mech.* **50**, 640–643 (1970).

[43] J. Stoer, "Einführung in die Numerische Mathematik I." Springer-Verlag, Berlin and New York, 1972.

[44] F. Stummel and K. Hainer, "Praktische Mathematik." Teubner Verlag, Stuttgart, 1971.
[45] T. Sunaga, Theory of an interval algebra and its applications to numerical analysis. "RAAG Memoirs II." Gakujutsu Bunken Fukyu-Kai Tokyo, 1958.
[46] J. F. Traub, "Iterative Methods for the Solution of Equations." Prentice-Hall, Englewood Cliffs, New Jersey, 1964.

Chapter 8 / METHODS FOR THE SIMULTANEOUS INCLUSION OF REAL ZEROS OF POLYNOMIALS

We shall in this chapter consider Newton-like interval methods that compute including intervals for all the real zeros of a real polynomial. The case where all zeros are real is treated first. Complex roots will be treated in Chapter 9. If all the zeros of the polynomial are real and simple, then we develop a single step method converging faster than quadratically. As an application we use this method to compute all the eigenvalues of a symmetric tridiagonal matrix.

We are given the real polynomial

$$p(x) = a^{(n)}x^n + a^{(n-1)}x^{n-1} + \cdots + a^{(0)}, \tag{1}$$

and we assume that

$$a^{(n)} = 1$$

in the sequel. The polynomial is furthermore assumed to have n real roots $\xi^{(1)}, \xi^{(2)}, \ldots, \xi^{(n)}$ that are collected together in the vector $(\xi^{(i)})$ where multiple roots are entered according to their multiplicities. Including intervals

$$\xi^{(j)} \in X^{(0,j)} = [x_1^{(0,j)}, x_2^{(0,j)}], \qquad 1 \leqslant j \leqslant n,$$

are assumed to be known for all the roots.

It is first assumed that these including intervals are pairwise disjoint; that is,

$$X^{(0,j)} \cap X^{(0,k)} = \varnothing, \qquad 1 \leqslant j < k \leqslant n. \tag{2}$$

The polynomial $p(x)$ may be written

$$p(x) = \prod_{j=1}^{n} (x - \xi^{(j)})$$

or

$$p(x) = (x - \xi^{(i)}) \prod_{j=1, j \neq i}^{n} (x - \xi^{(j)}),$$

from which

$$\xi^{(i)} = x - p(x) \Big/ \prod_{j=1, j \neq i}^{n} (x - \xi^{(j)})$$

follows. If one chooses $x = x^{(0,i)} \in X^{(0,i)}$, then we get

$$0 \notin \prod_{j=1, j \neq i}^{n} (x^{(0,i)} - X^{(0,j)})$$

and from (1.9) it follows that

$$\xi^{(i)} \in X^{(1,i)} = \left\{ x^{(0,i)} - p(x^{(0,i)}) \Big/ \prod_{j=1, j \neq i}^{n} (x^{(0,i)} - X^{(0,j)}) \right\} \cap X^{(0,i)}.$$

The interval expression on the right is therefore a new including interval $X^{(1,i)}$ for which

$$\xi^{(i)} \in X^{(1,i)} \subseteq X^{(0,i)}$$

holds. This relation gives rise to the following iteration scheme:

$$X^{(k+1,i)} = \left\{ x^{(k,i)} - p(x^{(k,i)}) \Big/ \prod_{j=1, j \neq i}^{n} (x^{(k,i)} - X^{(k,j)}) \right\} \cap X^{(k,i)}, \tag{3}$$

with

$$x^{(k,i)} \in X^{(k,i)}, \qquad 1 \leqslant i \leqslant n, \qquad k \geqslant 0.$$

The interval expression in the denominator is shortened to

$$Q^{(k,i)} = \prod_{j=1, j \neq i}^{n} (x^{(k,i)} - X^{(k,j)}).$$

The iteration scheme given by (3) is a total step method for the simultaneous inclusion of the polynomial roots $\xi^{(i)}$, $1 \leqslant i \leqslant n$. If we always use the last computed values of the inclusion intervals in $Q^{(k,i)}$, then we get

$$R^{(k,i)} = \prod_{j=1}^{i-1} (x^{(k,i)} - X^{(k+1,j)}) \prod_{j=i+1}^{n} (x^{(k,i)} - X^{(k,j)})$$

leading to the corresponding single step iteration. For this single step iteration, we now wish to make similar considerations as for (7.13). Depending on the sign of $p(x^{(k+1,i)})$ and $R^{(k,i)}$ the including interval $X^{(k+1,i)}$ is shrunk to $Y^{(k+1,i)}$. The sign function for intervals is defined by

$$\text{(4)} \qquad \operatorname{sign}(X) = \begin{cases} 1 & \text{if} \quad x_1 > 0 \\ -1 & \text{if} \quad x_2 < 0 \\ 0 & \text{otherwise.} \end{cases}$$

The intervals $Y^{(k+1,i)}$ that also contain the zeros $\xi^{(i)}$ are then defined as follows

$$Y^{(k+1,i)} = \begin{cases} [x_1^{(k+1,i)}, x^{(k+1,i)}] & \text{if} \quad \operatorname{sign}(R^{(k,i)}) \operatorname{sign}(p(x^{(k+1,i)})) > 0 \\ [x^{(k+1,i)}, x_2^{(k+1,i)}] & \text{if} \quad \operatorname{sign}(R^{(k,i)}) \operatorname{sign}(p(x^{(k+1,i)})) < 0 \\ X^{(k+1,i)} & \text{otherwise.} \end{cases}$$

Note that

$$\operatorname{sign}(R^{(0,i)}) = \operatorname{sign}(R^{(1,i)}) = \cdots, \qquad 1 \leqslant i \leqslant n,$$

is always true. Using these new including intervals we now compute the new expression in the denominator as

$$S^{(k+1,i)} = \prod_{j=1}^{i-1} (x^{(k+1,i)} - Y^{(k+2,j)}) \cdot \prod_{j=i+1}^{n} (x^{(k+1,i)} - Y^{(k+1,j)}).$$

Employing this we arrive at the following modified single step method:

$$\text{(5)} \quad \left\{ \begin{aligned} & Y^{(0,i)} = X^{(0,i)}, \; x^{(0,i)} \in X^{(0,i)}, \\ & X^{(k+1,i)} = \{x^{(k,i)} - p(x^{(k,i)})/S^{(k,i)}\} \cap X^{(k,i)}, \\ & \text{with} \\ & S^{(k,i)} = \prod_{j=1}^{i-1} (x^{(k,i)} - Y^{(k+1,j)}) \prod_{j=i+1}^{n} (x^{(k,i)} - Y^{(k,j)}), \\ & x^{(k+1,i)} \in X^{(k+1,i)}, \\ & Y^{(k+1,i)} = \begin{cases} [x_1^{(k+1,i)}, x^{(k+1,i)}] & \text{if} \quad \operatorname{sign}(S^{(k,i)}) \operatorname{sign}(p(x^{(k+1,i)})) > 0 \\ [x^{(k+1,i)}, x_2^{(k+1,i)}] & \text{if} \quad \operatorname{sign}(S^{(k,i)}) \operatorname{sign}(p(x^{(k+1,i)})) < 0 \\ X^{(k+1,i)} & \text{otherwise} \end{cases} \\ & 1 \leqslant i \leqslant n, \qquad k \geqslant 0. \end{aligned} \right.$$

Both methods (3) and (5) can be considered to be interval versions of known methods for the simultaneous determination of polynomial roots [7–9, 12]. The advantage of the interval versions of these methods is that they not only give including intervals for the roots, but they are also always convergent under the above assumptions. This is shown in the following theorem.

Theorem 1: Let the polynomial (1) be given with n simple real roots $\xi^{(i)}$, $1 \leqslant i \leqslant n$. Including intervals $X^{(0,i)} \ni \xi^{(i)}$, $1 \leqslant i \leqslant n$, are furthermore known for which (2) is valid. It then follows that the sequence $\{X^{(k,i)}\}_{k=0}^{\infty}$, $1 \leqslant i \leqslant n$, of iterates from (3) (resp., (5)) satisfies

$$\xi^{(i)} \in X^{(k,i)}, \qquad k \geqslant 0$$

and

$$X^{(0,i)} \supset X^{(1,i)} \supset X^{(2,i)} \supset \cdots \qquad \text{with} \quad \lim_{k \to \infty} X^{(k,i)} = \xi^{(i)},$$

or the sequence comes to rest at $[\xi^{(i)}, \xi^{(i)}]$ after a finite number of steps. ■

The statements of Theorem 1 for methods (3) and (5) follow in the same manner as the corresponding statements of Theorem 7.1, since both of the interval expressions $Q^{(k,i)}$ and $S^{(k,i)}$ either have the property (7.2) or they have the corresponding property with $m_2 < 0$ in the interval $X^{(k,i)}$.

By substituting

$$x^{(k,i)} = \tfrac{1}{2}(x_1^{(k,i)} + x_2^{(k,i)})$$

in both methods and considering the construction of (3) and (5), it follows immediately that the width of the inclusions for each zero is at least halved at each calculation of a new iterate.

Theorem 1 remains partially valid when the polynomial has multiple roots. If we collect these multiple roots together as

$$\xi^{(m)}, \xi^{(m+1)}, \ldots, \xi^{(n)},$$

then both methods (3) and (5) have to be changed so that the new calculations of the including intervals are only done for the indices $1 \leqslant i < m$. The assertions of Theorem 1 are then valid only for the simple zeros where the including intervals are recomputed in each step. The other intervals remain unchanged (see Alefeld and Herzberger [3]).

The iteration (3) may also be generalized in such a manner that in Theorem 1 the assumption (2) on the including intervals $X^{(0,i)}$, $1 \leqslant i \leqslant n$, may be replaced by a weaker condition. In order to do this one makes full use of the possibility that $x^{(k,i)} \in X^{(k,i)}$ rather than using a systematic selection procedure, for example, the arithmetic means of the bounds. Such a generalization was treated in Alefeld and Herzberger [2].

We shall now consider the behavior of the sequence

$$\{d(X^{(k,i)})\}_{k=0}^{\infty}, \qquad 1 \leqslant i \leqslant n,$$

of widths in more detail. In method (3) we therefore estimate

$$\begin{aligned} d(X^{(k+1,i)}) &\leqslant d(\{x^{(k,i)} - p(x^{(k,i)})/Q^{(k,i)}\}) \\ &= d(p(x^{(k,i)})/Q^{(k,i)}) = |p(x^{(k,i)})|\, d(1/Q^{(k,i)}), \end{aligned}$$

using (2.9), (2.10), and (2.14). Since

$$|p(x^{(k,i)})| = |p(x^{(k,i)}) - p(\xi^{(i)})| = |(x^{(k,i)} - \xi^{(i)})p'(\tilde{\eta}^{(k,i)})|$$
$$\leqslant d(X^{(k,i)})|p'(\tilde{\eta}^{(k,i)})| \leqslant d(X^{(k,i)})|p'(X^{(0,i)})|,$$

it follows that

$$d(X^{(k+1,i)}) \leqslant d(X^{(k,i)})|p'(X^{(0,i)})|\,d(1/Q^{(k,i)}).$$

Using Theorem 3.5 one gets the estimate

$$d(1/Q^{(k,i)}) \leqslant \gamma^{(k,i)}\,d(Q^{(k,i)}),$$

and since

$$Q^{(k,i)} \subseteq \prod_{j=1,\,j\neq i}^{n} (X^{(0,i)} - X^{(0,j)}),$$

one even gets

$$d\left(\frac{1}{Q^{(k,i)}}\right) \leqslant \gamma^{(i)}\,d(Q^{(k,i)}) = \gamma^{(i)}\,d\left(\prod_{j=1,\,j\neq i}^{n} (x^{(k,i)} - X^{(k,j)})\right)$$

with constants $\gamma^{(i)}$ that are only dependent on $X^{(0,j)}$, $1 \leqslant j \leqslant n$. Through repeated application of (2.12) one further obtains

$$d\left(\frac{1}{Q^{(k,i)}}\right) \leqslant \gamma^{(i)} \sum_{j=1,\,j\neq i}^{n} \eta^{(i,j)}\,d(X^{(k,j)})$$

with suitable constants $\eta^{(i,j)}$ that only depend upon $X^{(0,j)}$, $1 \leqslant j \leqslant n$, since $X^{(k,j)} \subseteq X^{(0,j)}$ and since we apply (2.3) and (2.9). By collecting these facts together we get the following inequality

(6) $$d(X^{(k+1,i)}) \leqslant |p'(X^{(0,i)})|\gamma^{(i)}\,d(X^{(k,i)}) \sum_{j=1,\,j\neq i}^{n} \eta^{(i,j)}\,d(X^{(k,j)}), \qquad 1 \leqslant i \leqslant n.$$

The same discussion may be carried through for (5) where the only additional consideration is the fact that $Y^{(k,i)} \subseteq X^{(k,i)}$ must be taken into account. This leads to

(7) $$d(X^{(k+1,i)}) \leqslant |p'(X^{(0,i)})|\gamma^{(i)}\,d(X^{(k,i)})\left(\sum_{j=1}^{i-1} \eta^{(i,j)}\,d(X^{(k+1,j)}) + \sum_{j=i+1}^{n} \eta^{(i,j)}\,d(X^{(k,j)})\right), \qquad 1 \leqslant i \leqslant n.$$

The next theorem will prove statements on the *R* order of the iterations (3) and (5) (see Appendix A).

Theorem 2: The assumptions and the notation are as in Theorem 1. For the *R* order of the iteration methods (3) and (5), it holds that

(8) $$O_R((3), (\xi^{(i)})) \geqslant 2$$

and

(9) $$O_R((5), (\xi^{(i)})) \geqslant 1 + \sigma^{(n)},$$

whereby $\sigma^{(n)} > 1$ is the unique positive root of the polynomial

$$\tilde{q}^{(n)}(y) = y^n - y - 1.$$

Proof: Of (8): We immediately get from (6) that

$$\begin{aligned} d(X^{(k+1,i)}) &\leqslant |p'(X^{(0,i)})|\gamma^{(i)}\left(\sum_{j=1, j\neq i}^{n} \eta^{(i,j)}\right)(d^{(k)})^2 \\ &\leqslant \max_{1\leqslant i\leqslant n} \left\{|p'(X^{(0,i)})|\gamma^{(i)}\left(\sum_{j=1, j\neq i}^{n} \eta^{(i,j)}\right)\right\}(d^{(k)})^2 \\ &\leqslant \gamma\,(d^{(k)})^2, \qquad 1 \leqslant i \leqslant n, \end{aligned}$$

where

$$d^{(k)} = \max_{1\leqslant i\leqslant n} \{d(X^{(k,i)})\}.$$

It then follows that

$$d^{(k+1)} = \max_{1\leqslant i\leqslant n} \{d(X^{(k+1,i)})\} \leqslant \gamma\,(d^{(k)})^2.$$

From Theorem 2 in Appendix A, it then follows that

$$O_R((3), (\xi^{(i)})) \geqslant 2.$$

Of (9): The proof of (9) requires more effort than the proof of (8). Let

$$\gamma = \max_{1\leqslant i,j\leqslant n} \{\eta^{(i,j)}|p'(X^{(0,i)})|\gamma^{(i)}\}.$$

Then we may rewrite (7)

$$d(X^{(k+1,i)}) \leqslant \gamma\, d(X^{(k,i)})\left(\sum_{j=1}^{i-1} d(X^{(k+1,j)}) + \sum_{j=i+1}^{n} d(X^{(k,j)})\right).$$

By using the substitution

$$d(X^{(k,i)}) = \frac{1}{(n-1)\gamma} h^{(k,i)}, \qquad 1 \leqslant i \leqslant n, \qquad \hat{\varepsilon} = \frac{1}{n-1},$$

it can be written

$$h^{(k+1,i)} \leqslant \hat{\varepsilon} h^{(k,i)}\left(\sum_{j=1}^{i-1} h^{(k+1,j)} + \sum_{j=i+1}^{n} h^{(k,j)}\right).$$

Assume without loss of generality that

$$h^{(0,i)} \leqslant h < 1, \qquad 1 \leqslant i \leqslant n.$$

We then have

$$h^{(k+1,i)} \leqslant h^{u^{(k+1,i)}}, \qquad 1 \leqslant i \leqslant n, \quad k \geqslant 0.$$

The vector $\mathfrak{u}^{(k+1)} = (u^{(k+1,1)}, \ldots, u^{(k+1,n)})^{\mathrm{T}}$ with integer components can be calculated from the rule

$$\mathfrak{u}^{(k+1)} = \mathscr{A}\mathfrak{u}^{(k)}, \qquad k \geqslant 0,$$

with the help of the starting vector $\mathfrak{u}^{(0)} = (1, 1, \ldots, 1)^{\mathrm{T}}$. (Vectors with real components, so-called *point vectors* are denoted by $\mathfrak{a}, \mathfrak{b}, \ldots$, in order that they may be distinguished from interval vectors. The corresponding is also done for "point matrices." These notations are introduced in a more thorough manner in Chapter 10.)

The matrix $\mathscr{A}$ then has the form

$$\mathscr{A} = \begin{pmatrix} 1 & 1 & & & & & \\ & 1 & 1 & & 0 & & \\ & & 1 & 1 & & & \\ & & & \ddots & \ddots & & \\ & 0 & & & & & \\ & & & & & 1 & 1 \\ 1 & 1 & 0 & & \cdots & 0 & 1 \end{pmatrix}.$$

The proof of this relation may be done using complete induction, and we omit it here. The matrix $\mathscr{A}$ is nonnegative and its directed graph (see Varga [17]) is clearly strongly connected. From this it follows that $\mathscr{A}$ is irreducible. From the theorem of Perron–Frobenius it then follows that $\mathscr{A}$ has an eigenvalue $\lambda^{(1)}$ equal to the spectral radius $\rho(\mathscr{A})$. Theorem 2.9 in Varga [17] states that $\mathscr{A}$ is also primitive. The remaining eigenvalues of $\mathscr{A}$ then satisfy the relation

$$\lambda^{(1)} = \rho(\mathscr{A}) > |\lambda^{(2)}| \geqslant \cdots \geqslant |\lambda^{(n)}|.$$

Since $\mathscr{A}$ is primitive it holds for a natural number $k^{(0)}$ that

$$\mathscr{A}^k = (a_{ij}^{(k)}) > \mathscr{O}, \qquad k \geqslant k^{(0)}.$$

It is, however, shown in Gröbner [10] that for matrices with the last two properties we have

$$\lim_{k \to \infty} (a_{ij}^{(k+1)}/a_{ij}^{(k)}) = \lambda^{(1)}.$$

For a given $\varepsilon > 0$ it is therefore true that

$$a_{ij}^{(k+1)}/a_{ij}^{(k)} \geqslant \rho(\mathscr{A}) - \varepsilon, \qquad k \geqslant k(\varepsilon) \geqslant k^{(0)},$$

or

$$a_{ij}^{(k+1)} \geqslant \alpha(\rho(\mathscr{A}) - \varepsilon), \qquad 1 \leqslant i, j \leqslant n,$$

where

$$\alpha = \min_{1 \leqslant i, j \leqslant n} a_{ij}^{(k)} > 0.$$

From this it then follows that

$$a_{ij}^{(k+2)} \geqslant a_{ij}^{(k+1)}(\rho(\mathscr{A}) - \varepsilon) \geqslant \alpha(\rho(\mathscr{A}) - \varepsilon)^2$$

or generally

$$a_{ij}^{(k+r)} \geqslant \alpha(\rho(\mathscr{A}) - \varepsilon)^r, \qquad 1 \leqslant i, \quad j \leqslant n, \quad r \geqslant 0.$$

If one uses this in the rule for calculating the vectors $u^{(k)}$ then one gets

$$u^{(k+r)} = \mathscr{A}^{k+r} u^{(0)} = \left(\sum_{j=1}^{n} a_{ij}^{(k+r)} \right)$$
$$\geqslant (n\alpha(\rho(\mathscr{A}) - \varepsilon)^r)e,$$

with $e = (1, 1, \ldots, 1)^{\mathrm{T}}$. We therefore get

$$h^{(k+r,i)} \leqslant h^{u^{(k+r,i)}} \leqslant h^{n\alpha(\rho(\mathscr{A}) - \varepsilon)^r},$$

$$1 \leqslant i \leqslant n, \qquad r \geqslant 0, \qquad k \geqslant k(\varepsilon) \geqslant k^{(0)}.$$

Expressed differently this means that

$$d(X^{(k+r,i)}) \leqslant (\hat{\varepsilon}/\gamma)h^{n\alpha(\rho(\mathscr{A}) - \varepsilon)^r}.$$

Let now

$$d^{(k)} = \max_{1 \leqslant i \leqslant n} \{d(X^{(k,i)})\}.$$

Then we get

$$d^{(k+r)} \leqslant (\hat{\varepsilon}/\gamma)h^{\alpha n(\rho(\mathscr{A}) - \varepsilon)^r}.$$

We may therefore conclude that the R factor satisfies

$$R_{\rho(\mathscr{A}) - \varepsilon}\{d^{(k)}\} = \limsup_{r \to \infty} (d^{(k+r)})^{[1/(\rho(\mathscr{A}) - \varepsilon)^r]}$$
$$\leqslant \limsup_{r \to \infty} \left(\frac{\hat{\varepsilon}}{\gamma} h^{\alpha n(\rho(\mathscr{A}) - \varepsilon)^r} \right)^{[1/(\rho(\mathscr{A}) - \varepsilon)^r]}$$
$$= h^{\alpha n} < 1.$$

From this it follows that

$$O_R((5), (\xi^{(i)})) \geqslant \rho(\mathscr{A}) - \varepsilon$$

for all $\varepsilon > 0$, and hence

$$O_R((5), (\xi^{(i)})) \geqslant \rho(\mathscr{A}).$$

Consider now the characteristic polynomial $q^{(n)}(\lambda)$ of the matrix $\mathscr{A}$

$$q^{(n)}(\lambda) = (\lambda - 1)^n - (\lambda - 1) - 1.$$

This can be written

$$\tilde{q}^{(n)}(\tau) = \tau^n - \tau - 1,$$

setting $\tau = \lambda - 1$.

The polynomial $\tilde{q}^{(n)}(\tau)$ has, according to the rule of Descartes, exactly one positive zero $\sigma^{(n)}$ for which

$$1 < \sigma^{(n)} < 2$$

since

$$\tilde{q}^{(n)}(1) = -1 < 0 \qquad \text{and} \qquad \tilde{q}^{(n)}(2) = 2^n - 3 \geqslant 1 > 0$$

is valid for $n \geqslant 2$. The spectral radius of $\mathscr{A}$ therefore satisfies

$$\rho(\mathscr{A}) = 1 + \sigma^{(n)} > 2$$

from which

$$O_R((5), (\xi^{(i)})) \geqslant 1 + \sigma^{(n)}. \quad \blacksquare$$

We now wish to present an application of the iteration (5). A real symmetric $n \times n$ matrix $\mathscr{A}' = (a_{ij})$ is given. The eigenvalues of the matrix $\mathscr{A}'$ are to be determined, that is, numbers λ for which

$$\mathscr{A}'x = \lambda x \qquad \text{with} \quad x \neq o$$

holds. In order to do this one applies a finite sequence of orthogonal similarity transformations of the form

$$\tilde{\mathscr{A}} = \mathscr{U}^{\mathrm{T}}\mathscr{A}\mathscr{U}$$

transforming the generally full matrix $\mathscr{A}'$ into a matrix $\mathscr{A}$ of the form

$$\mathscr{A} = \begin{pmatrix} a^{(1)} & b^{(1)} & & & 0 \\ b^{(1)} & a^{(2)} & b^{(2)} & & \\ & \ddots & \ddots & \ddots & \\ & 0 & & b^{(n-1)} & a^{(n)} \end{pmatrix}.$$

The eigenvalues of $\mathscr{A}$ (and therefore also those of $\mathscr{A}'$) are then calculated as the zeros of the characteristic polynomial

$$p(\lambda) = \det(\lambda\mathscr{I} - \mathscr{A})$$

of $\mathscr{A}$. The value of $p(\lambda)$ may be determined from the following recursion formula (see Schwarz [15]):

$$(10) \quad \begin{cases} f^{(0)}(\lambda) = 1, \qquad f^{(1)}(\lambda) = \lambda - a^{(1)}, \\ f^{(k)}(\lambda) = (\lambda - a^{(k)})f^{(k-1)}(\lambda) - (b^{(k-1)})^2 f^{(k-2)}(\lambda), \qquad 2 \leqslant k \leqslant n, \\ p(\lambda) = f^{(n)}(\lambda). \end{cases}$$

If now $\mathscr{A}$ has only simply eigenvalues and if disjoint including intervals are known for these from, for example, Gersgorin's theorem, then one may apply (5). The following example demonstrates this.

Example: (α) Consider the matrix

$$\mathscr{A} = \begin{pmatrix} 15 & 1 & & & & & & & \\ 1 & 10 & 1 & & & & 0 & & \\ & 1 & 7 & 1 & & & & & \\ & & 1 & 4 & 1 & & & & \\ & & & 1 & 0 & 1 & & & \\ & & & & 1 & -4 & 1 & & \\ & & & & & 1 & -7 & 1 & \\ & 0 & & & & & 1 & -10 & 1 \\ & & & & & & & 1 & -15 \end{pmatrix}.$$

Applying Gersgorin's theorem one obtains including intervals for the eigenvalues of $\mathscr{A}$:

$$\begin{aligned} X^{(0,1)} &= [+\,13.99999999995,\ +\,16.00000000005], \\ X^{(0,2)} &= [+\,7.999999999974,\ +\,12.00000000005], \\ X^{(0,3)} &= [+\,4.999999999981,\ +\,9.000000000015], \\ X^{(0,4)} &= [+\,1.999999999992,\ +\,6.000000000022], \\ X^{(0,5)} &= [-\,2.000000000008,\ +\,2.000000000008], \\ X^{(0,6)} &= [-\,6.000000000022,\ -\,1.999999999992], \\ X^{(0,7)} &= [-\,9.000000000015,\ -\,4.999999999981], \\ X^{(0,8)} &= [-\,12.00000000005,\ -\,7.999999999974], \\ X^{(0,9)} &= [-\,17.00000000005,\ -\,12.99999999995]. \end{aligned}$$

With these intervals one iterates according to (5) getting the following results:

$$\begin{aligned} X^{(5,1)} &= [+\,\underline{15.197093008}68, & &+\,\underline{15.197093008}72], \\ X^{(4,2)} &= [+\,\underline{10.131745154}64, & &+\,\underline{10.131745154}71], \\ X^{(4,3)} &= [+\,\underline{7.0019275809}04, & &+\,\underline{7.0019275809}60], \\ X^{(4,4)} &= [+\,\underline{3.920346203}678, & &+\,\underline{3.920346203}715], \\ X^{(5,5)} &= [-\,0.1096791595101 \times 10^{-10}, & &+\,0.1096791595101 \times 10^{-10}], \\ X^{(4,6)} &= [-\,\underline{3.920346203}719, & &-\,\underline{3.920346203}674], \\ X^{(4,7)} &= [-\,\underline{7.001927580}969, & &-\,\underline{7.001927580}895], \\ X^{(3,8)} &= [-\,\underline{10.131745154}73, & &-\,\underline{10.131745154}63], \\ X^{(3,9)} &= [-\,\underline{15.197093008}76, & &-\,\underline{15.197093008}66]. \end{aligned}$$

These intervals cannot be improved any further using the available program (see Appendix C). The digits coinciding in the upper and lower bounds are underlined.

(β) Consider further the matrix

$$\mathscr{A} = \begin{pmatrix} 12 & 1 & & & 0 \\ 1 & 9 & 1 & & \\ & 1 & 6 & 1 & \\ & & 1 & 3 & 1 \\ 0 & & & 1 & 0 \end{pmatrix}$$

(compare to Stoer and Bulirsch [16, p. 375]). Again using Gersgorin's theorem we get the following including intervals for the eigenvalues of $\mathscr{A}$:

$$X^{(0,1)} = [+\,10.99999999998,\ +\,13.00000000003],$$
$$X^{(0,2)} = [+\,6.999999999970,\ +\,11.00000000003],$$
$$X^{(0,3)} = [+\,3.999999999989,\ +\,8.000000000021],$$
$$X^{(0,4)} = [+\,0.9999999999945,\ +\,5.000000000019],$$
$$X^{(0,5)} = [-\,1.000000000004,\ +\,1.000000000004].$$

The following improved intervals were then calculated using the iteration procedure (5) (compare with remarks following Theorem 1):

$$X^{(1,1)} = [+\,\underline{12}.11013986010,\ +\,\underline{12}.55506993010],$$
$$X^{(1,2)} = [+\,\underline{9.0}06328989416,\ +\,\underline{9.0}48379503166],$$
$$X^{(1,3)} = [+\,5.999999999958,\ +\,6.000000000041],$$
$$X^{(1,4)} = [+\,\underline{2.9}79804773200,\ +\,\underline{2.9}87022580008],$$
$$X^{(1,5)} = [-\,0.\underline{3}230758693540,\ -\,0.\underline{3}162523763767],$$

$$X^{(2,1)} = [+\,\underline{12.31}617201370,\ +\,\underline{12.31}774922532],$$
$$X^{(2,2)} = [+\,\underline{9.0161}10401580,\ +\,\underline{9.0161}49094187],$$
$$X^{(2,3)} = [+\,5.999999999958,\ +\,6.000000000013],$$
$$X^{(2,4)} = [+\,\underline{2.98386}0239266,\ +\,\underline{2.98386}4788268],$$
$$X^{(2,5)} = [-\,0.\underline{3168759526}293,\ -\,0.\underline{31687595260}51],$$

$$X^{(3,1)} = [+\,\underline{12.31687595}112,\ +\,\underline{12.31687595}546],$$
$$X^{(3,2)} = [+\,\underline{9.0161363031}34,\ +\,\underline{9.0161363031}98],$$
$$X^{(3,3)} = X^{(2,3)}$$
$$X^{(3,4)} = [+\,\underline{2.9838636968}23,\ +\,\underline{2.9838636968}53],$$
$$X^{(3,5)} = [-\,0.\underline{3168759526}293,\ -\,0.\underline{31687595260}51],$$

$$X^{(4,1)} = [+\,\underline{12.316875952}58,\ +\,\underline{12.316875952}66],$$

$$X^{(4,2)} = [+\,\underline{9.0161363031}34,\ +\,\underline{9.0161363031}81],$$

$$X^{(4,3)} = X^{(3,3)}$$

$$X^{(4,4)} = X^{(3,4)}$$

$$X^{(4,5)} = [-\,0.\underline{3168759526}284,\ -\,0.\underline{3168759526}051].$$

REFERENCES

[1] O. Aberth, Iteration methods for finding all zeros of polynomial simultaneously. Technical Report, Texas A & M Univ. (1971); *Math. Comp.* **27**, 339–344 (1971).

[2] G. Alefeld and J. Herzberger, Über Simultanverfahren zur Bestimmung reeller Polynomwurzeln. *Z. Angew. Math. Mech.* **547**, 413–420 (1974).

[3] G. Alefeld and J. Herzberger, Über die Verbesserung von Schranken für die Eigenwerte symmetrischer Tridiagonalmatrizen. *Angew. Informat.* **17**, 27–35 (1974).

[4] G. Alefeld and J. Herzberger, On the convergence speed of some algorithms for the simultaneous approximation of polynomial roots. *SIAM J. Numer. Anal.* **11**, 237–243 (1974).

[5] W. Börsch-Supan, A posteriori error bounds for the zeros of polynomials. *Numer. Math.* **5**, 380–398 (1963).

[6] D. Braess and K. P. Hadeler, Simultaneous inclusion of the zeros of a polynomial. *Numer. Math.* **21**, 161–165 (1973).

[7] K. Dochev and P. Byrnev, Certain modifications of Newton's method for the approximate solution of algebraic equations. *Zh. Vych. Mat.* **4**, 915–920 (1964).

[8] E. Durand, "Solutions Numeriques des Equations Algébriques," Vol. I. Masson, Paris, 1960.

[9] L. W. Ehrlich, A modified Newton method for polynomials. *Comm. ACM* **10**, 107–108 (1967).

[10] W. Gröbner, "Matrizenrechnung." Bibliograph. Inst., Mannheim, 1966.

[11] J. Herzberger, Bemerkungen zu einem Verfahren von R. E. Moore. *Z. Angew. Math. Mech.* **53**, 356–358 (1973).

[12] I. O. Kerner, Ein Gesamtschrittverfahren zur Berechnung der Nullstellen von Polynomen. *Numer. Math.* **8**, 290–294 (1966).

[13] J. Petrić, M. Jovanović, and S. Stamatović, Algorithm for simultaneous determination of all roots of algebraic polynomial equations. *Mat. Vesnik* **9**, 325–332 (1972).

[14] J. W. Schmidt and H. Dressel, Fehlerabschätzungen bei Polynomgleichungen mit dem Fixpunktsatz von Brouwer. *Numer. Math.* **10**, 42–50 (1967).

[15] H. R. Schwarz, H. Rutishauser, and E. Stiefel, "Matrizen-Numerik." Teubner Verlag, Stuttgart, 1968.

[16] J. Stoer and R. Bulirsch, "Introduction to Numerical Analysis." Springer-Verlag, Berlin and New York, 1980.

[17] R. S. Varga, "Matrix Iterative Analysis." Prentice-Hall, Englewood Cliffs, New Jersey, 1962.

Chapter 9 / METHODS FOR THE SIMULTANEOUS INCLUSION OF COMPLEX ZEROS OF POLYNOMIALS

We shall in this chapter discuss a method for the simultaneous inclusion of the generally complex zeros of a polynomial due to Gargantini and Henrici [3]. Let the polynomial $p(z)$ be given as

$$(1) \qquad p(z) = a^{(n)}z^n + a^{(n-1)}z^{n-1} + \cdots + a^{(1)}z + a^{(0)}$$

with $a^{(i)} \in \mathbb{C}$, $0 \leqslant i \leqslant n$, $n \geqslant 2$. Furthermore assume that n intervals

$$W^{(0,i)} = \langle z^{(0,i)}, r^{(0,i)} \rangle \in K(\mathbb{C})$$

are given for which

$$(2) \qquad \zeta^{(i)} \in W^{(0,i)}, \qquad p(\zeta^{(i)}) = 0, \qquad 1 \leqslant i \leqslant n,$$

$$(3) \qquad W^{(0,i)} \cap W^{(0,j)} = \varnothing, \qquad 1 \leqslant i < j \leqslant n.$$

An element $Z \in K(\mathbb{C})$ is subsequently also represented by $Z = \langle m(Z), r(Z) \rangle$.

Let us consider the following iteration

$$(4) \qquad \begin{cases} z^{(k,i)} = m(W^{(k,i)}), \\ C^{(k,i)} = \displaystyle\sum_{j=1, j \neq i}^{n} \frac{1}{z^{(k,i)} - W^{(k,j)}}, \\ \vdots \end{cases}$$

$$\left\{\begin{aligned} &q(z^{(k,i)}) = \frac{p'(z^{(k,i)})}{p(z^{(k,i)})} \qquad \text{for} \quad p(z^{(k,i)}) \neq 0, \\ &W^{(k+1,i)} = \langle z^{(k+1,i)}, r^{(k+1,i)} \rangle = -\frac{1}{q(z^{(k,i)}) - C^{(k,i)}}, \end{aligned}\right.$$

$$1 \leqslant i \leqslant n, \qquad k \geqslant 0,$$

and let

(5) $$r^{(k)} = \max_{1 \leqslant i \leqslant n} \{r^{(k,i)}\},$$

(6) $$\rho^{(k)} = \min_{1 \leqslant i < j \leqslant n} \{\min\{|z| \,|\, z \in z^{(k,i)} - W^{(k,j)}\}\}.$$

For $i \neq j$ it follows from (3) that

(7) $$\min\{|z| \,|\, z \in z^{(0,i)} - W^{(0,j)}\} = |z^{(0,i)} - z^{(0,j)}| - r^{(0,j)} \geqslant \rho^{(0)}.$$

Furthermore, let $\eta^{(k)}$ be defined by

(8) $$\rho^{(k)} = (n-1)\eta^{(k)}.$$

The following is then true for iteration scheme (4).

Theorem 1: Let $p(z)$ be a polynomial (1) whose roots $\zeta^{(i)}$, $1 \leqslant i \leqslant n$, satisfy the conditions (2) and (3). With (5), (6), and (8) let

(9) $$6r^{(0)} \leqslant \eta^{(0)}.$$

(a) The iteration (4) is then always feasible and furthermore

$$\zeta^{(i)} \in W^{(k,i)}, \qquad 1 \leqslant i \leqslant n, \qquad k \geqslant 0.$$

(b) The inequality

$$r^{(k+1)} \leqslant \frac{1}{\rho^{(0)}(\eta^{(0)} - 4r^{(0)})}(r^{(k)})^3 \leqslant \frac{1}{12(n-1)} r^{(k)}, \qquad k \geqslant 0,$$

also holds.

Remark: From (b) it follows that $\lim_{k\to\infty} r^{(k)} = 0$. From (a) it therefore necessarily holds that

$$\lim_{k\to\infty} W^{(k,i)} = \zeta^{(i)}, \qquad 1 \leqslant i \leqslant n.$$

If one takes into account Definition 6.4c and 10.6a, then it follows using Theorem 2, Appendix A, from (b) that the R order of the iteration (4) satisfies $O_R((4), (\xi^{(i)})) \geqslant 3$.

Proof: Of (a): Since

$$|z^{(0,i)} - \zeta^{(i)}| \leqslant r^{(0,i)} \leqslant r^{(0)},$$

$$|z^{(0,i)} - \zeta^{(j)}| \geqslant |z^{(0,i)} - z^{(0,j)}| - |z^{(0,j)} - \zeta^{(j)}|$$

$$\geqslant |z^{(0,i)} - z^{(0,j)}| - r^{(0,j)} \geqslant \rho^{(0)},$$

it follows that

$$|q(z^{(0,i)})| = \left| \sum_{j=1}^{n} \frac{1}{z^{(0,i)} - \zeta^{(j)}} \right| \tag{10}$$

$$\geqslant \left| \frac{1}{z^{(0,i)} - \zeta^{(i)}} \right| - \sum_{j=1, j \neq i}^{n} \left| \frac{1}{z^{(0,i)} - \zeta^{(j)}} \right|$$

$$\geqslant \frac{1}{r^{(0)}} - \frac{1}{\eta^{(0)}} \qquad \text{for} \quad z^{(0,i)} \neq \zeta^{(i)}.$$

From

$$|z^{(0,i)} - z^{(0,j)}| - r^{(0,j)} \geqslant \rho^{(0)} > 0,$$

we have

$$0 \notin z^{(0,i)} - W^{(0,j)}$$

as well as

$$\frac{1}{z^{(0,i)} - W^{(0,j)}} \subset \left\langle 0, \frac{1}{\rho^{(0)}} \right\rangle,$$

$$C^{(0,i)} = \sum_{j=1, j \neq i}^{n} \frac{1}{z^{(0,i)} - W^{(0,j)}}$$

$$\subset \sum_{j=1, j \neq i}^{n} \left\langle 0, \frac{1}{\rho^{(0)}} \right\rangle = \left\langle 0, \frac{1}{\eta^{(0)}} \right\rangle,$$

$$q(z^{(0,i)}) - C^{(0,i)} \subset \langle q(z^{(0,i)}), 1/\eta^{(0)} \rangle. \tag{11}$$

Since

$$|q(z^{(0,i)})| - 1/\eta^{(0)} \geqslant 1/r^{(0)} - 2/\eta^{(0)} > 0,$$

clearly

$$0 \notin q(z^{(0,i)}) - C^{(0,i)}$$

and therefore

$$W^{(1,i)}, \qquad 1 \leqslant i \leqslant n,$$

is defined. Because of

$$\frac{p'(z^{(0,i)})}{p(z^{(0,i)})} = \sum_{j=1}^{n} \frac{1}{z^{(0,i)} - \zeta^{(j)}},$$

it follows from (2) and the inclusion monotonicity that

$$\zeta^{(i)} = z^{(0,i)} - p(z^{(0,i)}) \bigg/ \left[p'(z^{(0,i)}) - p(z^{(0,i)}) \sum_{j=1, j \neq i}^{n} \frac{1}{z^{(0,i)} - \zeta^{(j)}} \right]$$

$$\in z^{(0,i)} - \frac{1}{q(z^{(0,i)}) - C^{(0,i)}} = W^{(1,i)}, \qquad 1 \leqslant i \leqslant n.$$

This proves part (a) for $k = 1$.

Of (b): From

$$|z^{(0,i)} - z^{(0,j)}|^2 - (r^{(0,j)})^2 \geqslant (\rho^{(0)} + r^{(0,j)})^2 - (r^{(0,j)})^2 \geqslant (\rho^{(0)})^2,$$

we get that

$$r\left(\frac{1}{z^{(0,i)} - W^{(0,j)}}\right) = \frac{r^{(0,j)}}{|z^{(0,i)} - z^{(0,j)}|^2 - (r^{(0,j)})^2} \leqslant \frac{r^{(0)}}{(\rho^{(0)})^2}$$

and therefore

$$r(C^{(0,i)}) \leqslant \frac{n-1}{\rho^{(0)}} \cdot \frac{r^{(0)}}{\rho^{(0)}} = \frac{r^{(0)}}{\eta^{(0)}\rho^{(0)}}.$$

Using this inequality as well as (11) we now get

$$r(q(z^{(0,i)}) - C^{(0,i)}) = r(C^{(0,i)}),$$

$$|m(q(z^{(0,i)}) - C^{(0,i)})| \geqslant 1/r^{(0)} - 2/\eta^{(0)} + r(q(z^{(0,i)}) - C^{(0,i)})$$

$$= 1/r^{(0)} - 2/\eta^{(0)} + r(C^{(0,i)})$$

and therefore the inequality

$$r(W^{(1,i)}) = r\left(\frac{1}{q(z^{(0,i)}) - C^{(0,i)}}\right)$$

$$= \frac{r(q(z^{(0,i)}) - C^{(0,i)})}{|m(q(z^{(0,i)}) - C^{(0,i)})|^2 - (r(q(z^{(0,i)}) - C^{(0,i)}))^2}$$

$$\leqslant \frac{(r^{(0)})^3}{\rho^{(0)}(\eta^{(0)} - 4r^{(0)})},$$

that is

(12) $$r^{(1)} \leqslant \frac{(r^{(0)})^3}{\rho^{(0)}(\eta^{(0)} - 4r^{(0)})}.$$

Using (9) we get the inequality

$$r^{(1)} \leqslant \frac{1}{12(n-1)} r^{(0)}$$

from the above estimate. Let

$$\delta^{(0)} = \max_{1 \leqslant i \leqslant n} \{|z^{(0,i)} - z^{(1,i)}|\}.$$

We then get, using (6),

$$\rho^{(1)} \geqslant \rho^{(0)} - \delta^{(0)} - 2r^{(1)}. \tag{13}$$

In order to estimate $\delta^{(0)}$ we use (10), (11), and the relation

$$z^{(1,i)} - z^{(0,i)} \in \frac{1}{q(z^{(0,i)}) - C^{(0,i)}}$$

to obtain

$$|z^{(1,i)} - z^{(0,i)}| \leqslant \left| \frac{1}{\langle q(z^{(0,i)}), 1/\eta^{(0)} \rangle} \right| = \frac{1}{|q(z^{(0,i)})| - 1/\eta^{(0)}}$$
$$\leqslant \frac{r^{(0)}\eta^{(0)}}{\eta^{(0)} - 2r^{(0)}}$$

which finally gives

$$\delta^{(0)} \leqslant \frac{r^{(0)}\eta^{(0)}}{\eta^{(0)} - 2r^{(0)}}. \tag{14}$$

Using (12), (13), and (14), it follows from (9) that

$$\begin{aligned} \eta^{(1)} - 6r^{(1)} &= \rho^{(1)}/(n-1) - 6r^{(1)} \\ &\geqslant \eta^{(0)} - r^{(0)} \left(\frac{\eta^{(0)}}{\eta^{(0)} - 2r^{(0)}} + \frac{8(r^{(0)})^2}{\rho^{(0)}(\eta^{(0)} - 4r^{(0)})} \right) \\ &\geqslant \eta^{(0)} - 3r^{(0)} \geqslant 0; \end{aligned} \tag{15}$$

that is,

$$\eta^{(1)} \geqslant 6r^{(1)}.$$

With this it may be shown as above that

$$r^{(2)} \leqslant \frac{1}{\rho^{(1)}(\eta^{(1)} - 4r^{(1)})} (r^{(1)})^3 \leqslant \frac{1}{12(n-1)} r^{(1)}.$$

From (13), it follows analogously to (15) that

$$\eta^{(1)} - 4r^{(1)} \geqslant \eta^{(0)} - r^{(0)} \left(\frac{\eta^{(0)}}{\eta^{(0)} - 2r^{(0)}} + \frac{6(r^{(0)})^2}{\rho^{(0)}(\eta^{(0)} - 4r^{(0)})} \right) \geqslant 0 \tag{16}$$

as well as

$$(17) \qquad \eta^{(1)} \geqslant \eta^{(0)} - r^{(0)}\left(\frac{\eta^{(0)}}{\eta^{(0)} - 2r^{(0)}} + \frac{2(r^{(0)})^2}{\rho^{(0)}(\eta^{(0)} - 4r^{(0)})}\right) \geqslant 0.$$

Using both the above inequalities, we get from (9) that

$$\eta^{(1)}(\eta^{(1)} - 4r^{(1)}) \geqslant (\eta^{(0)})^2 - \eta^{(0)}r^{(0)}\left(\frac{2\eta^{(0)}}{\eta^{(0)} - 2r^{(0)}} + \frac{8(r^{(0)})^2}{\rho^{(0)}(\eta^{(0)} - 4r^{(0)})}\right)$$

$$\geqslant \eta^{(0)}(\eta^{(0)} - 4r^{(0)})$$

and therefore

$$r^{(2)} \leqslant \frac{1}{\rho^{(0)}(\eta^{(0)} - 4r^{(0)})}(r^{(1)})^3.$$

The remainder of the proof follows by complete induction. ■

We shall now show an application of the iteration (4). For this we consider the problem of the computation of the eigenvalues of a lower Hessenberg matrix using a sequence of inclusions. The values of the characteristic polynomial and its derivative, required in the iteration, may be computed using the method of Hyman (see Stoer and Bulirsch [6, p. 353]). As a concrete example we consider the matrix

$$\mathscr{H} = \begin{pmatrix} 12 + 16i & 1 & 0 & 0 \\ 0 & 9 + 12i & 1 & 0 \\ 0 & 0 & 6 + 8i & 1 \\ 1 & 0 & 0 & 3 + 4i \end{pmatrix},$$

where $i = \sqrt{-1}$. Using Gersgorin's theorem we get that the circular disks

$$W^{(0,1)} = \langle 12 + 16i, 1\rangle, \qquad W^{(0,2)} = \langle 9 + 12i, 1\rangle$$

$$W^{(0,3)} = \langle 6 + 8i, 1\rangle, \qquad W^{(0,4)} = \langle 3 + 4i, 1\rangle$$

each contain exactly one eigenvalue of $\mathscr{H}$. With the help of (4) we get the following improved inclusion sets $W^{(k,i)}$ for the eigenvalues of $\mathscr{H}$ where the representation

$$W^{(k,i)} = \langle m(W^{(k,i)}), r(W^{(k,i)})\rangle$$

with

$$m(W^{(k,i)}) = \operatorname{Re}(m(W^{(k,i)})) + i\operatorname{Im}(m(W^{(k,i)}))$$

is used (see Table 1). Further numerical examples are found in Gargantini and Henrici [3] and Speck [5].

Table 1

k	i	Re	Im	r
1	1	+ 11.99875131516	+ 15.99953080496	0.1001255×10^{-6}
	2	+ 9.003742419628	+ 12.00140833328	0.1494005×10^{-5}
	3	+ 5.996257580383	+ 7.998591666711	0.1493969×10^{-5}
	4	+ 3.001240654037	+ 4.000469195035	0.1000782×10^{-6}
2	1	+ 11.99875136181	+ 15.99953080159	0.1019500×10^{-9}
	2	+ 9.003742437190	+ 12.00140832752	$0.8760740 \times 10^{-10}$
	3	+ 5.996257562811	+ 7.998591672458	$0.3665239 \times 10^{-10}$
	4	+ 3.001248638204	+ 4.000469198423	$0.2555951 \times 10^{-10}$
3	1	+ 11.99875136181	+ 15.99953080159	0.1019496×10^{-9}
	2	+ 9.003742437190	+ 12.00140832752	$0.8760740 \times 10^{-10}$
	3	+ 5.996257562811	+ 7.998591672458	$0.3665353 \times 10^{-10}$
	4	+ 3.001248638204	+ 4.000469198423	$0.2556093 \times 10^{-10}$

Remarks: The iteration method investigated in Theorem 1 may be denoted a total step method. If one uses at every step the currently improved values of the approximations, then one obtains a method that may be described as a single step method. Similarly to Alefeld and Herzberger [1] it can be shown that this method has an order of convergence of the radius of the approximating circular disk to zero that is faster than cubic. Another class of methods for finding zeros of polynomials that employ circular arithmetic was introduced in Gargantini [2].

REFERENCES

[1] G. Alefeld and J. Herzberger, On the convergence speed of some algorithms for the simultaneous approximation of polynomial roots. *SIAM J. Numer. Anal.* **11**, 237–243 (1974).

[2] J. Gargantini, Further applications of circular arithmetic: Schröder-like algorithms with error bounds for finding zeros of polynomials. *SIAM J. Numer. Anal.* **15**, 497 – 510 (1978).

[3] J. Gargantini and P. Henrici, Circular arithmetic and the determination of polynomial zeros. *Numer. Math.* **18**, 305–320 (1972).

[4] P. Henrici, Circular arithmetic and the determination of polynomial zeros. Springer Lecture Notes 228, 86–92 (1971).

[5] P. T. Speck, Kreisarithmetik, Realisation und Anwendung auf einer Rechenanlage. Diplomarbeit, Inst. Angew. Math., Univ. Karlsruhe (1974).

[6] J. Stoer and R. Bulirsch, "Introduction to Numerical Analysis." Springer-Verlag, Berlin and New York, 1980.

Chapter 10 / INTERVAL MATRIX OPERATIONS

The set of $m \times n$ matrices over the real numbers is denoted by $M_{mn}(\mathbb{R})$ and the set of $m \times n$ matrices over the complex numbers by $M_{mn}(\mathbb{C})$. Elements from $M_{mn}(\mathbb{R})$, respectively, $M_{mn}(\mathbb{C})$, are denoted by $\mathscr{A}, \mathscr{B}, \mathscr{C}, \ldots, \mathscr{X}, \mathscr{Y}, \mathscr{Z}$. Column matrices, that is, real or complex vectors, are written as $\mathscr{a}, \mathscr{b}, \mathscr{c}, \ldots, \mathscr{x}, \mathscr{y}, \mathscr{z}$. The set of real column vectors is denoted by $V_n(\mathbb{R})$ and the set of complex column vectors as $V_n(\mathbb{C})$. In a similar manner we write $M_{mn}(I(\mathbb{R}))$ for the set of matrices over the real intervals and $M_{mn}(I(\mathbb{C}))$ for the set of matrices over the complex intervals where $I(\mathbb{C})$ may denote either $R(\mathbb{C})$ or $K(\mathbb{C})$. Elements from $M_{mn}(I(\mathbb{R}))$ (resp. $M_{mn}(I(\mathbb{C}))$) are written as $\mathscr{A}, \mathscr{B}, \mathscr{C}, \ldots, \mathscr{X}, \mathscr{Y}, \mathscr{Z}$ and we call them real (resp. complex) interval matrices. Column interval matrices, that is, real or complex interval vectors, are denoted by $\mathscr{a}, \mathscr{b}, \mathscr{c}, \ldots, \mathscr{x}, \mathscr{y}, \mathscr{z}$. The set of real interval column vectors is denoted by $V_n(I(\mathbb{R}))$ and the set of complex interval column vectors by $V_n(I(\mathbb{C}))$. Interval matrices and interval vectors are represented, as is usual for real or complex matrices, by their components in the form $\mathscr{A} = (A_{ij})$ for interval matrices and $\mathscr{a} = (A_i)$ for interval vectors. An interval matrix where all the components are point intervals is called a *point matrix*. A *point vector* is defined in a similar manner. We also mention the following obvious relationships $M_{mn}(I(\mathbb{R})) \subset M_{mn}(R(\mathbb{C}))$ and $V_n(I(\mathbb{R})) \subset V_n(R(\mathbb{C}))$.

Definition 1: Two $m \times n$ interval matrices $\mathscr{A} = (A_{ij})$ and $\mathscr{B} = (B_{ij})$ are equal, that is, $\mathscr{A} = \mathscr{B}$, iff there is equality between all corresponding components of the matrices. This may be written $\mathcal{A} = \mathcal{B} \Leftrightarrow A_{ij} = B_{ij}$, $1 \leqslant i \leqslant m$, $1 \leqslant j \leqslant n$. ■

We now introduce a partial order on the set of interval matrices.

Definition 2: Let $\mathcal{A} = (A_{ij})$ and $\mathcal{B} = (B_{ij})$ be two $m \times n$ interval matrices. Then it holds that

$$\mathcal{A} \subseteq \mathcal{B} \Leftrightarrow A_{ij} \subseteq B_{ij}, \qquad 1 \leqslant i \leqslant m, \qquad 1 \leqslant j \leqslant n. \quad ■$$

$\mathcal{A} \subset \mathcal{B}$ is similarly defined by means of the elements. If $\mathcal{A} = (a_{ij})$ is a point matrix, then we also write $\mathcal{A} \in \mathcal{B}$. Each interval matrix may be considered to be a set of point matrices. The relations $\subseteq$ and $\subset$ between sets of point matrices are defined in the usual set theoretic sense.

Our next goal is to define operations between interval matrices that formally correspond to the operations on point matrices.

Definition 3: (a) Let $\mathcal{A} = (A_{ij})$, $\mathcal{B} = (B_{ij})$ be two $m \times n$ interval matrices. Then

$$\mathcal{A} \pm \mathcal{B} := (A_{ij} \pm B_{ij})$$

defines an interval matrix addition, respectively, subtraction.

(b) Let $\mathcal{A} = (A_{ij})$ be an $m \times r$ interval matrix and let $\mathcal{B} = (B_{ij})$ be an $r \times n$ interval matrix. Then

$$\mathcal{A}\mathcal{B} := \left(\sum_{\nu=1}^{r} A_{i\nu} B_{\nu j} \right)$$

defines an interval matrix multiplication. In particular, for an $n \times r$ interval matrix $\mathcal{A} = (A_{ij})$ and an interval vector $\mathscr{u} = (U_i)$ with r components, we have

$$\mathcal{A}\mathscr{u} = \left(\sum_{\nu=1}^{r} A_{i\nu} U_{\nu} \right).$$

(c) Let $\mathcal{A} = (A_{ij})$ be an interval matrix and let X be an interval. Then

$$X\mathcal{A} = \mathcal{A}X := (XA_{ij}). \quad ■$$

It is assumed in the sequel that the interval matrices that occur in interval matrix operations have the correct number of rows and columns for the operations, and this will therefore not be explicitly stated. Furthermore, it is assumed that the interval operands have appropriate elements. If we have $\mathcal{A} \in M_{mn}(K(\mathbb{C}))$, then the product $\mathcal{A}\mathcal{B}$, for example, is defined only if $\mathcal{B} \in M_{nr}(K(\mathbb{C}))$.

The operations on interval matrices and vectors were formally introduced in Definition 3. For real interval operations we have the simple Definition 1.2. An analogous interpretation is not possible for the operations on interval matrices. It generally holds, however, that

$$\{\mathcal{A}\mathcal{B} \mid \mathcal{A} \in \mathcal{A}, \mathcal{B} \in \mathcal{B}\} \subseteq \{\mathcal{C} \mid \mathcal{C} \in \mathcal{A}\mathcal{B}\}.$$

The proof follows by using the inclusion monotonicity of the interval operations. Equality does not hold in general as shown by the following example. Let

$$\mathcal{A} = \begin{pmatrix} 1 & 1 \\ -1 & 1 \end{pmatrix}, \qquad u = \begin{pmatrix} [0, 1] \\ [0, 1] \end{pmatrix}.$$

Then we have

$$\mathcal{A}u = \begin{pmatrix} [0, 2] \\ [-1, 1] \end{pmatrix},$$

and if we choose

$$x = \begin{pmatrix} 2 \\ -1 \end{pmatrix} \in \mathcal{A}u,$$

then we see that there is no $y \in u$ such that $\mathcal{A}y = x$.

Let $\mathcal{A}, \mathcal{B} \in M_{mn}(I(\mathbb{R}))$ and $c \in V_n(\mathbb{R})$. Then the relations

$$(1) \qquad \begin{cases} \{\mathcal{A} \pm \mathcal{B} \mid \mathcal{A} \in \mathcal{A}, \mathcal{B} \in \mathcal{B}\} = \mathcal{A} \pm \mathcal{B}, \\ \{\mathcal{A}c \mid \mathcal{A} \in \mathcal{A}\} = \mathcal{A}c, \end{cases}$$

which we shall use later, are valid.

The set of interval matrices is closed with respect to the operations from Definition 3. The set of the real or complex matrices is isomorphic to the corresponding point matrices. This is the reason for using the same operation symbol in Definition 3 as for the corresponding real or complex operations.

We now state some useful properties for the above operations.

Theorem 4: Let $\mathcal{A}$, $\mathcal{B}$, $\mathcal{C}$ be interval matrices. Then

$$(2) \qquad \mathcal{A} + \mathcal{B} = \mathcal{B} + \mathcal{A},$$

$$(3) \qquad \mathcal{A} + (\mathcal{B} + \mathcal{C}) = (\mathcal{A} + \mathcal{B}) + \mathcal{C},$$

$$(4) \qquad \mathcal{A} + \mathcal{O} = \mathcal{O} + \mathcal{A} = \mathcal{A}, \qquad \mathcal{O} = \text{zero matrix},$$

$$(5) \qquad \mathcal{A}\mathcal{I} = \mathcal{I}\mathcal{A} = \mathcal{A}, \qquad \mathcal{I} = \text{unit matrix},$$

$$(6) \qquad \left.\begin{matrix} (\mathcal{A} + \mathcal{B})\mathcal{C} \subseteq \mathcal{A}\mathcal{C} + \mathcal{B}\mathcal{C}, \\ \mathcal{C}(\mathcal{A} + \mathcal{B}) \subseteq \mathcal{C}\mathcal{A} + \mathcal{C}\mathcal{B}, \end{matrix}\right\} \qquad \text{(subdistributivity)},$$

$$(7) \qquad (\mathcal{A} + \mathcal{B})\mathcal{C} = \mathcal{A}\mathcal{C} + \mathcal{B}\mathcal{C},$$

$$(8) \qquad \mathcal{C}(\mathcal{A} + \mathcal{B}) = \mathcal{C}\mathcal{A} + \mathcal{C}\mathcal{B},$$

$$
(9) \qquad \begin{cases}
\mathscr{A}(\mathscr{B}\mathscr{C}) \subseteq (\mathscr{A}\mathscr{B})\mathscr{C}, & \\
(\mathscr{A}\mathscr{B})\mathscr{C} \subseteq \mathscr{A}(\mathscr{B}\mathscr{C}) & \text{if} \quad \mathscr{C} = -\mathscr{C}, \\
\mathscr{A}(\mathscr{B}\mathscr{C}) = (\mathscr{A}\mathscr{B})\mathscr{C}, & \\
\mathscr{A}(\mathscr{B}\mathscr{C}) = (\mathscr{A}\mathscr{B})\mathscr{C} & \text{for} \quad \mathscr{A}, \mathscr{B}, \mathscr{C} \in M_{mn}(I(\mathbb{R})), \\
& \text{and} \quad \mathscr{B} = -\mathscr{B}, \quad \mathscr{C} = -\mathscr{C}.
\end{cases}
$$

Proof: Relations (2)–(8) are proved element by element using the formulas from Theorems 1.4 and 5.8. We prove (9) for square matrices. From the subdistributivity (1.8), as well as (5.6), we get

$$
\begin{aligned}
\mathscr{A}(\mathscr{B}\mathscr{C}) &= \left(\sum_{j=1}^{n} A_{ij} \left(\sum_{l=1}^{n} b_{jl} c_{lk} \right) \right) \\
&\subseteq \left(\sum_{j=1}^{n} \sum_{l=1}^{n} A_{ij} b_{jl} c_{lk} \right) \\
&= \left(\sum_{l=1}^{n} \left(\sum_{j=1}^{n} A_{ij} b_{jl} \right) c_{lk} \right) = (\mathscr{A}\mathscr{B})\mathscr{C},
\end{aligned}
$$

which proves the first part of (9).

Because of

$$
\begin{aligned}
(\mathscr{A}\mathscr{B})\mathscr{C} &= \left(\sum_{l=1}^{n} \left(\sum_{k=1}^{n} a_{ik} B_{kl} \right) C_{lj} \right) \\
&= \left(\sum_{l=1}^{n} \left| \sum_{k=1}^{n} a_{ik} B_{kl} \right| C_{lj} \right) \\
&\subseteq \left(\sum_{l=1}^{n} \left(\sum_{k=1}^{n} |a_{ik}|\,|B_{kl}| \right) C_{lj} \right) \\
&\subseteq \left(\sum_{l=1}^{n} \left(\sum_{k=1}^{n} |a_{ik}|\,|B_{kl}| C_{lj} \right) \right) \\
&= \left(\sum_{k=1}^{n} |a_{ik}| \left(\sum_{l=1}^{n} |B_{kl}| C_{lj} \right) \right) \\
&= \left(\sum_{k=1}^{n} a_{ik} \left(\sum_{l=1}^{n} B_{kl} C_{lj} \right) \right) = \mathscr{A}(\mathscr{B}\mathscr{C}),
\end{aligned}
$$

we have the second relation. From the equalities

$$\begin{aligned}(\mathscr{A}\mathscr{B})\mathscr{C} &= \left(\sum_{l=1}^{n}\left(\sum_{k=1}^{n} a_{ik}B_{kl}\right)c_{lj}\right)\\ &= \left(\sum_{l=1}^{n}\left(\sum_{k=1}^{n} a_{ik}B_{kl}c_{lj}\right)\right)\\ &= \left(\sum_{k=1}^{n} a_{ik}\left(\sum_{l=1}^{n} B_{kl}c_{lj}\right)\right),\end{aligned}$$

we get the third relation.

The last relation is proven using the third formula in (1.8) in the following manner

$$\begin{aligned}\mathscr{A}(\mathscr{B}\mathscr{C}) &= \left(\sum_{k=1}^{n} A_{ik}\left(\sum_{l=1}^{n} B_{kl}C_{lj}\right)\right)\\ &= \left(\sum_{k=1}^{n} A_{ik}\left(\sum_{l=1}^{n} |B_{kl}|C_{lj}\right)\right)\\ &= \left(\sum_{k=1}^{n} |A_{ik}|\left(\sum_{l=1}^{n} |B_{kl}|C_{lj}\right)\right)\\ &= \left(\sum_{k=1}^{n}\left(\sum_{l=1}^{n} |A_{ik}|\,|B_{kl}|C_{lj}\right)\right)\\ &= \left(\sum_{l=1}^{n}\left(\sum_{k=1}^{n} |A_{ik}|\,|B_{kl}|C_{lj}\right)\right)\\ &= \left(\sum_{l=1}^{n}\left(\sum_{k=1}^{n} |A_{ik}|\,|B_{kl}|\right)C_{lj}\right)\\ &= \left(\sum_{l=1}^{n}\left(\sum_{k=1}^{n} |A_{ik}|B_{kl}\right)C_{lj}\right)\\ &= \left(\sum_{l=1}^{n}\left(\sum_{k=1}^{n} A_{ik}B_{kl}\right)C_{lj}\right) = (\mathscr{A}\mathscr{B})\mathscr{C}. \quad \blacksquare\end{aligned}$$

The associative law is in general not valid for interval matrices. This is shown by the following

Example

$$\begin{bmatrix}[-1,1] & 1\\ -1 & [0,1]\end{bmatrix}\left\{\begin{bmatrix}1 & 1\\ 0 & 1\end{bmatrix}\begin{bmatrix}-1 & 0\\ 1 & -1\end{bmatrix}\right\} = \begin{bmatrix}1 & [-2,0]\\ [0,1] & [0,1]\end{bmatrix},$$

$$\left\{\begin{bmatrix}[-1,1] & 1\\ -1 & [0,1]\end{bmatrix}\begin{bmatrix}1 & 1\\ 0 & 1\end{bmatrix}\right\}\begin{bmatrix}-1 & 0\\ 1 & -1\end{bmatrix} = \begin{bmatrix}[-1,3] & [-2,0]\\ [0,1] & [0,1]\end{bmatrix}.$$

The fundamental property of inclusion monotonicity is also valid for interval matrix operations.

Theorem 5: Let $\mathscr{A}^{(k)}, \mathscr{B}^{(k)}, k = 1, 2$, be interval matrices. Furthermore, let X, Y be intervals and assume that

$$\mathscr{A}^{(k)} \subseteq \mathscr{B}^{(k)}, \qquad k = 1, 2, \qquad \text{and} \qquad X \subseteq Y.$$

The relations

$$\text{(10)} \qquad \begin{cases} \mathscr{A}^{(1)} * \mathscr{A}^{(2)} \subseteq \mathscr{B}^{(1)} * \mathscr{B}^{(2)}, \\ X\mathscr{A}^{(1)} \subseteq Y\mathscr{B}^{(1)}, \end{cases}$$

then hold for $* \in \{+, -, \cdot\}$. ■

The proof of (10) may be carried out component by component using (1.9) and Theorem 5.9. As a special case of (10) we have

$$\text{(10')} \qquad \begin{array}{l} \underset{\cdot}{\mathscr{A}} \in \mathscr{A}, \ \underset{\cdot}{\mathscr{B}} \in \mathscr{B} \Rightarrow \underset{\cdot}{\mathscr{A}} * \underset{\cdot}{\mathscr{B}} \in \mathscr{A} * \mathscr{B}, \qquad * \in \{+, -, \cdot\}, \\ x \in X, \ \underset{\cdot}{\mathscr{A}} \in \mathscr{A} \Rightarrow x\underset{\cdot}{\mathscr{A}} \in X\mathscr{A}. \end{array}$$

We now introduce the concepts of width and absolute value for interval matrices.

Definition 6: If $\mathscr{A} = (A_{ij})$ is an interval matrix, then

(a) the real nonnegative matrix

$$d(\mathscr{A}) := (d(A_{ij}))$$

is called the width matrix or width of $\mathscr{A}$; and

(b) the real, nonnegative matrix

$$|\mathscr{A}| := (|A_{ij}|)$$

is called the absolute value matrix or the absolute value of $\mathscr{A}$. ■

We now gather together some properties of the width and the absolute value of interval matrices. The partial order

$$\underset{\cdot}{\mathscr{X}} \leqslant \underset{\cdot}{\mathscr{Y}} \Leftrightarrow x_{ij} \leqslant y_{ij}, \qquad 1 \leqslant i \leqslant m, \qquad 1 \leqslant j \leqslant n,$$

is here used for real $m \times n$ matrices $\underset{\cdot}{\mathscr{X}}$ and $\underset{\cdot}{\mathscr{Y}}$. These properties are

$$\text{(11)} \qquad \mathscr{A} \subseteq \mathscr{B} \Rightarrow d(\mathscr{A}) \leqslant d(\mathscr{B}),$$

$$\text{(12)} \qquad d(\mathscr{A} \pm \mathscr{B}) = d(\mathscr{A}) + d(\mathscr{B}),$$

$$\text{(13)} \qquad d(\mathscr{A}) = \sup_{\underset{\cdot}{\mathscr{A}}, \underset{\cdot}{\mathscr{A}}' \in \mathscr{A}} |\underset{\cdot}{\mathscr{A}} - \underset{\cdot}{\mathscr{A}}'|,$$

$$\text{(14)} \qquad |\mathscr{A}| = \sup_{\underset{\cdot}{\mathscr{A}} \in \mathscr{A}} |\underset{\cdot}{\mathscr{A}}|,$$

$$\text{(15)} \qquad \mathscr{A} \subseteq \mathscr{B} \Rightarrow |\mathscr{A}| \leqslant |\mathscr{B}|,$$

$$(16)\qquad \begin{cases} |\mathscr{A}| \geqslant \underset{\cdot}{O} \quad \text{and} \quad |\mathscr{A}| = \underset{\cdot}{O} \Leftrightarrow \mathscr{A} = \underset{\cdot}{O}, \\ |\mathscr{A} + \mathscr{B}| \leqslant |\mathscr{A}| + |\mathscr{B}|, \\ |x\mathscr{A}| = |\mathscr{A}x| = |x|\,|\mathscr{A}|, \qquad x \in \mathbb{C}, \\ \mathscr{A} \in M_{mn}(I(\mathbb{R})), \quad \text{or} \quad \mathscr{A} \in M_{mn}(K(\mathbb{C})), \\ |\mathscr{A}\mathscr{B}| \leqslant |\mathscr{A}|\,|\mathscr{B}|, \end{cases}$$

$$(17)\qquad d(\mathscr{A}\mathscr{B}) \leqslant d(\mathscr{A})|\mathscr{B}| + |\mathscr{A}|\,d(\mathscr{B}),$$

$$(18)\qquad d(\mathscr{A}\mathscr{B}) \geqslant |\mathscr{A}|\,d(\mathscr{B}), \qquad d(\mathscr{A}\mathscr{B}) \geqslant d(\mathscr{A})|\mathscr{B}|,$$

$$(19)\qquad \begin{cases} d(a\mathscr{B}) = |a|\,d(\mathscr{B}), \qquad a \in \mathbb{C}, \\ d(\underset{\cdot}{\mathscr{A}}\mathscr{B}) = |\underset{\cdot}{\mathscr{A}}|\,d(\mathscr{B}), \qquad d(\mathscr{B}\underset{\cdot}{\mathscr{A}}) = d(\mathscr{B})|\underset{\cdot}{\mathscr{A}}|. \end{cases}$$

For real interval matrices $\mathscr{A}$ and $\mathscr{B}$ we have

$$(20)\qquad \underset{\cdot}{O} \in \mathscr{A} \Rightarrow |\mathscr{A}| \leqslant d(\mathscr{A}) \leqslant 2|\mathscr{A}|,$$

$$(21)\qquad \mathscr{A} = (-1)\mathscr{A} \Rightarrow \mathscr{A}\mathscr{B} = \mathscr{A}|\mathscr{B}|,$$

$$(22)\qquad \underset{\cdot}{O} \in \mathscr{A}, \quad 0 \notin B_{ij} \quad \text{for} \quad \mathscr{B} = (B_{ij}) \Rightarrow d(\mathscr{A}\mathscr{B}) = d(\mathscr{A})|\mathscr{B}|.$$

The proof of these properties follows in a componentwise manner using the properties given in Chapter 2 for $I(\mathbb{R})$ and Chapter 6 for $I(\mathbb{C})$.

We also remark that properties (20)–(22) are not all valid for complex interval matrices. Consider, for example, (21) for an $m \times m$ matrix with elements from $R(\mathbb{C})$ with $m = 1$; that is, an interval $A = A_1 + iA_2 \in R(\mathbb{C})$. The assumption $A = (-1)A$ is equivalent to

$$A_1 = (-1)A_1, \qquad A_2 = (-1)A_2,$$

for $A = A_1 + iA_2$. Using (2.18) we get with $B = B_1 + iB_2$ that

$$\begin{aligned} AB &= (A_1B_1 - A_2B_2) + i(A_1B_2 + A_2B_1) \\ &= A_1|B_1| + A_2|B_2| + i(A_1|B_2| + A_2|B_1|). \end{aligned}$$

On the other hand we have

$$A|B| = A_1(|B_1| + |B_2|) + iA_2(|B_1| + |B_2|).$$

The two intervals are, for example, for $B_1 = 0$, in general different. Since we do not need the relations (20)–(22) for complex intervals we do not discuss the cases for which they are valid.

We now introduce the concept of the distance matrix of two interval matrices.

Definition 7: Let $\mathscr{A} = (A_{ij})$ and $\mathscr{B} = (B_{ij})$ be two interval matrices. The real nonnegative matrix

$$q(\mathscr{A}, \mathscr{B}) := (q(A_{ij}, B_{ij}))$$

is then called the distance matrix or the distance between $\mathscr{A}$ and $\mathscr{B}$. ■

The relations

$$q(\mathscr{A},\mathscr{B}) = \mathscr{O} \Leftrightarrow \mathscr{A} = \mathscr{B},$$

$$q(\mathscr{A},\mathscr{B}) \leqslant q(\mathscr{A},\mathscr{C}) + q(\mathscr{B},\mathscr{C}),$$

are clearly valid for the distance between interval matrices as well as

(23) $$q(\mathscr{A} + \mathscr{C}, \mathscr{B} + \mathscr{C}) = q(\mathscr{A},\mathscr{B}),$$

(24) $$q(\mathscr{A} + \mathscr{B}, \mathscr{C} + \mathscr{D}) \leqslant q(\mathscr{A},\mathscr{C}) + q(\mathscr{B},\mathscr{D}),$$

(25) $$q(\mathscr{A}\mathscr{B}, \mathscr{A}\mathscr{C}) \leqslant |\mathscr{A}| q(\mathscr{B},\mathscr{C}).$$

The proofs of the last properties are carried out by using the corresponding properties in $I(\mathbb{R})$ (see Chapter 2) or $I(\mathbb{C})$ (see Chapter 6) in a componentwise manner. From the distance concept for interval matrices established in Definition 7 one may, using a monotone matrix norm $\|\cdot\|$, define a metric for the set of interval matrices (see, for example, Mayer [3]) using $\|q(\mathscr{A},\mathscr{B})\|$. The set of all $m \times n$ interval matrices may also be considered as an $m \cdot n$ fold product $I(\mathbb{C}) \times I(\mathbb{C}) \times \cdots \times I(\mathbb{C})$ of the complete metric space $I(\mathbb{C})$. From known theorems in topology (see, for example, Thron [8]) we have that this product space is again a complete space. In product spaces we have that the convergence in the space is equivalent to the convergence of the individual components. Consequently we have the following statements:

(26) A sequence $\{\mathscr{A}^{(k)}\}_{k=0}^{\infty}$ of $m \times n$ interval matrices converges to $\mathscr{A}$, written $\lim_{k\to\infty} \mathscr{A}^{(k)} = \mathscr{A} \Leftrightarrow \lim_{k\to\infty} A_{ij}^{(k)} = A_{ij}$, $1 \leqslant i \leqslant m$, $1 \leqslant j \leqslant n$.

Corollary 8: Every sequence of $m \times n$ interval matrices $\{\mathscr{A}^{(k)}\}_{k=0}^{\infty}$, for which

$$\mathscr{A}^{(0)} \supseteq \mathscr{A}^{(1)} \supseteq \mathscr{A}^{(2)} \supseteq \cdots$$

holds, converges to an interval matrix $\mathscr{A} = (A_{ij})$, where

$$A_{ij} = \bigcap_{k=0}^{\infty} A_{ij}^{(k)}, \qquad 1 \leqslant i \leqslant m, \quad 1 \leqslant j \leqslant n. \quad ■$$

The proof of this corollary follows from (26), Definition 2, as well as the analogous statement of Corollary 8 for intervals.

Corollary 9: The operations introduced in Definition 3 are continuous. ■

The proof follows from the fact that the continuity of the operations between the individual elements imply the continuity of the operations as a whole. From Definition 3 and because of Theorems 2.4 and 6.6, the elements of the result of the operation are continuously dependent on the operands.

The following relation is true because of (2.21) and (5.17)

(27) $$\mathscr{X} \subseteq \mathscr{Y} \Rightarrow \tfrac{1}{2}(d(\mathscr{Y}) - d(\mathscr{X})) \leqslant q(\mathscr{X},\mathscr{Y}) \leqslant d(\mathscr{Y}) - d(\mathscr{X}).$$

In the same manner as in Chapter 2 for elements from $I(\mathbb{R})$ and in Chapter 6 for elements from $R(\mathbb{C})$ we now wish to introduce yet another binary operation on $M_{mn}(I(\mathbb{R}))$ as well as on $M_{mn}(R(\mathbb{C}))$. Since $M_{mn}(I(\mathbb{R})) \subset M_{mn}(R(\mathbb{C}))$ it suffices to define the operation for $M_{mn}(R(\mathbb{C}))$. Let $\mathscr{A} = (A_{ij})$, $\mathscr{B} = (B_{ij}) \in M_{mn}(R(\mathbb{C}))$. Then the set theoretic intersection

$$\mathscr{A} \cap \mathscr{B} = \{\mathscr{C} \mid \mathscr{C} \in \mathscr{A}, \mathscr{C} \in \mathscr{B}\}$$

denotes the intersection of $\mathscr{A}$ and $\mathscr{B}$. The intersection of two interval matrices $\mathscr{A}$ and $\mathscr{B}$ is in $M_{mn}(R(\mathbb{C}))$ iff the set theoretic intersection is nonempty. In this case we have

$$\mathscr{A} \cap \mathscr{B} = (A_{ij} \cap B_{ij}),$$

where $A_{ij} \cap B_{ij}$, $1 \leqslant i \leqslant n$, $1 \leqslant j \leqslant m$, is formed according to (2.23) (resp. (6.19)).

Corresponding to Corollaries 2.12 and 6.7 we get

Corollary 10: Let $\mathscr{A}, \mathscr{B}, \mathscr{C}, \mathscr{D} \in M_{mn}(R(\mathbb{C}))$. Then we have

$$\mathscr{A} \subseteq \mathscr{C}, \mathscr{B} \subseteq \mathscr{D} \Rightarrow \mathscr{A} \cap \mathscr{B} \subseteq \mathscr{C} \cap \mathscr{D} \qquad \text{(inclusion monotonicity)},$$

and the intersection is a continuous operation when it can be executed in $M_{mn}(R(\mathbb{C}))$. ■

As in the Corollaries 2.12 and 6.7 the proof again follows from the fact that continuity of the operations on the individual elements implies continuity of the operation as a whole.

We now consider triply indexed arrays whose elements are intervals in order to generalize the bilinear operators from $V_n(\mathbb{C}) \times V_n(\mathbb{C})$ to $V_n(\mathbb{C})$. The set of all such arrays is denoted by $M_{n^3}(I(\mathbb{C}))$. We therefore have

$$\mathscr{B} = (B_{ijk}) \in M_{n^3}(I(\mathbb{C})),$$

where

$$B_{ijk} \in I(\mathbb{C}), \qquad 1 \leqslant i,j,k \leqslant n.$$

Equality, inclusion, and addition are defined element by element as for interval matrices. The concepts of width, distance, and absolute value are similarly defined. For example,

$$|\mathscr{B}| := (|B_{ijk}|)$$

is a bilinear operator from $V_n(\mathbb{R}) \times V_n(\mathbb{R})$ to $V_n(\mathbb{R})$. The set of all bilinear operators from $V_n(\mathbb{R}) \times V_n(\mathbb{R})$ to $V_n(\mathbb{R})$ is called $M_{n^3}(\mathbb{R})$.

Definition 11: Let $\mathscr{B} = (B_{ijk}) \in M_{n^3}(I(\mathbb{C}))$, $x = (X_i)$, $y = (Y_i) \in V_n(I(\mathbb{C}))$ and $\mathscr{A} \in M_{nn}(I(\mathbb{C}))$. Then

(a) $\mathscr{C} := \mathscr{B}x \in M_{nn}(I(\mathbb{C}))$ with

$$C_{ij} = \sum_{k=1}^{n} B_{ijk} X_k, \qquad 1 \leqslant i,j \leqslant n,$$

and

(b) $\mathscr{z} := (\mathscr{B}\mathscr{x})\mathscr{y} \in V_n(I(\mathbb{C}))$ with

$$Z_i = \sum_{j=1}^{n} C_{ij} Y_j = \sum_{j=1}^{n} \left(\sum_{k=1}^{n} B_{ijk} X_k \right) Y_j, \qquad 1 \leqslant i \leqslant n.$$

We also write $\mathscr{B}\mathscr{x}\mathscr{y}$ for $(\mathscr{B}\mathscr{x})\mathscr{y}$.

(c) Furthermore

$$\mathscr{C} := \mathscr{A}\mathscr{B} \in M_{n^3}(I(\mathbb{C}))$$

with

$$C_{ijk} = \sum_{\nu=1}^{n} A_{i\nu} B_{\nu jk}, \qquad 1 \leqslant i, j, k \leqslant n. \quad \blacksquare$$

The following theorem contains some relations needed later.

Theorem 12: Let

$$\mathscr{A} = (a_{ij}) \in M_{nn}(\mathbb{C}), \qquad \mathscr{B} = (B_{ijk}) \in M_{n^3}(I(\mathbb{C})),$$
$$\mathscr{x} = (X_i), \qquad \mathscr{y} = (Y_i) \in V_n(I(\mathbb{C})).$$

Then

(a) $(\mathscr{A}\mathscr{B})\mathscr{x}\mathscr{y} \subseteq \mathscr{A}(\mathscr{B}\mathscr{x}\mathscr{y})$.

If $\mathscr{x} = -\mathscr{x}$, then

(b) $\mathscr{B}\mathscr{x}\mathscr{x} = \frac{1}{2}|\mathscr{B}|\, d(\mathscr{x})\mathscr{x}$

and

(c) $d(\mathscr{B}\mathscr{x}\mathscr{x}) = \frac{1}{2}|\mathscr{B}|\, d(\mathscr{x})\, d(\mathscr{x})$.

Proof: Of (a): Let $\mathscr{C} = \mathscr{A}\mathscr{B} = (C_{ijk})$ and $\mathscr{z} = (Z_i) = \mathscr{B}\mathscr{x}\mathscr{y}$. Then using Definition 11 and subdistributivity, we get

$$\begin{aligned}
(\mathscr{A}\mathscr{B})\mathscr{x}\mathscr{y} &= \left(\sum_{j=1}^{n} \left(\sum_{k=1}^{n} C_{ijk} X_k \right) Y_j \right) = \left(\sum_{j=1}^{n} \left(\sum_{k=1}^{n} \left(\sum_{\nu=1}^{n} a_{i\nu} B_{\nu jk} \right) X_k \right) Y_j \right) \\
&\subseteq \left(\sum_{j=1}^{n} \left(\sum_{k=1}^{n} \left(\sum_{\nu=1}^{n} a_{i\nu} B_{\nu jk} X_k \right) \right) Y_j \right) = \left(\sum_{j=1}^{n} \left(\sum_{\nu=1}^{n} \left(\sum_{k=1}^{n} a_{i\nu} B_{\nu jk} X_k \right) \right) Y_j \right) \\
&= \left(\sum_{j=1}^{n} \left(\sum_{\nu=1}^{n} a_{i\nu} \left(\sum_{k=1}^{n} B_{\nu jk} X_k \right) \right) Y_j \right) \subseteq \left(\sum_{j=1}^{n} \left(\sum_{\nu=1}^{n} a_{i\nu} \left(\sum_{k=1}^{n} B_{\nu jk} X_k \right) Y_j \right) \right) \\
&= \left(\sum_{\nu=1}^{n} \left(\sum_{j=1}^{n} a_{i\nu} \left(\sum_{k=1}^{n} B_{\nu jk} X_k \right) Y_j \right) \right) = \left(\sum_{\nu=1}^{n} a_{i\nu} \left(\sum_{j=1}^{n} \left(\sum_{k=1}^{n} B_{\nu jk} X_k \right) Y_j \right) \right) \\
&= \left(\sum_{\nu=1}^{n} a_{i\nu} Z_\nu \right) = \mathscr{A}(\mathscr{B}\mathscr{x}\mathscr{y}).
\end{aligned}$$

University Library
GOVERNORS STATE UNIVERISTY

Of (b): From Definition 11b it follows using the symmetry of $\mathscr{x}$ for the ith component Z_i and from $\mathscr{z} = \mathscr{B}\mathscr{x}\mathscr{x}$ that

$$Z_i = \sum_{j=1}^{n}\left(\sum_{k=1}^{n} B_{ijk}X_k\right)X_j = \sum_{j=1}^{n}\left|\sum_{k=1}^{n} B_{ijk}X_k\right|X_j$$

$$= \sum_{j=1}^{n}\left|\left(\sum_{k=1}^{n} |B_{ijk}|X_k\right)\right|X_j = \sum_{j=1}^{n}\left(\sum_{k=1}^{n} \tfrac{1}{2}|B_{ijk}|\, d(X_k)\right)X_j.$$

Therefore $\mathscr{B}\mathscr{x}\mathscr{x} = \frac{1}{2}|\mathscr{B}|\, d(\mathscr{x})\mathscr{x}$.

Of (c): Using the representation given in (b) we get

$$d(Z_i) = \sum_{j=1}^{n}\left(\sum_{k=1}^{n} \tfrac{1}{2}|B_{ijk}|\, d(X_k)\right)d(X_j);$$

that is,

$$d(\mathscr{B}\mathscr{x}\mathscr{x}) = \tfrac{1}{2}|\mathscr{B}|\, d(\mathscr{x})\, d(\mathscr{x}). \quad \blacksquare$$

Remarks: Interval matrix operations were used frequently prior to the systematic investigation of their properties in Apostolatos and Kulisch [1]. The statements regarding triply indexed arrays with interval elements were proven by Platzöder [6].

REFERENCES

[1] N. Apostolatos and U. Kulisch, Grundzüge einer Intervallrechnung für Matrizen und einige Anwendungen. *Elektron. Rech.* **10**, 73–83 (1968).

[2] U. Kulisch, Grundzüge der Intervallrechnung. *In* "Überblicke Mathematik 2." Bibliograph. Inst. Mannheim, 1969.

[3] O. Mayer, Über die in der Intervallrechnung auftretenden Räume und einige Anwendungen. Ph.D. Thesis, Univ. Karlsruhe (1968).

[4] O. Mayer, Algebraische und metrische Strukturen in der Intervallrechnung und einige Anwendungen. *Computing* **5**, 144–162 (1970).

[5] R. E. Moore, "Interval Analysis." Prentice-Hall, Englewood Cliffs, New Jersey, 1966.

[6] L. Platzöder, Einige Beiträge über die Existenz von Lösungen nichtlinearer Gleichungssysteme und Verfahren zu ihrer Berechnung. Ph.D. Thesis, Techn. Univ. Berlin (1981).

[7] L. B. Rall, "Computational Solution of Nonlinear Operator Equations." Wiley, New York, 1969.

[8] W. J. Thron, "Topological Structures." Holt, New York, 1966.

[9] Ch. Ullrich, Rundungsinvariante Strukturen mit äußeren Verknüpfungen. Ph.D. Thesis, Univ. Karlsruhe (1972).

Chapter 11 / FIXED POINT ITERATION FOR NONLINEAR SYSTEMS OF EQUATIONS

We now consider functions $f(x)$ of the vector variable $x = (x_1, \ldots, x_n)^T \in V_n(\mathbb{C})$ that takes on values in $\mathbb{C}$. We furthermore assume that the function f is constructed from the basic arithmetic operations as well as the standard functions sine, cosine, etc. This means that the function may also be evaluated as an interval function. The function f is furthermore assumed to depend on m parameters $a_1, a_2, \ldots, a_m \in \mathbb{C}$; that is, we may write f as

$$f(x) = f(x_1, x_2, \ldots, x_n; a_1, a_2, \ldots, a_m).$$

Let now n functions $f_i(x)$, $1 \leqslant i \leqslant n$, be given in the above form. The relation

$$y = f(x) = (f_i(x))$$

defines a map from $V_n(\mathbb{C})$ to $V_n(\mathbb{C})$. The relation

$$y = f(x) = (f_i(x))$$

furthermore defines a map on the set of interval vectors having n components; that is, it is defined by

$$f: V_n(I(\mathbb{C})) \to V_n(I(\mathbb{C})) \qquad \text{with} \quad f(x) = (f_i(x)).$$

In the following we wish to use interval arithmetic in order to include the solution set of the set

$$f(x) = o$$

of equations where

$$f(x) = (f_i(x))$$

as well as

$$f_i(x) = f_i(x_1, \ldots, x_n; a_{i1}, \ldots, a_{im_i}), \qquad 1 \leq i \leq n,$$

assuming that the parameters a_{ij} are allowed to vary independently over the (generally complex) intervals.

An important possibility for solving this problem is to solve the system using an iteration method. We note that the given equation may always be transformed to the form

$$x = f(x).$$

The interval evaluation of the right-hand side of this equation for an arbitrary interval vector $x^{(0)}$ gives an interval vector $x^{(1)}$. Continuing in this manner one gets the iteration method

$$x^{(k+1)} = f(x^{(k)}), \qquad k \geq 0.$$

The following questions then immediately arise regarding (a) when the sequence $\{x^{(k)}\}_{k=0}^{\infty}$ exists, (b) when the sequence converges, (c) when the limit x^* is unique, and (d) what relationship the limit x^* has with the solution of the above problem.

First, we give an elementary fixed point theorem that depends on the inclusion monotonicity of the interval evaluation of functions (see also Chapter 3).

Theorem 1: Let the map

$$f: V_n(\mathbb{C}) \to V_n(\mathbb{C}) \qquad \text{with} \quad f(x) = (f_i(x))$$

be given where the functions $f_i(x)$ are of the form described above. Consider an iteration method in $V_n(I(\mathbb{C}))$ given by

$$x^{(k+1)} = f(x^{(k)}), \qquad k \geq 0,$$

which satisfies

$$x^{(1)} \subseteq x^{(0)}.$$

Then the following is true:

(1) The sequence of iterates $\{x^{(k)}\}_{k=0}^{\infty}$ satisfies $\lim_{k\to\infty} x^{(k)} = x$ with $x = f(x)$.

(2) Each vector $x \in x^{(0)}$ satisfying the equation $x = f(x)$ is contained in x; that is,

$$\{x \mid x \in x^{(0)}, x = f(x)\} \subseteq x,$$

holds.

Proof: Of (1): The assumption is that $x^{(1)} \subseteq x^{(0)}$. Since the interval evaluation satisfies the property of inclusion monotonicity, it follows that

$$x^{(2)} = f(x^{(1)}) \subseteq f(x^{(0)}) = x^{(1)} \subseteq x^{(0)}.$$

Using complete induction it can be shown that

$$\cdots \subseteq x^{(3)} \subseteq x^{(2)} \subseteq x^{(1)} \subseteq x^{(0)}.$$

From Corollary 10.8 it follows that the sequence converges towards an element $x \in V_n(I(\mathbb{C}))$. The continuity of the interval evaluation then implies that

$$x = \lim_{k \to \infty} x^{(k)} = \lim_{k \to \infty} f(x^{(k)}) = f(x).$$

Of (2): Let $x \in x^{(0)}$ with $x = f(x)$. Again from the inclusion monotonicity of interval evaluations it follows that

$$x = f(x) \in f(x^{(0)}) = x^{(1)}$$

and in general by complete induction that

$$x \in x^{(k)}, \qquad k \geqslant 0,$$

and therefore also $x \in x$. ■

The fixed point that exists under the assumption in Theorem 1 is not necessarily unique. This is shown in the following simple example.

Example: Consider the equation

$$X = X \cdot X \cdot X$$

in $R(\mathbb{C})$. This equation is clearly satisfied by the following members of $R(\mathbb{C})$:

$$X = [-1, 1], [1, 1], [-1, -1], [0, 1], [-1, 0], [0, 0], i[-1, 1].$$

We now prove another fixed point theorem that is based on a slightly modified iteration as well as on somewhat different conditions.

Theorem 2: Let the map

$$f: V_n(\mathbb{C}) \to V_n(\mathbb{C}) \qquad \text{with} \quad f(x) = (f_i(x))$$

be given where the functions $f_i(x)$ again are of the form described above. Consider the iteration

$$x^{(k+1)} = f(x^{(k)}) \cap x^{(k)}, \qquad k \geqslant 0,$$

in $V_n(R(\mathbb{C}))$. Assume the existence of at least one element $\tilde{x} \in x^{(0)}$ satisfying the equation $\tilde{x} = f(\tilde{x})$. Then the following statements are true:

(3) The sequence $\{x^{(k)}\}_{k=0}^{\infty}$ of iterates satisfies $\lim_{k \to \infty} x^{(k)} = x$ with $x = f(x) \cap x$.

(4) Each vector $\underset{\cdot}{x} \in x^{(0)}$ satisfying the equation $\underset{\cdot}{x} = \underset{\cdot}{f}(\underset{\cdot}{x})$ is contained in x; that is,

$$\{\underset{\cdot}{x} \mid \underset{\cdot}{x} \in x^{(0)}, \underset{\cdot}{x} = \underset{\cdot}{f}(\underset{\cdot}{x})\} \subseteq x.$$

Proof: The proof is similar to the proof of Theorem 1. Since $\tilde{\underset{\cdot}{x}} \in x^{(0)}$ we get the relation $\tilde{\underset{\cdot}{x}} \in f(x^{(0)}) \cap x^{(0)} = x^{(1)}$. Using complete induction we get $\tilde{\underset{\cdot}{x}} \in x^{(k)}$, $k \geqslant 0$. Since the intersection is always nonempty one gets the sequence of intervals

$$x^{(0)} \supseteq x^{(1)} \supseteq \cdots$$

that according to Corollary 10.8 converges to a limiting value x. Since the interval evaluation and the process of taking the intersection always are continuous we have $x = f(x) \cap x$ for the limiting value as well as $\hat{\underset{\cdot}{x}} \in x$ for all $\hat{\underset{\cdot}{x}} = \underset{\cdot}{f}(\hat{\underset{\cdot}{x}}) \in x^{(0)}$. ■

We may make the same remark as for Theorem 1 with respect to the uniqueness of the fixed point in Theorem 2. These two fairly elementary fixed point theorems were treated in detail in Nickel [8]. We now bring two fixed point theorems that we shall use later for concrete iteration procedures. These fixed point theorems allow us also to make statements about the uniqueness of the fixed point in contrast to the rather general statements of Theorems 1 and 2. First we have

Definition 3: (See Schröder [10].) Let

$$\underset{\cdot}{f}: V_n(\mathbb{C}) \to V_n(\mathbb{C}), \qquad \underset{\cdot}{f}(\underset{\cdot}{x}) = (f_i(\underset{\cdot}{x}))$$

be a map in the form described above. $\underset{\cdot}{f}$ is called a $\underset{\cdot}{\mathscr{P}}$ contraction if there exists a nonnegative matrix $\underset{\cdot}{\mathscr{P}}$ such that

$$q(\underset{\cdot}{f}(x), \underset{\cdot}{f}(y)) \leqslant \underset{\cdot}{\mathscr{P}} q(x, y) \qquad \text{for all} \quad x, y \in V_n(I(\mathbb{C}))$$

with

$$\rho(\underset{\cdot}{\mathscr{P}}) < 1. \quad ■$$

Here ρ denotes the spectral radius of $\underset{\cdot}{\mathscr{P}}$ and q denotes the distance of two interval vectors as defined in Chapter 10. We first prove the following theorem based on this definition.

Theorem 4: If $\underset{\cdot}{f}: V_n(\mathbb{C}) \to V_n(\mathbb{C})$ is a $\underset{\cdot}{\mathscr{P}}$ contraction, then $x = \underset{\cdot}{f}(x)$ has a unique fixed point $x^* \in V_n(I(\mathbb{C}))$. Furthermore, for every $x^{(0)} \in \dot{V}_n(I(\mathbb{C}))$ the iteration

$$x^{(k+1)} = \underset{\cdot}{f}(x^{(k)}), \qquad k \geqslant 0,$$

converges to x^*.

Proof: Since $\rho(\mathscr{P}) < 1$ and since $\mathscr{P} \geqslant \mathscr{O}$ it follows that $(\mathscr{I} - \mathscr{P})^{-1}$ exists and that

$$(\mathscr{I} - \mathscr{P})^{-1} = \sum_{j=0}^{\infty} \mathscr{P}^j \geqslant \sum_{j=0}^{m-1} \mathscr{P}^j \geqslant \mathscr{O}.$$

For arbitrary k and $m \geqslant 1$, we then get

$$\begin{aligned} q(x^{(k+m)}, x^{(k)}) &\leqslant \sum_{j=0}^{m-1} \mathscr{P}^j q(x^{(k+1)}, x^{(k)}) \\ &\leqslant (\mathscr{I} - \mathscr{P})^{-1} \mathscr{P}^k q(x^{(1)}, x^{(0)}). \end{aligned}$$

Since $\lim_{k \to \infty} \mathscr{P}^k = \mathscr{O}$, each component of $\{x^{(k)}\}_{k=0}^{\infty}$ and therefore $\{x^{(k)}\}_{k=0}^{\infty}$ itself forms a Cauchy sequence. Since $V_n(I(\mathbb{C}))$ is complete and since a $\mathscr{P}$ contraction is continuous, we get

$$\lim_{k \to \infty} x^{(k)} = x^* \qquad \text{and} \qquad x^* = f(x^*).$$

The uniqueness of the fixed point follows from

$$q(x^*, y^*) = q(f(x^*), f(y^*)) \leqslant \mathscr{P} q(x^*, y^*)$$

and $(\mathscr{I} - \mathscr{P})^{-1} \geqslant \mathscr{O}$. ∎

The content of this theorem is a special case of the more general result proved by Schröder [10].

A further fixed point theorem that will be used later is

Theorem 5: Let

$$f: V_n(\mathbb{C}) \to V_n(\mathbb{C})$$

and

$$g: V_n(\mathbb{C}) \times V_n(\mathbb{C}) \to V_n(\mathbb{C}),$$

where f and g have the form described above and where

$$g(x, x) = f(x) \qquad \text{for all} \quad x \in V_n(I(\mathbb{C})),$$

$$\begin{aligned} &q(g(x, z), g(y, z)) \leqslant \mathscr{Q} q(x, y), \\ &q(g(z, x), g(z, y)) \leqslant \mathscr{R} q(x, y) \qquad \text{for all} \quad x, y, z \in V_n(I(\mathbb{C})), \end{aligned}$$

with

$$\mathscr{Q} \geqslant \mathscr{O}, \qquad \mathscr{R} \geqslant \mathscr{O}, \qquad \rho(\mathscr{Q}) < 1, \qquad \rho((\mathscr{I} - \mathscr{Q})^{-1} \mathscr{R}) < 1.$$

Then $f(x) = x$ has a unique fixed point $x^* \in V_n(I(\mathbb{C}))$ and for every $x^{(0)}$ there exists a unique sequence $\{x^{(k)}\}_{k=0}^{\infty}$ that satisfies the equation

$$x^{(k+1)} = g(x^{(k+1)}, x^{(k)}), \qquad k \geqslant 0.$$

Furthermore $\lim_{k \to \infty} x^{(k)} = x^*$.

Proof: The matrix $(\mathcal{I} - \mathcal{Q})^{-1}\mathcal{R}$ may be considered to be an iteration matrix corresponding to the matrix $\mathcal{I} - \mathcal{Q} - \mathcal{R}$. Since $\mathcal{P} = \mathcal{Q} + \mathcal{R} \geqslant \mathcal{O}$ it follows that $\rho(\mathcal{P}) < 1$ using the fact that $\rho((\mathcal{I} - \mathcal{Q})^{-1}\mathcal{R}) < 1$. (See also Apostolatos and Kulisch [2] and Varga [11].) For arbitrary $x, y \in V_n(I_n(\mathbb{C}))$ using (10.23) and (10.24), it follows that

$$q(f(x), f(y)) \leqslant q(g(x, x), g(x, y)) + q(g(x, y), g(y, y))$$
$$\leqslant (\mathcal{R} + \mathcal{Q})q(x, y);$$

that is, f is a $\mathcal{P}$ contraction. By the previous theorem we have that $f(x) = x$ has a unique fixed point $x^* \in V_n(I_n(\mathbb{C}))$. From the assumptions we have that for a fixed $z \in V_n(I_n(\mathbb{C}))$ the map $g(\cdot, z)$ is a $\mathcal{Q}$ contraction. By applying the previous theorem it follows that $g(\cdot, x^{(k)})$ has exactly one fixed point $x^{(k+1)}$. This means that the existence of the sequence $\{x^{(k)}\}_{k=0}^{\infty}$ has been shown for arbitrary $x^{(0)} \in V_n(I(\mathbb{C}))$. From

$$q(x^{(k+1)}, x^*) \leqslant q(g(x^{(k+1)}, x^{(k)}), g(x^*, x^{(k)})) + q(g(x^*, x^{(k)}), g(x^*, x^*))$$
$$\leqslant \mathcal{Q}q(x^{(k+1)}, x^*) + \mathcal{R}q(x^{(k)}, x^*)$$

or

$$q(x^{(k+1)}, x^*) \leqslant (\mathcal{I} - \mathcal{Q})^{-1}\mathcal{R}q(x^{(k)}, x^*) \leqslant ((\mathcal{I} - \mathcal{Q})^{-1}\mathcal{R})^{k+1}q(x^{(0)}, x^*),$$

it follows that $\lim_{k\to\infty} x^{(k)} = x^*$ since $\rho((\mathcal{I} - \mathcal{Q})^{-1}\mathcal{R}) < 1$. This proves the theorem. ■

The contents of Theorem 5 were originally proved for maps from $V_n(\mathbb{R})$ to $V_n(\mathbb{R})$ (see Ortega and Rheinholdt [9]).

In connection with the last two theorems we now wish to show a connection between the unique fixed point and the potential solutions of the equation

$$x = f(x), \qquad x \in V_n(\mathbb{C}).$$

We formulate

Corollary 6: Let

$$f: V_n(\mathbb{C}) \to V_n(\mathbb{C}), \qquad f(x) = (f_i(x))$$

be a map and let

$$f_i(x) = f_i(X_1, X_2, \ldots, X_n; A_{i1}, \ldots, A_{im_i}), \qquad 1 \leqslant i \leqslant n.$$

Assume that the assumptions of Theorem 4 or 5 are satisfied and let x^* be the proven unique fixed point of $x = f(x)$. Then it holds that

$$\{x \mid x = f(x), a_{ij} \in A_{ij}, 1 \leqslant i \leqslant n, 1 \leqslant j \leqslant m_i\} \subseteq x^*.$$

Proof: Consider the equation

$$x = f(x), \qquad x \in V_n(\mathbb{C}),$$

for a fixed choice of $a_{ij} \in A_{ij}$, $1 \leqslant i \leqslant n$, $1 \leqslant j \leqslant m_i$, and assume that it is satisfied by an element $x^* \in V_n(\mathbb{C})$. We may then start the iteration in Theorem 4, respectively, Theorem 5, with $x^{(0)} = x^*$ and we get as a limiting value the fixed point x^*. As in the proof of (2) in Theorem 1 we can use inclusion monotonicity to show that

$$x^* \in x^{(k)}, \qquad k \geqslant 0,$$

always holds. It therefore follows that $x^* \in x^*$. ■

We now wish to consider the practical calculation of the Lipschitz constants of interval evaluations. We shall see that the Lipschitz constants of interval evaluations will majorize the Lipschitz constants of the underlying functions. This means that each of the systems $x = f(x)$ considered in Corollary 6 satisfies the conditions of Theorems 4 and 5 when restricted to $V_n(\mathbb{C})$, and they have therefore a unique solution x^*.

In the following we consider the practical calculation of Lipschitz constants for interval evaluations. Through this it will be possible to check whether the conditions of Theorem 4, respectively, Theorem 5, are satisfied in specific cases. We limit ourselves to the space $V_n(I(\mathbb{R}))$ for simplicity. Similar formulas may be developed for $V_n(I(\mathbb{C}))$ without difficulty. As a preparation we prove Lemma 7 that states a property of the metric q that follows from the fact that it is a Hausdorff metric (compare with Moore [4]).

Lemma 7: Let $Y, Z \in I(\mathbb{R})$. Then if $\alpha \geqslant 0$, it follows that

$q(Y, Z) \leqslant \alpha \Leftrightarrow$ For every $y \in Y$ there exists a $z \in Z$ with $|y - z| \leqslant \alpha$ and for every $z \in Z$ there exists a $y \in Y$ with $|z - y| \leqslant \alpha$.

Proof: We have $q(Y, Z) = \max\{|y_1 - z_1|, |y_2 - z_2|\}$, if $Y = [y_1, y_2]$ and $Z = [z_1, z_2]$. The above equivalence is first proved in the direction $\Rightarrow$: Let $q(Y, Z) \leqslant \alpha$ and fix the choice of $y \in Y$. If $y \in Z$, then we can choose $z = y$, which is the first statement on the right. If, on the other hand, $y \notin Z$, then if $z_1 > y$, we get with $z = z_1$ that

$$\alpha \geqslant |z_1 - y_1| = z_1 - y_1 \geqslant z_1 - y = |z_1 - y|$$

and for the case $z_2 < y$ we get with $z = z_2$ that

$$\alpha \geqslant |z_2 - y_2| = y_2 - z_2 \geqslant y - z_2 = |y - z_2|.$$

Since this is symmetric in y and z, it follows analogously that the second statement on the right is valid for a fixed $z \in Z$.

$\Leftarrow$: Let first $y_1 \leqslant z_1$. We then fix $y = y_1$, and there exists a $z \geqslant z_1$ such that

$$\alpha \geqslant |z - y_1| = z - y_1 \geqslant z_1 - y_1 = |z_1 - y_1|.$$

If, on the other hand, we have $z_1 < y_1$, then we choose a fixed $z = z_1$ and there exists a $y \geqslant y_1$ for which

$$\alpha \geqslant |y - z_1| = y - z_1 \geqslant y_1 - z_1 = |y_1 - z_1|.$$

It therefore always follows that $|y_1 - z_1| \leqslant \alpha$. One proves $|y_2 - z_2| \leqslant \alpha$ in the same manner. Finally one has $q(Y, Z) \leqslant \alpha$. ■

We shall now use this lemma in order to derive a statement regarding the Lipschitz constant of interval evaluations. This is contained in

Theorem 8: Let f be a real function of the real variable x. From the expression $f(x; a_1, \ldots, a_m)$ belonging to the function f (see Chapter 3) the expression $\tilde{f}(x_1, x_2, \ldots, x_n; a_1, \ldots, a_m)$ is formed in such a manner that each occurrence of the variable x is replaced by a new variable x_i, $1 \leqslant i \leqslant n$. Let this new expression satisfy a Lipschitz condition

$$|\tilde{f}(x_1, \ldots, y_i, \ldots, x_n; a_1, \ldots, a_m) - \tilde{f}(x_1, \ldots, z_i, \ldots, x_n; a_1, \ldots, a_m)| \leqslant l_i |y_i - z_i|$$

for each i with $1 \leqslant i \leqslant n$ choosing fixed

$$x_k \in X, \qquad 1 \leqslant k \leqslant n, \qquad i \neq k, \qquad \text{and} \qquad a_k \in A_k, \qquad 1 \leqslant k \leqslant m.$$

If the interval evaluation $f(X; A_1, \ldots, A_m)$ exists, then for all intervals $Y \subseteq X$ and $Z \subseteq X$ the following Lipschitz condition

$$q(f(Y; A_1, \ldots, A_m), f(Z; A_1, \ldots, A_m)) \leqslant \left(\sum_{i=1}^{n} l_i \right) q(Y, Z) \tag{5}$$

is satisfied.

Proof: From the construction of the expression $\tilde{f}(x_1, \ldots, x_n; a_1, \ldots, a_m)$ it follows that (compare to Chapter 3)

$$f(X; A_1, \ldots, A_m) = \tilde{f}(X, \ldots, X; A_1, \ldots, A_m) = W(\tilde{f}, X, \ldots, X; A_1, \ldots, A_m).$$

Let $u \in f(Y; A_1, \ldots, A_m)$ be chosen arbitrarily, then we have

$$u = \tilde{f}(y_1, \ldots, y_n; a_1, \ldots, a_m) \quad \text{for} \quad y_i \in Y, \quad 1 \leqslant i \leqslant n, \quad a_k \in A_k, \quad 1 \leqslant k \leqslant m.$$

From Lemma 7 we have that for every y_i there exists an element $z_i \in Z$ for which

$$|y_i - z_i| \leqslant q(Y, Z).$$

If we consider the function value

$$\tilde{f}(z_1, \ldots, z_n; a_1, \ldots, a_m) \in f(Z; A_1, \ldots, A_m),$$

then we get by repeated application of the triangular inequality while remembering the above assumptions that the inequality

$$\begin{aligned} &|\mathfrak{f}(y_1, \ldots, y_n; a_1, \ldots, a_m) - \mathfrak{f}(z_1, \ldots, z_n; a_1, \ldots, a_m)| \\ &\quad \leqslant |\mathfrak{f}(y_1, \ldots, y_n; a_1, \ldots, a_m) \\ &\qquad - \mathfrak{f}(z_1, y_2, \ldots, y_n; a_1, \ldots, a_m)| + \cdots \\ &\qquad + |\mathfrak{f}(z_1, \ldots, z_{n-1}, y_n; a_1, \ldots, a_m) \\ &\qquad - \mathfrak{f}(z_1, \ldots, z_n; a_1, \ldots, a_m)| \\ &\quad \leqslant \sum_{i=1}^{n} l_i |y_i - z_i| \leqslant \left(\sum_{i=1}^{n} l_i \right) q(Y, Z) = \alpha \end{aligned}$$

is valid. An analogous inequality may be derived for an arbitrary $v \in f(Z, A_1, \ldots, A_m)$. From Lemma 7 we then get the inequality (5). ■

The assumptions in Theorem 8 are almost always satisfied in practice. The function expressions that are considered here are put together using the basic arithmetic operations as well as a number of standard functions. They are therefore even almost always differentiable over the domain of definition.

As an application of the above theorem we now give some concrete examples.

Examples: (a) $f(x; a) = ax$. The interval evaluation of this function for $A \in I(\mathbb{R})$ and arbitrary $Y, Z \in I(\mathbb{R})$ satisfies the inequality

$$q(AY, AZ) \leqslant |A| q(Y, Z)$$

according to (5) (compare also with formula (2.7)).

(b) $f(x; a_0, \ldots, a_n) = \sum_{k=0}^{n} a_k x^k$.

The interval evaluation of this function satisfies with $A_k \in I(\mathbb{R})$, $0 \leqslant k \leqslant n$, and arbitrary $Y, Z \in I(\mathbb{R})$ the inequality

$$q\left(\sum_{k=0}^{n} A_k Y^k, \sum_{k=0}^{n} A_k Z^k \right) \leqslant \left(\sum_{k=1}^{n} k |A_k| \, |X|^{k-1} \right) q(Y, Z),$$

according to (5). Here X is the smallest interval containing $Y \cup Z$. This inequality is independent of whether X^k is calculated as $X \cdot \cdots \cdot X$ or as a unary operation according to Definition 1.3.

A formally analogous formula may be derived for rational expressions. This is left as a simple exercise for the reader.

(c) $f(x; a) = x/2 - 2e^{ax}$.

The interval evaluation of this function satisfies with $A \in I(\mathbb{R})$ and arbitrary

$Y, Z \in I(\mathbb{R})$ the inequality

$$q(Y/2 - 2e^{AY}, Z/2 - 2e^{AZ}) \leqslant (\tfrac{1}{2} + 2|A|e^{|A||X|})q(Y, Z),$$

according to (5). Here again X is the smallest interval containing $Y \cup Z$.

Theorem 8 may be carried over to functions of several variables without any difficulties. One then obtains formulas for the interval evaluation of maps $\not{f}$ as described in the introductory parts of this chapter. One only has to apply the inequality componentwise. This follows without any difficulties from the already proven results and we therefore omit the details. Such formulas then allow us to calculate the matrix $\mathscr{P}$ that occurs in Definition 3 in practical cases. We give here a simple example of this.

Example: We are given the n functions

$$f_i(x_1, \ldots, x_n; a_{i1}, \ldots, a_{in}) = \sum_{v=1}^{n} \sin(a_{iv}x_v), \qquad 1 \leqslant i \leqslant n.$$

The interval evaluations exist for arbitrary intervals $X_v \in I(\mathbb{R})$, $1 \leqslant v \leqslant n$, as well as $A_{ij} \in I(\mathbb{R})$, $1 \leqslant i, j \leqslant n$. Each of these functions furthermore, satisfies the assumptions of Theorem 8 for given intervals A_{ij}, $1 \leqslant i, j \leqslant n$, with

$$|f_i(x_1, \ldots, x_v, \ldots, x_n; a_{i1}, \ldots, a_{in}) - f_i(x_1, \ldots, y_v, \ldots, x_n; a_{i1}, \ldots, a_{in})|$$
$$\leqslant |A_{iv}|\,|x_v - y_v|, \qquad x_j \in X_j, \quad 1 \leqslant j \leqslant n, \quad j \neq v, \quad x_v, y_v \in X_v.$$

If we now use Theorem 8 on each component then we get

$$\not{f}(\mathscr{x}) = (f_i(\mathscr{x})) = \left(\sum_{v=1}^{n} \sin(A_{iv}X_v)\right),$$

$$q(\not{f}(\mathscr{x}), \not{f}(\mathscr{y})) \leqslant \mathscr{P}q(\mathscr{x}, \mathscr{y}), \qquad \text{with} \quad \mathscr{P} = (|A_{ij}|).$$

If now $\rho(\mathscr{P}) < 1$, then the equation

$$\mathscr{x} = \not{f}(\mathscr{x}) = (f_i(\mathscr{x}))$$

has exactly one fixed point according to Theorem 4, which may be determined by iteration. It may be shown using Theorem 5 that a single step method also converges.

In Theorem 4 we assumed for the sake of simplicity that the map $\not{f}$ was defined on the whole of the space $V_n(\mathbb{C})$, that $\not{f}$ had an interval evaluation for all members of $V_n(I(\mathbb{C}))$, and that $\not{f}$ is a $\mathscr{P}$ contraction. Simple examples [see, for instance, Example (b)] show, however, that the matrix $\mathscr{P}$ may be dependent of $\mathscr{x}$ and $\mathscr{y}$ in Definition 3 and that the condition $\rho(\mathscr{P}) < 1$ may not be satisfied for all $\mathscr{x}, \mathscr{y}$. For this case we have

Theorem 9: Suppose $\not{f}: \vartheta \subseteq V_n(\mathbb{C}) \to V_n(\mathbb{C})$ is a $\mathscr{P}$ contraction for all $\mathscr{x}, \mathscr{y}$ from the closed set $I(\vartheta) = \{\mathscr{z} \in V_n(I(\mathbb{C})) | \mathscr{z} \subseteq \vartheta\}$. If $\not{f}(\mathscr{x}) \in I(\vartheta)$ for all $\mathscr{x} \in I(\vartheta)$, then

$x = f(x)$ has a unique fixed point $x^* \in I(\vartheta)$ and the iteration $x^{(k+1)} = f(x^{(k)})$, $k \geqslant 0$, converges for every starting vector $x^{(0)} \in I(\vartheta)$ to this fixed point x^*. ■

The proof may be carried through analogously to the proof of Theorem 4. We remark that from the inequality

$$q(x^{(k+m)}, x^{(k)}) \leqslant \sum_{j=1}^{m} \mathscr{P}^j q(x^{(k)}, x^{(k-1)})$$
$$\leqslant (\mathscr{I} - \mathscr{P})^{-1} \mathscr{P} q(x^{(k)}, x^{(k-1)})$$

that may be derived from Theorem 4 we get the error estimate

$$q(x^{(k)}, x^*) \leqslant (\mathscr{I} - \mathscr{P})^{-1} \mathscr{P} q(x^{(k)}, x^{(k-1)})$$

for both Theorems 4 and 9 by letting $m \to \infty$ in the previous inequality.

We mention another lemma that we shall use later.

Lemma 10: Let $f\colon x \subseteq \vartheta \subseteq V_n(\mathbb{R}) \to V_n(\mathbb{R})$ be a continuous map. Let also $\mathscr{Y}\colon x \subseteq \vartheta \subseteq V_n(\mathbb{R}) \to V_n(\mathbb{R})$ be continuous and assume that for all $x \in x$ the inverse $\mathscr{Y}(x)^{-1}$ exists. If $p(x) \in x$ for $x \in x$ holds for the map $p\colon x \to V_n(\mathbb{R})$ where

$$p(x) := x - \mathscr{Y}(x) f(x),$$

then f has a zero in the interval x. ■

The simple proof of this assertion is given by using the fixed point theorem of Brouwer: p maps the compact and convex set $\{x \mid x \in x\} \subset V_n(\mathbb{R})$ into itself and it therefore has a fixed point. Since $\mathscr{Y}(x)$ is nonsingular f has a zero in x.

Remarks: Lemma 10 as well as the proof of this lemma is given in Alefeld [1] and it gives a generalization of the assertions for constant $\mathscr{Y}$ proven by Moore in [5]. See also Moore [6].

This lemma is important since one can test the assumptions of the lemma using a finite number of arithmetic operations for an interval vector through the majorizing of the map $\mathscr{Y}(x)$ by an interval expression depending on an interval vector x.

The details of the above remarks are described in later chapters, in particular in Chapter 19.

The direct verification of the assumptions of Brouwer's fixed point theorem is in contrast a very difficult task.

REFERENCES

[1] G. Alefeld, Intervallanalytische Methoden bei nichtlinearen Gleichungen. *Jahrbuch Überb. Math.* 63–78 (1979).

[2] N. Apostolatos and U. Kulisch, Über die Konvergenz des Relaxationsverfahrens bei nichtnegativen und diagonaldominanten Matrizen. *Computing* **2**, 17–24 (1967).
[3] J. Herzberger, Über metrische Eigenschaften von Mengensystemen und einige Anwendungen. Ph.D. Thesis, Univ. Karlsruhe (1969).
[4] R. E. Moore, "Interval Analysis." Prentice-Hall, Englewood Cliffs, New Jersey, 1966.
[5] R. E. Moore, A test for existence of solutions to non-linear systems. *SIAM J. Numer. Anal.* **14**, 611–615 (1977).
[6] R. E. Moore, "Methods and Applications of Interval Analysis," SIAM Studies in Applied Mathematics. SIAM, Philadelphia, Pennsylvania, 1979.
[7] S. B. Nadler, Multi-valued contraction mappings. *Pacific J. Math.* **30**, 475–488 (1969).
[8] K. Nickel, The contraction mapping fixpoint theorem in interval analysis. MRC Technical Summary Rep. No. 1334, Madison, Wisconsin (1973).
[9] J. M. Ortega and W. C. Rheinholdt, Numerical solution of nonlinear problems. *In* "Studies in Numerical Analysis," Vol. 2, pp. 122–143. Society for Industrial and Applied Mathematics, Philadelphia, Pennsylvania, 1970.
[10] J. Schröder, Das Iterationsverfahren bei allgemeinem Abstandsbegriff. *Math. Z.* **66**, 111–116 (1956).
[11] R. Varga, "Matrix Iterative Analysis." Prentice-Hall, Englewood Cliffs, New Jersey, 1962.

Chapter 12 / SYSTEMS OF LINEAR EQUATIONS AMENABLE TO ITERATION

We assume that the system of linear equations considered here already is of the form

(1) $$x = \mathscr{A} x + b$$

with $\mathscr{A} = (a_{ij})$ and $b = (b_i)$.

Let it be known that the elements a_{ij} of the matrix $\mathscr{A}$ lie in intervals A_{ij} and the components b_i of the vector b lie in intervals B_i. We are interested in the set of solutions of (1) that is obtained when the input data is allowed to vary over the given intervals. We therefore define an $n \times n$ interval matrix $\mathscr{A} = (A_{ij})$ containing the interval coefficients as well as an interval vector $b = (B_i)$ containing the interval right-hand side components. Let us now consider the map

$$f\colon V_n(\mathbb{C}) \to V_n(\mathbb{C})$$

defined by

$$f(x) = \mathscr{A} x + b.$$

First we have

Theorem 1: The iteration

(2) $$x^{(k+1)} = \mathscr{A} x^{(k)} + b, \qquad k \geqslant 0,$$

converges to a unique fixed point x^* of the equation

$$x = \mathscr{A}x + b,$$

for every $x^{(0)} \in V_n(I(\mathbb{C}))$ iff $\rho(|\mathscr{A}|) < 1$.

Proof: We show that f is an $|\mathscr{A}|$ contraction. Let $x, y \in V_n(I(\mathbb{C}))$. Using (10.23) as well as (10.25) we get

$$q(f(x), f(y)) = q(\mathscr{A}x + b, \mathscr{A}y + b) \leqslant |\mathscr{A}| q(x, y).$$

From Theorem 11.4 it follows that the condition $\rho(|\mathscr{A}|) < 1$ suffices for the convergence and uniqueness of the fixed point. Conversely, let the iteration $x^{(k+1)} = \mathscr{A}x^{(k)} + b$, $k \geqslant 0$, converge for each $x^{(0)} \in V_n(I(\mathbb{C}))$ toward the fixed point x^*. We have to show that $\rho(|\mathscr{A}|) < 1$. The theorem of Perron and Frobenius (see Varga [21]) shows that the real nonnegative matrix $|\mathscr{A}|$ has a nonnegative eigenvector for the eigenvalue $\lambda = \rho(|\mathscr{A}|)$. From the convergence of the iteration $x^{(k+1)} = \mathscr{A}x^{(k)} + b$, $k \geqslant 0$, toward x^* for each starting vector $x^{(0)}$ it follows that the sequence $\{d(x^{(k)})\}_{k=0}^{\infty}$ converges to $d(x^*)$. We now choose $x^{(0)}$ such that $d(x^{(0)})$ is an eigenvector for the eigenvalue $\lambda = \rho(|\mathscr{A}|)$ of $|\mathscr{A}|$ and such that at least one component of $d(x^{(0)})$ is larger than the corresponding component of $d(x^*)$. It then follows from (2), using (10.12) and (10.18) and assuming that $\lambda = \rho(|\mathscr{A}|) \geqslant 1$, that

$$\begin{aligned} d(x^{(1)}) &= d(\mathscr{A}x^{(0)} + b) = d(\mathscr{A}x^{(0)}) + d(b) \\ &\geqslant d(\mathscr{A}x^{(0)}) \geqslant |\mathscr{A}|\, d(x^{(0)}) = \lambda\, d(x^{(0)}) \geqslant d(x^{(0)}), \\ d(x^{(2)}) &\geqslant |\mathscr{A}|\, d(x^{(1)}) \geqslant \lambda |\mathscr{A}|\, d(x^{(0)}) = \lambda^2\, d(x^{(0)}) \geqslant d(x^{(0)}). \end{aligned}$$

In general we obtain

$$d(x^{(k+1)}) \geqslant |\mathscr{A}|\, d(x^{(k)}) \geqslant \cdots \geqslant \lambda^{k+1}\, d(x^{(0)}) \geqslant d(x^{(0)}).$$

Taking limits as $k \to \infty$ we get

$$d(x^*) \geqslant d(x^{(0)}),$$

which contradicts the choice of $x^{(0)}$. We therefore necessarily have $\rho(|\mathscr{A}|) < 1$. ■

We now wish to establish the connection of this theorem with the task defined in the introduction to this section. (See also Corollary 11.6.)

Theorem 2: Let $\mathscr{A}$ be an interval matrix for which $\rho(|\mathscr{A}|) < 1$. It then follows that for the fixed point x^* (that exists and is unique according to Theorem 1) of the equation $x^* = \mathscr{A}x^* + b$ the relation

$$\{y = (I - A)^{-1} b \mid A \in \mathscr{A},\ b \in b\} \subseteq \{x \mid x \in x^*\}$$

holds.

If $\mathscr{A} = (A_{ij}) \in M_{nn}(I(\mathbb{R}))$, $\mathscr{b} \in V_n(I(\mathbb{R}))$, and if the inequality $i(A_{ij}) \geqslant 0$ holds for $A_{ij} = [i(A_{ij}), s(A_{ij})]$, then $\mathscr{x}^*$ is optimal in the following sense: There is no interval vector $\mathscr{x} \in V_n(I(\mathbb{R}))$ which satisfies $\mathscr{x} \subseteq \mathscr{x}^*$, $\mathscr{x} \neq \mathscr{x}^*$, and for which

$$\{\underset{\cdot}{\mathscr{y}} = (\underset{\cdot}{\mathscr{I}} - \underset{\cdot}{\mathscr{A}})^{-1}\underset{\cdot}{\mathscr{b}} \mid \underset{\cdot}{\mathscr{A}} \in \mathscr{A}, \underset{\cdot}{\mathscr{b}} \in \mathscr{b}\} \subseteq \{\underset{\cdot}{\mathscr{x}} \mid \underset{\cdot}{\mathscr{x}} \in \mathscr{x}\}$$

holds.

Proof: We show first that the linear system of equations

$$\underset{\cdot}{\mathscr{y}} = \underset{\cdot}{\mathscr{A}}\underset{\cdot}{\mathscr{y}} + \underset{\cdot}{\mathscr{b}}$$

has the solution

$$\underset{\cdot}{\mathscr{y}} = (\underset{\cdot}{\mathscr{I}} - \underset{\cdot}{\mathscr{A}})^{-1}\underset{\cdot}{\mathscr{b}}$$

for $\underset{\cdot}{\mathscr{A}} \in \mathscr{A}$ and $\underset{\cdot}{\mathscr{b}} \in \mathscr{b}$. Since $\underset{\cdot}{\mathscr{A}} \in \mathscr{A}$, we have that $|\underset{\cdot}{\mathscr{A}}| \leqslant |\mathscr{A}|$ and from the theorem of Perron and Frobenius (see Varga [21]) we get

$$\rho(\underset{\cdot}{\mathscr{A}}) \leqslant \rho(|\underset{\cdot}{\mathscr{A}}|) \leqslant \rho(|\mathscr{A}|) < 1.$$

It therefore follows that $\underset{\cdot}{\mathscr{I}} - \underset{\cdot}{\mathscr{A}}$ is nonsingular and hence that the assertion is true.

We now consider the iteration

$$\mathscr{x}^{(k+1)} = \mathscr{A}\mathscr{x}^{(k)} + \mathscr{b}, \qquad k \geqslant 0,$$

with $\mathscr{x}^{(0)} = \underset{\cdot}{\mathscr{y}} = \underset{\cdot}{\mathscr{A}}\underset{\cdot}{\mathscr{y}} + \underset{\cdot}{\mathscr{b}}$. From inclusion monotonicity it follows that

$$\underset{\cdot}{\mathscr{y}} = \underset{\cdot}{\mathscr{A}}\underset{\cdot}{\mathscr{y}} + \underset{\cdot}{\mathscr{b}} \in \mathscr{A}\mathscr{x}^{(0)} + \mathscr{b} = \mathscr{x}^{(1)},$$

or in general

$$\underset{\cdot}{\mathscr{y}} = \underset{\cdot}{\mathscr{A}}\underset{\cdot}{\mathscr{y}} + \underset{\cdot}{\mathscr{b}} \in \mathscr{A}\mathscr{x}^{(k)} + \mathscr{b} = \mathscr{x}^{(k+1)}.$$

Since $\rho(|\mathscr{A}|) < 1$ it follows that $\lim_{k\to\infty} \mathscr{x}^{(k)} = \mathscr{x}^*$ and therefore also that $\underset{\cdot}{\mathscr{y}} \in \mathscr{x}^*$. Since $\mathscr{x}^*$ is independent from the starting vector it follows that the first part of the theorem holds.

For the proof of the second part of the theorem we form a vector $\underset{\cdot}{\mathscr{u}} \in V_n(\mathbb{R})$ from the n lower bounds of the components of $\mathscr{x}^*$. The upper bounds are similarly combined into a vector $\underset{\cdot}{\mathscr{v}} \in V_n(\mathbb{R})$. It then follows from $\mathscr{x}^* = \mathscr{A}\mathscr{x}^* + \mathscr{b}$ using the rules of interval arithmetic that

$$\underset{\cdot}{\mathscr{u}} = \underset{\cdot}{\mathscr{A}}^*\underset{\cdot}{\mathscr{u}} + \underset{\cdot}{\mathscr{l}} \qquad \text{and} \qquad \underset{\cdot}{\mathscr{v}} = \underset{\cdot}{\mathscr{A}}^{**}\underset{\cdot}{\mathscr{v}} + \underset{\cdot}{\mathscr{s}}$$

where $\underset{\cdot}{\mathscr{u}}_i = (u_i)$, $\underset{\cdot}{\mathscr{v}}_i = (v_i)$, and

$$\underset{\cdot}{\mathscr{A}}^* = (a_{ij}^*), \qquad a_{ij}^* = \begin{cases} i(A_{ij}), & u_j > 0 \\ s(A_{ij}), & u_j \leqslant 0, \end{cases}$$

$$\underset{\cdot}{\mathscr{A}}^{**} = (a_{ij}^{**}), \qquad a_{ij}^{**} = \begin{cases} s(A_{ij}), & v_j > 0 \\ i(A_{ij}), & v_j \leqslant 0, \end{cases}$$

$$\underset{\cdot}{\mathscr{l}} = (i(B_i)), \qquad \underset{\cdot}{\mathscr{s}} = (s(B_i)).$$

These equations imply that u and v are members of the set

$$\{y = (\mathcal{I} - \mathcal{A})^{-1}b \mid \mathcal{A} \in \mathcal{A},\ b \in b\}.$$

This finishes the proof. ■

The iteration method considered in Theorem 1 may be called a total step method (T) analogously to the corresponding method for "point systems of equations." The analogous single step method (S) is obtained by decomposing the interval matrix $\mathcal{A}$ in

$$\mathcal{A} = \mathcal{L} + \mathcal{D} + \mathcal{U},$$

where $\mathcal{L}$ is a strictly lower triangular matrix, $\mathcal{D}$ a diagonal matrix, and $\mathcal{U}$ a strictly upper triangular matrix. The single step method is then defined by the iteration

$$x^{(k+1)} = \mathcal{L}x^{(k+1)} + (\mathcal{D} + \mathcal{U})x^{(k)} + b, \qquad k \geqslant 0. \tag{3}$$

The convergence of the single step method is considered in the following.

Theorem 3: The iteration

$$x^{(k+1)} = \mathcal{L}x^{(k+1)} + (\mathcal{D} + \mathcal{U})x^{(k)} + b, \qquad k \geqslant 0,$$

with an arbitrary starting vector $x^{(0)} \in V_n(I(\mathbb{C}))$ converges to a unique fixed point x^* iff

$$\rho((\mathcal{I} - |\mathcal{L}|)^{-1}(|\mathcal{D}| + |\mathcal{U}|)) < 1.$$

Proof: We prove this theorem by appealing to Theorem 11.5, and we therefore define

$$f\colon V_n(\mathbb{C}) \to V_n(\mathbb{C})$$

with

$$f(x) = \mathcal{L}x + (\mathcal{D} + \mathcal{U})x + b$$

and

$$g\colon V_n(\mathbb{C}) \times V_n(\mathbb{C}) \to V_n(\mathbb{C})$$

with

$$g(x, y) = \mathcal{L}x + (\mathcal{D} + \mathcal{U})y + b.$$

We then have

$$g(x, x) = f(x) \qquad \text{for all} \quad x \in V_n(I(\mathbb{C})),$$

and from (10.23) and (10.25) it follows that

$$\begin{aligned} q(g(x, z), g(y, z)) &= q(\mathcal{L}x + (\mathcal{D} + \mathcal{U})z + b, \mathcal{L}y + (\mathcal{D} + \mathcal{U})z + b) \\ &\leqslant |\mathcal{L}|q(x, y), \end{aligned}$$

and

$$q(g(z, x), g(z, y)) = q(\mathcal{L}z + (\mathcal{D} + \mathcal{U})x + b, \mathcal{L}z + (\mathcal{D} + \mathcal{U})y + b)$$
$$\leqslant (|\mathcal{D}| + |\mathcal{U}|)q(x, y)$$

for all $x, y, z \in V_n(I(\mathbb{C}))$. Since $|\mathcal{L}|$ is a strictly lower triangular matrix we have $\rho(|\mathcal{L}|) = 0$. Therefore, using $\mathcal{Q} := |\mathcal{L}|$, $\mathcal{R} := |\mathcal{D}| + |\mathcal{U}|$, we have satisfied all the conditions of Theorem 11.5 and the condition

$$\rho((\mathcal{I} - |\mathcal{L}|)^{-1}(|\mathcal{D}| + |\mathcal{U}|)) < 1$$

is sufficient. The proof of the necessity of this condition for the convergence for every starting vector is carried out as in the corresponding proof in Theorem 1 for the total step method. ■

The theorem of Stein and Rosenberg as well as a generalization of this (see Varga [21] and Apostolatos and Kulisch [6]) states that with $\mathcal{A} = \mathcal{L} + \mathcal{D} + \mathcal{U}$ it follows that

$$\rho(|\mathcal{A}|) < 1 \qquad \text{iff} \qquad \rho((\mathcal{I} - |\mathcal{L}|)^{-1}(|\mathcal{D}| + |\mathcal{U}|)) < 1.$$

Since the conditions on the convergence of the total step method as well as the single step method are both necessary and sufficient, we get the following

Theorem 4: The total step method (2) converges for an arbitrary starting value $x^{(0)} \in V_n(I(\mathbb{C}))$ to a unique fixed point iff the single step method (3) converges to a unique fixed point for an arbitrary starting value $x^{(0)} \in V_n(I(\mathbb{C}))$. ■

This result differs essentially from the corresponding result for point systems of equations (see, for example, Varga [21]): The convergence or divergence of the total step method does not necessarily mean the convergence or divergence of the single step method.

Since the multiplication of interval matrices with interval vectors is in general not distributive it is not obvious that for the case $\rho(|\mathcal{A}|) < 1$ the fixed point of the total step iteration, namely,

$$x^* = \mathcal{A}x^* + b$$

is identical to the fixed point of the single step iteration, namely,

$$\tilde{x}^* = \mathcal{L}\tilde{x}^* + (\mathcal{D} + \mathcal{U})\tilde{x}^* + b.$$

We have, however, from the particular form of the matrices $\mathcal{L}$, $\mathcal{D}$, and $\mathcal{U}$ that

$$\tilde{x}^* = \mathcal{L}\tilde{x}^* + (\mathcal{D} + \mathcal{U})\tilde{x}^* + b = (\mathcal{L} + \mathcal{D} + \mathcal{U})\tilde{x}^* + b = \mathcal{A}\tilde{x}^* + b.$$

This is shown using the definition of interval matrix and interval vector operations using the particular form of the matrices $\mathcal{L}$, $\mathcal{D}$, and $\mathcal{U}$. From this it follows that $x^* = \tilde{x}^*$ and we get

Corollary 5: If $\rho(|\mathscr{A}|) < 1$, then both the total step and the single step iterations converge to the fixed point x^* of the equation

$$x = \mathscr{A}x + b. \quad \blacksquare$$

We now consider the next method, the symmetric single step method (SS), where the matrix $\mathscr{A}$ is decomposed into

$$\mathscr{A} = \mathscr{L} + \mathscr{U}$$

with $\mathscr{L}$ being a strictly lower and $\mathscr{U}$ being a strictly upper triangular matrix. The method (SS) is then defined by

$$\text{(SS)} \qquad \begin{cases} x^{(k+1/2)} = \mathscr{L}x^{(k+1/2)} + \mathscr{U}x^{(k)} + b, \\ x^{(k+1)} = \mathscr{L}x^{(k+1/2)} + \mathscr{U}x^{(k+1)} + b, \qquad k \geqslant 0. \end{cases}$$

If the diagonal elements of $\mathscr{A}$ do not all vanish then we can instead consider the iteration

$$\text{(SS}') \qquad \begin{cases} x^{(k+1/2)} = \mathscr{L}x^{(k+1/2)} + \mathscr{D}x^{(k)} + \mathscr{U}x^{(k)} + b, \\ x^{(k+1)} = \mathscr{L}x^{(k+1/2)} + \mathscr{D}x^{(k+1/2)} + \mathscr{U}x^{(k+1)} + b, \qquad k \geqslant 0. \end{cases}$$

For this iteration it is also possible to prove similar statements as in the following for the method (SS).

For point matrices and point vectors the method (SS) was first considered by Aitken [1]. In Niethammer [18] it was pointed out that in the case of point systems of equations the method (SS) can be carried out with the same amount of work as the single step method except for the first iteration step. This will be considered in more detail in Chapter 14, where we shall consider a modification of the method (SS) for which the same result holds. The convergence of the method (SS) is given by a corollary of the following more general result.

Theorem 6: The interval matrix $\mathscr{A} \in M_{nn}(I(\mathbb{C}))$ is decomposed into the sum $\mathscr{A} = \mathscr{M} + \mathscr{N}$ of two interval matrices $\mathscr{M}$ and $\mathscr{N}$ for which $\rho(|\mathscr{M}|) < 1$ as well as $\rho(|\mathscr{N}|) < 1$. The following is then true for an arbitrary interval vector $b \in V_n(I(\mathbb{C}))$.

(a) For each interval vector $x^{(0)} \in V_n(I(\mathbb{C}))$ there exists a unique sequence $\{x^{(k)}\}_{k=0}^{\infty}$ which satisfies the iteration method

$$\text{(V)} \qquad \begin{cases} x^{(k+1/2)} = \mathscr{M}x^{(k+1/2)} + \mathscr{N}x^{(k)} + b, \\ x^{(k+1)} = \mathscr{M}x^{(k+1/2)} + \mathscr{N}x^{(k+1)} + b, \qquad k = 0, 1, 2, \ldots. \end{cases}$$

(b) If $\rho((\mathscr{I} - |\mathscr{N}|)^{-1}|\mathscr{M}|(\mathscr{I} - |\mathscr{M}|)^{-1}|\mathscr{N}|) < 1$, then the equation $x = \mathscr{A}x + b$ has a unique fixed point x^*. If furthermore

$$\mathscr{A}x^* = (\mathscr{M} + \mathscr{N})x^* = \mathscr{M}x^* + \mathscr{N}x^*,$$

then the sequence $\{x^{(k)}\}_{k=0}^{\infty}$ calculated according to (V) converges for all interval vectors $x^{(0)}$ to x^*. (As we have seen, the distributive law is not satisfied for interval matrices. See Chapter 10, formula (6).)

(c) Conversely, if the equation $x = \mathcal{A}x + b$ has exactly one fixed point x^* and if the method (V) converges for all $x^{(0)}$ towards x^*, then it follows that

$$\mathcal{A}x^* = (\mathcal{M} + \mathcal{N})x^* = \mathcal{M}x^* + \mathcal{N}x^*$$

and

$$\rho((\mathcal{I} - |\mathcal{N}|)^{-1}|\mathcal{M}|(\mathcal{I} - |\mathcal{M}|)^{-1}|\mathcal{N}|) < 1.$$

Proof: Of (a): For an arbitrary interval vector z using (10.23) and (10.25) it holds for all vectors x and y that

$$q(\mathcal{M}x + \mathcal{N}z + b, \mathcal{M}y + \mathcal{N}z + b) = q(\mathcal{M}x, \mathcal{M}y) \leqslant |\mathcal{M}|q(x, y).$$

Since $\rho(|\mathcal{M}|) < 1$ we have from Theorem 1 that the equation

$$x^{(k+1/2)} = \mathcal{M}x^{(k+1/2)} + \mathcal{N}x^{(k)} + b$$

has a unique fixed point $x^{(k+1/2)}$ for each k. Correspondingly one may show that the equation

$$x^{(k+1)} = \mathcal{M}x^{(k+1/2)} + \mathcal{N}x^{(k+1)} + b$$

also has a unique fixed point $x^{(k+1)}$. This proves the existence and the uniqueness of the sequence $\{x^{(k)}\}_{k=0}^{\infty}$ given the starting vector $x^{(0)}$.

Of (b): We show that $\rho(|\mathcal{A}|) < 1$. Since $\rho(|\mathcal{M}|) < 1$, $\rho(|\mathcal{N}|) < 1$, the inverses

$$(\mathcal{I} - |\mathcal{M}|)^{-1} \qquad \text{and} \qquad (\mathcal{I} - |\mathcal{N}|)^{-1}$$

exist and are nonnegative according to Theorem 3.8 of Varga [21]. Therefore the real matrix

$$(\mathcal{I} - |\mathcal{N}|)^{-1}|\mathcal{M}|(\mathcal{I} - |\mathcal{M}|)^{-1}|\mathcal{N}| = (\mathcal{I} - |\mathcal{N}|)^{-1}(\mathcal{I} - |\mathcal{M}|)^{-1}|\mathcal{M}||\mathcal{N}|$$

is also nonnegative. Applying again Theorem 3.8 of Varga [21] we get

$$\begin{aligned} \mathcal{O} &\leqslant (\mathcal{I} - (\mathcal{I} - |\mathcal{N}|)^{-1}(\mathcal{I} - |\mathcal{M}|)^{-1}|\mathcal{M}||\mathcal{N}|)^{-1} \\ &= (\mathcal{I} - (|\mathcal{M}| + |\mathcal{N}|))^{-1}(\mathcal{I} - |\mathcal{M}|)(\mathcal{I} - |\mathcal{N}|) \end{aligned}$$

and by using

$$(\mathcal{I} - |\mathcal{M}|)^{-1} \geqslant \mathcal{O}, \qquad (\mathcal{I} - |\mathcal{N}|)^{-1} \geqslant \mathcal{O}$$

finally

$$(\mathcal{I} - (|\mathcal{M}| + |\mathcal{N}|))^{-1} \geqslant \mathcal{O}.$$

It therefore follows that the inequality

$$\rho(|\mathcal{M}| + |\mathcal{N}|) < 1$$

holds by Theorem 3.8 of Varga [21]. From

$$|\mathscr{A}| = |\mathscr{M} + \mathscr{N}| \leqslant |\mathscr{M}| + |\mathscr{N}|$$

(see Chapter 10, Eq. (15)) and because of the theorem of Perron and Frobenius [21, Theorem 2.7] it follows that

$$\rho(|\mathscr{A}|) \leqslant \rho(|\mathscr{M}| + |\mathscr{N}|) < 1.$$

From Theorem 1 the equation $x = \mathscr{A}x + b$ has therefore a unique fixed point x^*. From

$$x^* = \mathscr{A}x^* + b = \mathscr{M}x^* + \mathscr{N}x^* + b$$

it follows using (10.23)–(10.25) and (V) that

$$\begin{aligned} q(x^{(k+1)}, x^*) &= q(\mathscr{M}x^{(k+1/2)} + \mathscr{N}x^{(k+1)} + b, \mathscr{M}x^* + \mathscr{N}x^* + b) \\ &\leqslant q(\mathscr{M}x^{(k+1/2)} + \mathscr{N}x^{(k+1)}, \mathscr{M}x^{(k+1/2)} + \mathscr{N}x^*) \\ &\quad + q(\mathscr{M}x^{(k+1/2)} + \mathscr{N}x^*, \mathscr{M}x^* + \mathscr{N}x^*) \\ &\leqslant |\mathscr{N}|q(x^{(k+1)}, x^*) + |\mathscr{M}|q(x^{(k+1/2)}, x^*), \end{aligned}$$

or since $(\mathscr{I} - |\mathscr{N}|)^{-1} \geqslant \mathscr{O}$, that

$$q(x^{(k+1)}, x^*) \leqslant (\mathscr{I} - |\mathscr{N}|)^{-1}|\mathscr{M}|q(x^{(k+1/2)}, x^*).$$

Correspondingly one gets

$$q(x^{(k+1/2)}, x^*) \leqslant (\mathscr{I} - |\mathscr{M}|)^{-1}|\mathscr{N}|q(x^{(k)}, x^*);$$

that is, finally

$$\begin{aligned} q(x^{(k+1)}, x^*) &\leqslant (\mathscr{I} - |\mathscr{N}|)^{-1}|\mathscr{M}|(\mathscr{I} - |\mathscr{M}|)^{-1}|\mathscr{N}|q(x^{(k)}, x^*) \\ &\leqslant \{(\mathscr{I} - |\mathscr{N}|)^{-1}|\mathscr{M}|(\mathscr{I} - |\mathscr{M}|)^{-1}|\mathscr{N}|\}^{k+1}q(x^{(0)}, x^*). \end{aligned}$$

Since the spectral radius of the expression in curly brackets is less than 1 it follows that $\lim_{k\to\infty} x^{(k)} = x^*$.

Of (c): Let the equation $x = \mathscr{A}x + b$ have a unique fixed point x^*. From the inequality

$$q(x^{(k+1/2)}, x^*) \leqslant (\mathscr{I} - |\mathscr{M}|)^{-1}|\mathscr{N}|q(x^{(k)}, x^*)$$

that may be derived as in the proof of (b), it follows that the sequence $\{x^{(k+1/2)}\}_{k=0}^{\infty}$ converges to x^* for arbitrary $x^{(0)}$. It then follows from the first part of (V) as $k \to \infty$ that

$$x^* = \mathscr{M}x^* + \mathscr{N}x^* + b;$$

that is,

$$o = q(x^*, x^*) = q(\mathscr{M}x^* + \mathscr{N}x^* + b, \mathscr{A}x^* + b) = q(\mathscr{M}x^* + \mathscr{N}x^*, \mathscr{A}x^*)$$

or

$$\mathscr{A}x^* = (\mathscr{M} + \mathscr{N})x^* = \mathscr{M}x^* + \mathscr{N}x^*.$$

We still have to prove

$$\rho((\mathscr{I} - |\mathscr{N}|)^{-1}|\mathscr{M}|(\mathscr{I} - |\mathscr{M}|)^{-1}|\mathscr{N}|) < 1.$$

In order to do this we proceed as in Theorem 1.

From the theorem of Perron and Frobenius [21, Theorem 2.7] we know that the matrix

$$(\mathscr{I} - |\mathscr{N}|)^{-1}|\mathscr{M}|(\mathscr{I} - |\mathscr{M}|)^{-1}|\mathscr{N}|$$

has a nonnegative eigenvector for the nonnegative eigenvalue

$$\tilde{\lambda} = \rho((\mathscr{I} - |\mathscr{N}|)^{-1}|\mathscr{M}|(\mathscr{I} - |\mathscr{M}|)^{-1}|\mathscr{N}|).$$

We now choose $x^{(0)}$ such that $d(x^{(0)})$ is an eigenvector for the eigenvalue $\tilde{\lambda}$ and such that at least one component of $d(x^{(0)})$ is larger than the corresponding component of $d(x^*)$. It then follows from (V) using (10.12) and (10.18) that

$$d(x^{(k+1/2)}) \geqslant |\mathscr{M}|\, d(x^{(k+1/2)}) + |\mathscr{N}|\, d(x^{(k)})$$

or

$$d(x^{(k+1/2)}) \geqslant (\mathscr{I} - |\mathscr{M}|)^{-1}|\mathscr{N}|\, d(x^{(k)}),$$

as well as

$$d(x^{(k+1)}) \geqslant |\mathscr{M}|\, d(x^{(k+1/2)}) + |\mathscr{N}|\, d(x^{(k+1)})$$

or

$$d(x^{(k+1)}) \geqslant (\mathscr{I} - |\mathscr{N}|)^{-1}|\mathscr{M}|\, d(x^{(k+1/2)}).$$

Finally we get

$$\begin{aligned} d(x^{(k+1)}) &\geqslant (\mathscr{I} - |\mathscr{N}|)^{-1}|\mathscr{M}|(\mathscr{I} - |\mathscr{M}|)^{-1}|\mathscr{N}|\, d(x^{(k)}) \\ &\geqslant \{(\mathscr{I} - |\mathscr{N}|)^{-1}|\mathscr{M}|(\mathscr{I} - |\mathscr{M}|)^{-1}|\mathscr{N}|\}^{k+1}\, d(x^{(0)}) \\ &= \tilde{\lambda}^{k+1}\, d(x^{(0)}). \end{aligned}$$

The convergence of the iteration (V) to x^* implies the convergence of the sequence $\{d(x^{(k)})\}_{k=0}^{\infty}$ to $d(x^*)$. The assumption $\tilde{\lambda} \geqslant 1$ leads to the inequality

$$d(x^{(k+1)}) \geqslant \tilde{\lambda}^{k+1} d(x^{(0)}) \geqslant d(x^{(0)}), \qquad k \geqslant 0,$$

and for $k \to \infty$ to $d(x^*) \geqslant d(x^{(0)})$, which contradicts the choice of $x^{(0)}$. Therefore $\tilde{\lambda} < 1$, which proves the theorem. ∎

We now choose

$$\mathscr{M} := \mathscr{L}, \qquad \mathscr{N} := \mathscr{U}$$

for Theorem 6. Then we have $\rho(|\mathcal{M}|) = \rho(|\mathcal{N}|) = 0$. The equality

$$\mathcal{A}x = \mathcal{L}x + \mathcal{U}x$$

holds because of the particular form of the matrices $\mathcal{L}$ and $\mathcal{U}$ for all interval vectors. One therefore gets the following corollary from Theorem 6.

Corollary 7: The symmetric single step method (SS) converges to the unique fixed point x^* of the equation $x = \mathcal{A}x + \ell$ for each starting vector $x^{(0)}$ iff the spectral radius of

$$(\mathcal{I} - |\mathcal{U}|)^{-1}|\mathcal{L}|(\mathcal{I} - |\mathcal{L}|)^{-1}|\mathcal{U}|$$

is less than one. ■

We already showed in the proof of Theorem 6, part (b), that the inequality

$$\rho((\mathcal{I} - |\mathcal{N}|)^{-1}|\mathcal{M}|(\mathcal{I} - |\mathcal{M}|)^{-1}|\mathcal{N}|) < 1$$

implies $\rho(|\mathcal{A}|) < 1$. If we furthermore assume that

$$|\mathcal{A}| = |\mathcal{M}| + |\mathcal{N}|,$$

then we now wish to show that the converse is true under the assumptions that $\rho(|\mathcal{M}|) < 1$ and $\rho(|\mathcal{N}|) < 1$ from Theorem 6. Let therefore

$$\rho(|\mathcal{A}|) = \rho(|\mathcal{M}| + |\mathcal{N}|) < 1.$$

According to Theorem 3.8 in Varga [21] the inverse of $\mathcal{I} - |\mathcal{A}|$ exists and we have $(\mathcal{I} - |\mathcal{A}|)^{-1} \geqslant \mathcal{O}$. Consider now the decomposition of $\mathcal{I} - |\mathcal{A}|$ into

$$\mathcal{I} - |\mathcal{A}| = (\mathcal{I} - |\mathcal{M}|)(\mathcal{I} - |\mathcal{N}|) - |\mathcal{M}|\,|\mathcal{N}|.$$

Since $\rho(|\mathcal{M}|) < 1$, $\rho(|\mathcal{N}|) < 1$, it follows that the inverses of $\mathcal{I} - |\mathcal{M}|$ and $\mathcal{I} - |\mathcal{N}|$ exist and that they are nonnegative. The product $(\mathcal{I} - |\mathcal{N}|)^{-1} \times (\mathcal{I} - |\mathcal{M}|)^{-1}$ is therefore also nonnegative. The decomposition of $\mathcal{I} - |\mathcal{A}|$ is therefore regular since $|\mathcal{M}|\,|\mathcal{N}|$ is nonnegative [21, Definition 3.5]. Using Theorem 3.13 of Varga [21] it therefore follows that the relation

$$\rho((\mathcal{I} - |\mathcal{N}|)^{-1}(\mathcal{I} - |\mathcal{M}|)^{-1}|\mathcal{M}|\,|\mathcal{N}|) = \rho((\mathcal{I} - |\mathcal{N}|)^{-1}|\mathcal{M}|(\mathcal{I} - |\mathcal{M}|)^{-1}|\mathcal{N}|) < 1$$

is valid.

In summary using Theorem 6, part (b), one gets

Theorem 8: Let the interval matrix $\mathcal{A}$ be decomposed into the sum $\mathcal{A} = \mathcal{M} + \mathcal{N}$ of two interval matrices for which

$$|\mathcal{A}| = |\mathcal{M}| + |\mathcal{N}|, \qquad \text{and} \qquad \rho(|\mathcal{M}|) < 1, \qquad \rho(|\mathcal{N}|) < 1$$

hold. Then

$$\rho((\mathcal{I} - |\mathcal{N}|)^{-1}|\mathcal{M}|(\mathcal{I} - |\mathcal{M}|)^{-1}|\mathcal{N}|) < 1$$

iff

$$\rho(|\mathscr{A}|) < 1. \quad ■$$

The equality $|\mathscr{A}| = |\mathscr{M}| + |\mathscr{N}|$ is satisfied, for example, when $\mathscr{A} = (A_{ij})$ is decomposed into $\mathscr{A} = \mathscr{M} + \mathscr{N}$ with $\mathscr{M} = (M_{ij})$, $\mathscr{N} = (N_{ij})$ in such a manner that for each $1 \leqslant i, j \leqslant n$ at least one of the components M_{ij} and N_{ij} is equal to zero.

This condition is satisfied, for example, by the decomposition that leads to the symmetric single step method (SS). From Theorem 8 one immediately gets

Corollary 9: The symmetric single step method converges to the unique fixed point x^* of the equation $x = \mathscr{A}x + b$ for each starting vector $x^{(0)} \in V_n(I(\mathbb{C}))$ iff $\rho(|\mathscr{A}|) < 1$, that is, when the total step method (and therefore also the single step method) converges to x^* for each $x^{(0)} \in V_n(I(\mathbb{C}))$. ■

We shall now discuss the speed of convergence of a sequence $\{x^{(k)}\}_{k=0}^{\infty}$ of interval vectors converging to x^* that are generated by an iteration procedure

$$(4) \qquad x^{(k+1)} = f(x^{(k)}), \qquad k \geqslant 0.$$

Definition 10: Let $x^* = f(x^*)$ and let $\mathfrak{C}$ be the set of all sequences $\{x^{(k)}\}_{k=0}^{\infty}$ that are calculated using (4) for which $\lim_{k\to\infty} x^{(k)} = x^*$. The quantity

$$\alpha = \sup\left\{ \limsup_{k\to\infty} \|q(x^{(k)}, x^*)\|^{1/k} \;\middle|\; \{x^{(k)}\}_{k=0}^{\infty} \in \mathfrak{C} \right\}$$

is then called the asymptotic convergence factor of the iteration (4) at the point x^*. ■

Let $\{x^{(k)}\}_{k=0}^{\infty}$ be a sequence converging to x^* and let

$$\beta = \limsup_{k\to\infty} \|q(x^{(k)}, x^*)\|^{1/k}.$$

Since $\lim_{k\to\infty} \|q(x^{(k)}, x^*)\| = 0$ it follows that $0 \leqslant \beta \leqslant 1$ and hence $0 \leqslant \alpha \leqslant 1$. From the definition of β we have that for every $\varepsilon > 0$ there is a k_0, such that

$$(5) \qquad \|q(x^{(k)}, x^*)\| \leqslant (\beta + \varepsilon)^k, \qquad k \geqslant k_0.$$

If $\beta < 1$, then one can choose $\varepsilon > 0$ so that $\beta + \varepsilon < 1$. The inequality (5) then shows that $\|q(x^{(k)}, x^*)\|$ converges asymptotically to zero with a speed of convergence at least as great as a geometric sequence with the quotient $\beta + \varepsilon$. The supremum over all sequences from $\mathfrak{C}$ characterizes the asymptotically worst choice of $x^{(0)}$. Definition 10 is an immediate generalization of the corresponding definition of the asymptotic convergence factor for sequences of point vectors (see, for example, Ortega [19] or Ortega and Rheinboldt [20]).

In the following it is important that α is norm independent. For this it suffices to show that β is norm independent. Let $\|\cdot\|$ and $\|\cdot\|'$ be two vector norms. From the theorem of norm equivalence there exist two real numbers $d \geqslant c > 0$ such that $c\|x\| \leqslant \|x\|' \leqslant d\|x\|$ for all point vectors. One therefore gets

$$\begin{aligned}\limsup_{k\to\infty} \|q(x^{(k)}, x^*)\|^{1/k} &\leqslant \lim_{k\to\infty} \left(\frac{1}{c}\right)^{1/k} \limsup_{k\to\infty} \|q(x^{(k)}, x^*)\|'^{1/k} \\ &\leqslant \lim_{k\to\infty} \left(\frac{d}{c}\right)^{1/k} \limsup_{k\to\infty} \|q(x^{(k)}, x^*)\|^{1/k} \\ &= \limsup_{k\to\infty} \|q(x^{(k)}, x^*)\|^{1/k}.\end{aligned}$$

Before we apply Definition 10 to the methods considered in this chapter we prove a theorem relating to point matrices.

Consider a positive point vector $h = (h_i)$, $h_i > 0$, and the diagonal point matrix $\mathcal{H} = \operatorname{diag}(1/h_i)$. A monotone vector norm is then defined by

$$\|x\| = \max_{1 \leqslant i \leqslant n} (|x_i|/h_i);$$

that is, from

$$|x| \leqslant |y|$$

it follows that

$$\|x\| \leqslant \|y\|.$$

(In particular from $o \leqslant x \leqslant y$ it follows that the inequality $\|x\| \leqslant \|y\|$ holds). From

$$\|\mathcal{A}\| = \sup_{\|x\|=1} \|\mathcal{A}x\|$$

one then obtains a matrix norm subordinate to the vector norm $\|x\|$ (see Collatz [9, page 135]) and we have

$$\|\mathcal{A}\| = \max_{1 \leqslant i \leqslant n} \left(\frac{1}{h_i} \sum_{j=1}^{n} |a_{ij}| h_j\right).$$

We now prove

Theorem 11: For every point matrix $\mathcal{A} \geqslant \mathcal{O}$ and every $\varepsilon > 0$ there exists a monotone vector norm $\|\cdot\|$ so that

$$\|\mathcal{A}\| = \sup_{\|x\|=1} \|\mathcal{A}x\| \leqslant \rho(\mathcal{A}) + \varepsilon$$

holds.

Proof: Let $\mathcal{A} \geqslant \mathcal{O}$ be irreducible (see [21]). Then $\mathcal{A}$ has a positive eigenvector $c = (c_i)$, $c_i > 0$, to the eigenvalue $\lambda = \rho(\mathcal{A})$ (see Varga [21, Theorem 2.1]).

From the equation $\mathcal{A}\mathfrak{c} = \lambda\mathfrak{c}$ it follows that

$$\rho(\mathcal{A}) = \lambda = \frac{1}{c_i}\sum_{j=1}^{n} a_{ij}c_j = \|\mathcal{A}\| = \sup_{\|\mathfrak{x}\|=1} \|\mathcal{A}\mathfrak{x}\|$$

with

$$\|\mathfrak{x}\| = \max_{1 \leqslant i \leqslant n} \frac{|x_i|}{c_i};$$

that is,

$$\|\mathcal{A}\| \leqslant \rho(\mathcal{A}) + \varepsilon.$$

If $\mathcal{A} \geqslant \mathcal{O}$ is reducible, then we define an irreducible matrix $\tilde{\mathcal{A}} \geqslant \mathcal{O}$ by

$$\tilde{\mathcal{A}} = (\tilde{a}_{ij}) \qquad \text{with} \quad \tilde{a}_{ij} = \begin{cases} a_{ij} & \text{if} \quad a_{ij} > 0 \\ a > 0 & \text{if} \quad a_{ij} = 0. \end{cases}$$

Clearly $\tilde{\mathcal{A}} \geqslant \mathcal{A} \geqslant \mathcal{O}$. If now $\tilde{\mathfrak{c}} = (\tilde{c}_i)$, the positive eigenvector of $\tilde{\mathcal{A}}$ to the eigenvalue $\lambda = \rho(\tilde{\mathcal{A}})$, then from $|\lambda| \leqslant \|\mathcal{A}\|$ (true for all matrix norms) it follows that

$$\rho(\tilde{\mathcal{A}}) = \frac{1}{\tilde{c}_i}\sum_{j=1}^{n} \tilde{a}_{ij}\tilde{c}_j = \|\tilde{\mathcal{A}}\| \geqslant \max_{1 \leqslant i \leqslant n} \frac{1}{\tilde{c}_i}\sum_{j=1}^{n} a_{ij}\tilde{c}_j = \|\mathcal{A}\| \geqslant \rho(\mathcal{A})$$

with $\|\mathfrak{x}\| = \max_{1 \leqslant i \leqslant n}|x_i|/\tilde{c}_i$ and $\|\mathcal{A}\| = \sup_{\|\mathfrak{x}\|=1} \|\mathcal{A}\mathfrak{x}\|$. Since the spectral radius $\rho(\mathcal{A})$ is a continuous function of the entries of $\mathcal{A}$ (see, for example, Ortega [19]) we have that for each $\varepsilon > 0$ there exists a $\delta = \delta(\varepsilon) > 0$ so that

$$\rho(\tilde{\mathcal{A}}) - \rho(\mathcal{A}) \leqslant \varepsilon$$

holds for all $a \leqslant \delta(\varepsilon)$. Since $\rho(\tilde{\mathcal{A}}) \geqslant \|\mathcal{A}\|$ we then have

$$\|\mathcal{A}\| - \rho(\mathcal{A}) \leqslant \rho(\tilde{\mathcal{A}}) - \rho(\mathcal{A}) \leqslant \varepsilon$$

or

$$\|\mathcal{A}\| \leqslant \rho(\mathcal{A}) + \varepsilon. \quad \blacksquare$$

From these preparations we may now easily prove the following theorem.

Theorem 12: Let the equation

$$\mathfrak{x} = \mathcal{A}\mathfrak{x} + \mathfrak{b}$$

be given with an interval vector $\mathfrak{b}$ and an interval matrix $\mathcal{A}$ for which $\rho(|\mathcal{A}|) < 1$. It then follows that the asymptotic convergence factor α_T for the total step method satisfies

$$\alpha_T \leqslant \rho(|\mathcal{A}|),$$

α_S for the single step method satisfies

$$\alpha_S \leqslant \rho((\mathcal{I} - |\mathcal{L}|)^{-1}(|\mathcal{D}| + |\mathcal{U}|))$$

and α_{SS} for the symmetric single step method satisfies

$$\alpha_{SS} \leqslant \rho((\mathcal{I} - |\mathcal{U}|)^{-1}|\mathcal{L}|(\mathcal{I} - |\mathcal{L}|)^{-1}|\mathcal{U}|).$$

Proof: We show the proof for the total step method. If $\rho(|\mathcal{A}|) < 1$, then the total step method converges according to Theorem 1 for each starting vector $x^{(0)}$ to the unique fixed point x^* of $x = \mathcal{A}x + b$. If one chooses $x^{(0)}$ arbitrarily, then using the properties of the distance q one gets

$$q(x^{(1)}, x^*) \leqslant |\mathcal{A}| q(x^{(0)}, x^*),$$

$$q(x^{(2)}, x^*) \leqslant |\mathcal{A}|^2 q(x^{(0)}, x^*),$$

and in general

$$q(x^{(k)}, x^*) \leqslant |\mathcal{A}|^k q(x^{(0)}, x^*), \qquad k \geqslant 1.$$

By using the monotone vector norm that exists according to Theorem 11 and its subordinate matrix norm it follows that

$$\|q(x^{(k)}, x^*)\| \leqslant (\rho(|\mathcal{A}|) + \varepsilon)^k \|q(x^{(0)}, x^*)\|;$$

that is,

$$\limsup_{k\to\infty} \|q(x^{(k)}, x^*)\|^{1/k} \leqslant \rho(|\mathcal{A}|) + \varepsilon.$$

Since $\varepsilon > 0$ was arbitrary and since α_T is independent of the norm, one gets the result $\alpha_T \leqslant \rho(|\mathcal{A}|)$, which proves the theorem in this case. The other cases are handled in a similar manner. ■

It is an open question if the equality always holds in the estimates for $\alpha_T, \alpha_S, \alpha_{SS}$. In order to show this for the total step method, for example, one has to give a starting vector $x^{(0)}$ for which

$$\limsup_{k\to\infty} \|q(x^{(k)}, x^*)\| = \rho(|\mathcal{A}|)$$

holds.

This may be shown rather easily in particular cases. The existence has not been proved in the general case.

Remarks: The explicit application of interval analysis to the iterative solution of systems of equations was first done in [7]. There it was shown for a real interval matrix $\mathcal{A}$ and a real interval vector b that the total step method (2) is convergent to a unique fixed point if $\| |\mathcal{A}| \| < 1$. The necessary and sufficient conditions for Theorems 1 and 3 were given for real interval matrices in Mayer [12] and for interval matrices with elements from $R(\mathbb{C})$ in Alefeld [2]. The results for interval matrices with elements from $K(\mathbb{C})$ in Theorems 1 and 3 are new. Theorem 11, necessary for the proof of Theorem 12, may be found in Mayer [12].

The results for the symmetric single step method (SS) were proved in Alefeld [5]. Similar results to those of Theorem 2, second part, are found in the paper by Barth and Nuding [8]; see also Apostolatos and Kulisch [7] for this. Without going into details we note that, corresponding to what was done in this chapter, one may also investigate solutions of the fixed point equation

$$\mathscr{X} = \mathscr{A}\mathscr{X} + \mathscr{B}, \qquad \mathscr{A}, \mathscr{B}, \mathscr{X} \in M_{nn}(I(\mathbb{C}));$$

see Apostolatos and Kulisch [7]. For example, one has a unique fixed point $\mathscr{X}^*$ for the iteration

$$\mathscr{X}^{(k+1)} = \mathscr{A}\mathscr{X}^{(k)} + \mathscr{B}, \qquad k \geqslant 0,$$

for each starting matrix $\mathscr{X}^{(0)} \in M_{nn}(I(\mathbb{C}))$ if $\rho(|\mathscr{A}|) < 1$. We then have

$$\{\mathscr{Y} = (\mathscr{I} - \mathscr{A})^{-1}\mathscr{B} \mid \mathscr{A} \in \mathscr{A},\ \mathscr{B} \in \mathscr{B}\} \subseteq \mathscr{X}^*.$$

We also wish to make some remarks about the condition $\rho(|\mathscr{A}|) < 1$, which is a condition necessary and sufficient for the convergence of the total step method in Theorem 1. If we consider the particular case when $\mathscr{A}$ is a point matrix and $\mathscr{b}$ a point vector, then one gets that the iteration

$$(6) \qquad x^{(k+1)} = \mathscr{A}x^{(k)} + \mathscr{b}, \qquad k \geqslant 0,$$

converges to the solution

$$x = (\mathscr{I} - \mathscr{A})^{-1}\mathscr{b}$$

for every interval vector $x^{(0)}$ iff $\rho(|\mathscr{A}|) < 1$. On the other hand the iteration

$$(7) \qquad x^{(k+1)} = \mathscr{A}x^{(k)} + \mathscr{b}, \qquad k \geqslant 0,$$

converges for all $x^{(0)} \in V_n(\mathbb{C})$ to

$$x = (\mathscr{I} - \mathscr{A})^{-1}\mathscr{b}$$

if $\rho(\mathscr{A}) < 1$ (see Varga [21]).

If we carry out the iteration (6) in $V_n(K(\mathbb{C}))$, then with $\mathscr{A} = (a_{rs})$, $a_{rs} = a_{rs}^{(1)} + ia_{rs}^{(2)}$, $1 \leqslant r,\ s \leqslant n$, according to Definition 10.6b, considering Definition 6.4b, we have that $|\mathscr{A}|$ satisfies

$$|\mathscr{A}| = |\mathscr{A}|_2 := \left(\sqrt{(a_{rs}^{(1)})^2 + (a_{rs}^{(2)})^2}\right).$$

If we carry out the iteration (6) in $V_n(R(\mathbb{C}))$, on the other hand, then from Definition 10.6b, considering also Definition 6.2, we have that $|\mathscr{A}|$ satisfies

$$|\mathscr{A}| = |\mathscr{A}|_1 := (|a_{rs}^{(1)}| + |a_{rs}^{(2)}|).$$

Because of

$$\sqrt{a_1^2 + a_2^2} \leqslant |a_1| + |a_2|$$

we have

$$|\mathscr{A}|_2 \leqslant |\mathscr{A}|_1.$$

From

$$\mathscr{O} \leqslant |\mathscr{A}|_2 \leqslant |\mathscr{A}|_1$$

it follows using Theorem 2.8 of Varga [21] that

$$\rho(\mathscr{A}) \leqslant \rho(|\mathscr{A}|_2) \leqslant \rho(|\mathscr{A}|_1). \tag{8}$$

This inequality shows that one needs in general stronger assumptions for (6) as for (7). The reason is that for (6) the necessary and sufficient criterion guarantees the convergence for all interval vectors. The set of interval vectors contains as a proper subset the set of all point vectors that are allowed as starting vectors in (7). This was discussed in Mayer [12].

The second part of inequality (8) shows that when (6) is carried out in $V_n(K(\mathbb{C}))$ it requires at most as strong a condition for $\mathscr{A}$ as for the corresponding iteration in $V_n(R(\mathbb{C}))$. This is also true for systems of equations with proper interval coefficients. From this point of view the arithmetic in $K(\mathbb{C})$ has a certain advantage over the one in $R(\mathbb{C})$.

REFERENCES

[1] A. C. Aitken, Studies in practical mathematics V. On the iterative solution of linear equations. *Proc. Roy. Soc. Edinburgh Sect. A* **63**, 52–60 (1960).

[2] G. Alefeld, Intervallrechnung über den komplexen Zahlen und einige Anwendungen. Ph.D. Thesis, Univ. Karlsruhe (1968).

[3] G. Alefeld, Anwendungen des Fixpunktsatzes für Pseudometrische Räume in der Intervallrechnung. *Numer. Math.* **17**, 33–39 (1971).

[4] G. Alefeld, Eigenschaften und Anwendungsmöglichkeiten einer Intervallarithmetik über den komplexen Zahlen. *Z. Angew. Math. Mech.* **50**, 455–465 (1970).

[5] G. Alefeld, Das symmetrische Einzelschrittverfahren bei linearen Gleichungen mit Intervallen als Koeffizienten. *Computing* **18**, 329–340 (1977).

[6] N. Apostolatos and U. Kulisch, Über die Konvergenz des Relaxationsverfahrens bei nichtnegativen und diagonaldominanten Matrizen. *Computing* **2**, 17–24 (1967)

[7] N. Apostolatos and U. Kulisch, Grundzüge einer Intervallrechnung für Matrizen und einige Anwendungen. *Elektron. Rech.* **10**, 73–83 (1968).

[8] W. Barth and E. Nuding, Optimale Lösung von Intervallgleichungssystemen. *Computing* **12**, 117–125 (1974).

[9] L. Collatz, "Funktionalanalysis und numerische Mathematik." Berlin and New York, Springer-Verlag, 1964.

[10] J. Herzberger, Über metrische Eigenschaften von Mengensystemen und einige Anwendungen. Ph.D. Thesis, Univ. Karlsruhe (1969).

[11] U. Kulisch, Grundzüge der Intervallrechnung. In "Überblicke Mathematik 2." Bibliograph. Inst. Mannheim (1969).

[12] O. Mayer, Über die in der Intervallrechnung auftretenden Räume und einige Anwendungen. Ph.D. Thesis, Univ. Karlsruhe (1968).

[13] O. Mayer, Algebraische und metrische Strukturen in der Intervallrechnung und einige Anwendungen. *Computing* **5**, 144–162 (1970).
[14] O. Mayer, Über die Bestimmung von Einschließungsmengen für die Lösungen linearer Gleichungssysteme mit fehlerbehafteten Koeffizienten. *Elektron. Datenverarbeit.* **12**, 164–167 (1970).
[15] O. Mayer, Über die intervallmäßige Durchführung einiger Iterationsverfahren. *Z. Angew. Math. Mech.* **50**, T 65–T 66 (1970).
[16] O. Mayer, Über eine Klasse komplexer Intervallgleichungssysteme mit iterationsfähiger Gestalt. *Computing* **6**, 104–106 (1970).
[17] O. Mayer, Über intervallmäßige Iterationsverfahren bei linearen Gleichungssystemen und allgemeineren Intervallgleichungssystemen. *Z. Angew. Math. Mech.* **51**, 117–124 (1971).
[18] W. Niethammer, Relaxation bei komplexen Matrizen. *Math. Z.* **86**, 34–40 (1964).
[19] J. M. Ortega, "Numerical Analysis: A Second Course." Academic Press, New York, 1972.
[20] J. M. Ortega and W. C. Rheinboldt, "Iterative Solution of Nonlinear Equations in Several Variables." Academic Press, New York, 1970.
[21] R. Varga, "Matrix Iterative Analysis." Prentice-Hall, Englewood Cliffs, New Jersey, 1962.

Chapter 13 / RELAXATION METHODS

We have already discussed the total step, the single step, and the symmetric single step method. There also exist many further methods for solving a linear point system of equations of the form

$$x = \mathcal{A} x + b,$$

where the asymptotic factor of convergence may be made smaller by the introduction of one or more parameters (see Varga [8]). Most of these methods can easily be carried over to iteration methods using interval vectors. As an example we consider the relaxation method for the single step case.

We again decompose the matrix $\mathcal{A}$ into $\mathcal{A} = \mathcal{L} + \mathcal{D} + \mathcal{U}$ as in the single step method where $\mathcal{L}$ is a strictly lower triangular matrix, $\mathcal{U}$ a strictly upper triangular matrix, and $\mathcal{D}$ a diagonal matrix. We then form

$$\begin{cases} \tilde{X}_i^{(k+1)} = \sum_{j=1}^{i-1} A_{ij} X_j^{(k+1)} + \sum_{j=i}^{n} A_{ij} X_j^{(k)} + B_i, \\ X_i^{(k+1)} = (1-\omega) X_i^{(k)} + \omega \tilde{X}_i^{(k+1)}, \qquad 1 \leqslant i \leqslant n, \quad k \geqslant 0, \end{cases}$$

starting with an arbitrary interval vector $x^{(0)}$. Using vector notation this may be written

$$x^{(k+1)} = (1-\omega) x^{(k)} + \omega\{\mathcal{L} x^{(k+1)} + (\mathcal{D} + \mathcal{U}) x^{(k)} + b\}, \qquad k \geqslant 0,$$

where $\omega > 0$ is a paramcter. As in the case of the total step or the single step method it may be shown that

$$\rho((\mathcal{I} - \omega|\mathcal{L}|)^{-1}\{|1-\omega|\mathcal{I} + \omega(|\mathcal{D}| + |\mathcal{U}|)\}) < 1$$

is a necessary and sufficient condition for the convergence of the method to a unique fixed point for an arbitrary starting vector.

It can furthermore be shown that if $\rho(|\mathscr{A}|) < 1$, then the above condition is satisfied for all ω, for which

$$0 < \omega < 2/(1 + \rho(|\mathscr{A}|))$$

(see, for example, Alefeld [1]). If the total step method converges and if ω satisfies the above inequality condition, then the relaxation method also converges to a fixed point $\tilde{x}^*$ satisfying the equation

$$\tilde{x}^* = (1 - \omega)\tilde{x}^* + \omega(\mathscr{A}\tilde{x}^* + b).$$

This fixed point is in general not equal to the fixed point x^* satisfying the equation $x = \mathscr{A}x + b$. In order to show this we first note that for all real a, b and proper intervals Z (i.e., $d(Z) > 0$) the relation

$$(a + b)Z = aZ + bZ$$

holds if $ab > 0$.

Let $0 < \omega < 1$. Then we have

$$(1 - \omega)x^* + \omega(\mathscr{A}x^* + b) = (1 - \omega)x^* + \omega x^* = (1 - \omega + \omega)x^* = x^*;$$

that is, $x^* = \tilde{x}^*$ since the fixed point $\tilde{x}^*$ calculated using the relaxation method is unique.

If, on the other hand, we have $\omega > 1$, then it follows that

$$\begin{aligned} \mathscr{A}x^* + b = x^* &= (1 - \omega + \omega)x^* \subseteq (1 - \omega)x^* + \omega x^* \\ &= (1 - \omega)x^* + \omega(\mathscr{A}x^* + b) \\ &= (1 - \omega)x^* + \omega\{\mathscr{L}x^* + (\mathscr{D} + \mathscr{U})x^* + b\}. \end{aligned}$$

Starting the relaxation method with $\tilde{x}^{(0)} = x^*$ it then follows from this inclusion relation using inclusion monotonicity that

$$x^* \subseteq (1 - \omega)\tilde{x}^{(0)} + \omega\{\mathscr{L}\tilde{x}^{(1)} + (\mathscr{D} + \mathscr{U})\tilde{x}^{(0)} + b\} =: \tilde{x}^{(1)}.$$

By complete induction it follows that $x^* \subseteq \tilde{x}^{(k)}$, $k \geqslant 0$; that is, $x^* \subseteq \tilde{x}^*$. There are simple examples where the inclusion is proper. If, therefore, $\omega > 1$, then one must take into account that the fixed point of $x^* = \mathscr{A}x^* + b$ is made "larger" by the use of the relaxation method. This is an unwanted effect since the problem was to find an interval vector that encloses the set

$$\{y = (\mathscr{I} - A)^{-1}b \mid A \in \mathscr{A},\ b \in \mathscr{b}\}$$

and gives a fairly good inclusion.

In the particular case of $\mathscr{A}$ being a point matrix and b a point vector we always have $x^* = \tilde{x}^*$. The set of point vectors is a subset of the set of interval vectors. If one therefore chooses a point vector, then all the iterates and the fixed point will also be point vectors. We therefore have for the total step method that

$$x^* = \mathcal{A}x^* + b$$

and for the relaxation method that

$$\tilde{x}^* = (1 - \omega)\tilde{x}^* + \omega(\mathcal{A}\tilde{x}^* + b),$$

from which it follows that $\tilde{x}^* = x^*$. For the case of a point matrix and a point vector, both methods therefore converge to the solution

$$x^* = (\mathcal{I} - \mathcal{A})^{-1}b$$

of the equation $x = \mathcal{A}x + b$.

We now wish to investigate this case in some further detail. The sequence $\{d(x^{(k)})\}_{k=0}^{\infty}$ of widths of the sequence $\{x^{(k)}\}_{k=0}^{\infty}$ converges to the zero vector since $\lim_{k\to\infty} x^{(k)} = x^* = (\mathcal{I} - \mathcal{A})^{-1}b$. It therefore seems reasonable to characterize the speed of convergence by the quantity $\tilde{\alpha}$ given in the following definition.

Definition 1: Let $x^* = f(x^*)$ and let $\mathfrak{C}$ be the set of all sequences $\{x^{(k)}\}_{k=0}^{\infty}$ that may be calculated using the iteration method

$$x^{(k+1)} = f(x^{(k)}), \qquad k \geq 0,$$

and for which $\lim_{k\to\infty} x^{(k)} = x^*$ holds. The quantity

$$\tilde{\alpha} = \sup\left\{ \limsup_{k\to\infty} \|d(x^{(k)})\|^{1/k} \mid \{x^{(k)}\}_{k=0}^{\infty} \in \mathfrak{C} \right\}$$

is then called the asymptotic convergence factor of the iteration method at the point x^*. ■

$\tilde{\alpha}$ may be characterized as the asymptotically worst case for the choice of $x^{(0)}$ as was done with α (according to Definition 12.10). In the same way as for α, one shows that $\tilde{\alpha}$ is independent of the norm used.

We now state the following theorem.

Theorem 2: Let the equation

$$x = \mathcal{A}x + b$$

be given with a point matrix $\mathcal{A}$ for which $\rho(|\mathcal{A}|) < 1$ and a point vector b. The asymptotic convergence factor $\tilde{\alpha}_{\mathrm{T}}$ of the total step method (12.2) then satisfies

$$\tilde{\alpha}_{\mathrm{T}} = \rho(|\mathcal{A}|),$$

and the asymptotic convergence factor $\tilde{\alpha}_{\mathrm{R}}$ of the relaxation method satisfies

$$\tilde{\alpha}_{\mathrm{R}} = \rho((\mathcal{I} - \omega|\mathcal{L}|)^{-1}\{|1 - \omega|\mathcal{I} + \omega(|\mathcal{D}| + |\mathcal{U}|)\})$$

for

$$0 < \omega < 2/(1 + \rho(|\mathscr{A}|)).$$

Proof: We carry out the proof for the relaxation method. Starting with an arbitrary interval vector $\mathscr{x}^{(0)}$ we get from the iteration procedure

$$\mathscr{x}^{(k+1)} = (1 - \omega)\mathscr{x}^{(k)} + \omega\{\mathscr{L}\mathscr{x}^{(k+1)} + (\mathscr{D} + \mathscr{U})\mathscr{x}^{(k)} + \mathscr{b}\}, \qquad k \geqslant 0,$$

using the rules (10.12) and (10.19) that

$$d(\mathscr{x}^{(k+1)}) = |1 - \omega|\, d(\mathscr{x}^{(k)}) + \omega|\mathscr{L}|\, d(\mathscr{x}^{(k+1)}) + \omega(|\mathscr{D}| + |\mathscr{U}|)\, d(\mathscr{x}^{(k)})$$

or

$$\begin{aligned} d(\mathscr{x}^{(k+1)}) &= (\mathscr{I} - \omega|\mathscr{L}|)^{-1}\{|1 - \omega|\mathscr{I} + \omega(|\mathscr{D}| + |\mathscr{U}|)\}\, d(\mathscr{x}^{(k)}) \\ &= \{(\mathscr{I} - \omega|\mathscr{L}|)^{-1}(|1 - \omega|\mathscr{I} + \omega(|\mathscr{D}| + |\mathscr{U}|)\}^{k+1}\, d(\mathscr{x}^{(0)}). \end{aligned}$$

From that we immediately get

$$\tilde{\alpha}_{\mathrm{R}} \leqslant \rho((\mathscr{I} - \omega|\mathscr{L}|)^{-1}\{|1 - \omega|\mathscr{I} + \omega(|\mathscr{D}| + |\mathscr{U}|)\}).$$

If one now chooses a particular $\mathscr{x}^{(0)}$ so that $d(\mathscr{x}^{(0)})$ is an eigenvector of the nonnegative matrix $(\mathscr{I} - \omega|\mathscr{L}|)^{-1}\{|1 - \omega|\mathscr{I} + \omega(|\mathscr{D}| + |\mathscr{U}|)\}$ for the eigenvalue λ that is equal to the spectral radius of this matrix, then it follows from the equation for $d(\mathscr{x}^{(k+1)})$ that

$$d(\mathscr{x}^{(k+1)}) = \lambda^{k+1}\, d(\mathscr{x}^{(0)})$$

from which the assertion follows. The proof for the total step method may be carried out similarly. ■

The above theorem now allows us to give a precise statement about the asymptotically fastest method in the sense of Definition 1.

Theorem 3: Under the assumptions of Theorem 2 it holds that

$$\begin{aligned} \min\{\tilde{\alpha}_{\mathrm{R}} \mid 0 < \omega < 2/(1 + \rho(|\mathscr{A}|))\} &= \tilde{\alpha}_{\mathrm{R}}|_{\omega=1} \\ &= \rho((\mathscr{I} - |\mathscr{L}|)^{-1}(|\mathscr{D}| + |\mathscr{U}|)) = \tilde{\alpha}_{\mathrm{S}}, \end{aligned}$$

as well as

$$\tilde{\alpha}_{\mathrm{S}} \leqslant \tilde{\alpha}_{\mathrm{T}}.$$

Proof: Consider the real point matrix

$$\mathscr{P} = ((1 - |1 - \omega|)/\omega)\mathscr{I} - |\mathscr{A}|$$

partitioned into $\mathscr{P} = \mathscr{M}_\omega - \mathscr{N}_\omega$ with

$$\mathscr{M}_\omega = (1/\omega)(\mathscr{I} - \omega|\mathscr{L}|), \qquad \mathscr{N}_\omega = (1/\omega)\{|1 - \omega|\mathscr{I} + \omega(|\mathscr{D}| + |\mathscr{U}|)\}.$$

The above splitting is regular (see Varga [8]) since $\mathscr{M}_\omega^{-1}$ exists and since

$\mathcal{M}_\omega^{-1} \geqslant \mathcal{O}$, $\mathcal{N}_\omega \geqslant \mathcal{O}$. If ω satisfies $0 < \omega < 2/(1 + \rho(|\mathcal{A}|))$ then $\mathcal{P}^{-1} \geqslant \mathcal{O}$ and because of $\mathcal{N}_\omega \geqslant \mathcal{N}_1$ it follows from an obvious generalization of Theorem 3.15 in Varga [8, p. 90] that

$$\rho(\mathcal{M}_1^{-1}\mathcal{N}_1) = \rho((\mathcal{I} - |\mathcal{L}|)^{-1}(|\mathcal{D}| + |\mathcal{U}|)) \leqslant \rho(\mathcal{M}_\omega^{-1}\mathcal{N}_\omega) < 1,$$

which proves the first part of the theorem. The second part of the theorem follows from a theorem of Stein and Rosenberg and its generalization (see Varga [8] and Apostolatos and Kulisch [3]), which says that if $\rho(|\mathcal{A}|) < 1$ it follows that

$$\rho((\mathcal{I} - |\mathcal{L}|)^{-1}(|\mathcal{D}| + |\mathcal{U}|)) \leqslant \rho(|\mathcal{A}|). \quad ■$$

The previous theorem says that in the case of a point system of equations one cannot asymptotically (in the sense of Definition 1) accelerate the convergence of the single step method using the relaxation method. The single step method furthermore converges at least as fast as the total step method.

We shall now discuss the connection between the above iteration method in the space of interval vectors and a principle of inclusion of solutions derived in a different manner (see Collatz [5, p. 275 ff], Albrecht [2], and Schröder [7]).

We consider the linear system of equations

$$x = \mathcal{A}x + b,$$

where $\mathcal{A}$ is a real point matrix and b a real point vector. We introduce the natural (that is, elementwise (resp. componentwise)) order in the set of point matrices (resp. point vectors). A point matrix $\mathcal{P}$ is called isotone (resp. antitone) when $x \geqslant o$ implies $\mathcal{P}x \geqslant o$ (resp. $\leqslant o$). The matrix $\mathcal{A}$ is then decomposed into an isotone and an antitone point matrix

$$\mathcal{A} = \mathcal{T}_1 + \mathcal{T}_2.$$

Starting with point vectors $v^{(0)}$ and $w^{(0)}$ for which $v^{(0)} \leqslant w^{(0)}$ the iteration method

$$(1) \qquad \begin{cases} v^{(k+1)} = \mathcal{T}_1 v^{(k)} + \mathcal{T}_2 w^{(k)} + b, \\ w^{(k+1)} = \mathcal{T}_1 w^{(k)} + \mathcal{T}_2 v^{(k)} + b, \qquad k \geqslant 0, \end{cases}$$

then calculates two sequences $\{v^{(k)}\}_{k=0}^\infty$ and $\{w^{(k)}\}_{k=0}^\infty$ of point vectors. If now $v^{(0)} \leqslant v^{(1)} \leqslant w^{(1)} \leqslant w^{(0)}$, then one can show using complete induction that

$$v^{(0)} \leqslant v^{(1)} \leqslant \cdots \leqslant v^{(k)} \leqslant v^{(k+1)} \leqslant w^{(k+1)} \leqslant w^{(k)} \leqslant \cdots \leqslant w^{(1)} \leqslant w^{(0)}.$$

The sequences $\{v^{(k)}\}_{k=0}^\infty$ and $\{w^{(k)}\}_{k=0}^\infty$ are therefore convergent, and from simple deliberations it follows that a solution x^* of the equation $x = \mathcal{A}x + b$ exists and is enclosed by the limiting elements v^* and w^*. If $\rho(|\mathcal{A}|) = \rho(\mathcal{T}_1 - \mathcal{T}_2) < 1$, then $v^* = w^* = x^*$.

If one considers the iteration method

$$(2) \qquad x^{(k+1)} = \mathcal{A}x^{(k)} + b, \qquad k \geqslant 0,$$

in addition to the iteration method for $\{v^{(k)}\}_{k=0}^{\infty}$ and $\{w^{(k)}\}_{k=0}^{\infty}$ using $x^{(0)} = ([v_i^{(0)}, w_i^{(0)}])$, then one sees immediately that the bounds of the components of the sequence $\{x^{(k)}\}_{k=0}^{\infty}$ of interval vectors coincide with the components of the sequences $\{v^{(k)}\}_{k=0}^{\infty}$ and $\{w^{(k)}\}_{k=0}^{\infty}$ when one considers the rules for multiplication of interval matrices with interval vectors.

The previous considerations for the iteration method (2) show that one may dispense with the assumption that $v^{(0)} \leqslant v^{(1)} \leqslant w^{(1)} \leqslant w^{(0)}$ for method (1) if only $v^{(0)} \leqslant x^* \leqslant w^{(0)}$ in order that the inclusion of the solution x^* is guaranteed. The sequences $\{v^{(k)}\}_{k=0}^{\infty}$ and $\{w^{(k)}\}_{k=0}^{\infty}$ then do not necessarily converge monotonically. Monotonic behavior may be reintroduced by taking intersections after each iteration step (see Chapter 14).

Remarks: Similar statements as for the relaxation method are valid for the symmetric relaxation method (SR)

$$x^{(k+1/2)} = (1-\omega)x^{(k)} + \omega\{\mathcal{L}x^{(k+1/2)} + \mathcal{U}x^{(k)} + b\},$$
$$x^{(k+1)} = (1-\omega)x^{(k+1/2)} + \omega\{\mathcal{L}x^{(k+1/2)} + \mathcal{U}x^{(k+1)} + b\}, \qquad k \geqslant 0,$$

that reduces to the symmetric single step method (SS) of Chapter 12 for $\omega = 1$. Necessary and sufficient conditions for the convergence of this method for an arbitrary starting interval vector to a unique fixed point are given by the condition

$$\rho((\mathcal{I} - \omega|\mathcal{U}|)^{-1}(|1-\omega|\mathcal{I} + \omega|\mathcal{L}|)(\mathcal{I} - \omega|\mathcal{L}|)^{-1}(|1-\omega|\mathcal{I} + \omega|\mathcal{U}|)) < 1.$$

If, furthermore, $\rho(|\mathcal{A}|) < 1$, then this is the case if

$$0 < \omega < 2/(1 + \rho(|\mathcal{A}|)).$$

The proof may be performed in the same manner as for the relaxation method.

If $\mathcal{A}$ is a point matrix, then as in Theorem 2 it can be shown that the asymptotic convergence factor defined by Definition 1 is

$$\tilde{\alpha}_{\mathrm{SR}} = \rho((\mathcal{I} - \omega|\mathcal{U}|)^{-1}(|1-\omega|\mathcal{I} + \omega|\mathcal{L}|)(\mathcal{I} - \omega|\mathcal{L}|)^{-1}(|1-\omega|\mathcal{I} + \omega|\mathcal{U}|)),$$

and furthermore, corresponding to Theorem 3, we have

$$\tilde{\alpha}_{\mathrm{SS}} \leqslant \tilde{\alpha}_{\mathrm{SR}}$$

for $0 < \omega < 2/(1 + \rho(|\mathcal{A}|))$.

We show that $\tilde{\alpha}_{\mathrm{SS}} \leqslant \tilde{\alpha}_{\mathrm{S}}$ is valid. The matrix

$$(\mathcal{I} - |\mathcal{L}|)^{-1}|\mathcal{U}|$$

is nonnegative and always reducible since it has only zeros in the first column. If one adds a positive matrix $\Delta\mathcal{U}$ (in general, not an upper triangular matrix), then one is able to make $(\mathcal{I} - |\mathcal{L}|)^{-1}(|\mathcal{U}| + \Delta\mathcal{U})$ irreducible. Using the theorem of Perron and Frobenius (see Varga [8]) one gets

$$(\mathcal{I} - |\mathcal{L}|)^{-1}(|\mathcal{U}| + \Delta\mathcal{U})x = \lambda x,$$

where λ is the spectral radius of the matrix on the left and where x has only positive components. A simple transformation (valid for $0 < \lambda < 1$ and sufficiently small $\Delta\mathcal{U}$) gives

$$x = (\mathcal{I} - (1/\lambda)(|\mathcal{U}| + \Delta\mathcal{U}))^{-1}|\mathcal{L}|x \geqslant (\mathcal{I} - (|\mathcal{U}| + \Delta\mathcal{U}))^{-1}|\mathcal{L}|x.$$

One finally obtains

$$(\mathcal{I} - (|\mathcal{U}| + \Delta\mathcal{U}))^{-1}|\mathcal{L}|(\mathcal{I} - |\mathcal{L}|)^{-1}(|\mathcal{U}| + \Delta\mathcal{U})x \leqslant \lambda x.$$

This inequality is also valid for $\lambda = 0$. If one now applies Theorem 1.11 in Berman and Plemmons [4, p. 28], then one gets that the spectral radius of the matrix on the left is not larger than λ. Since this is true for all $\Delta\mathcal{U}$ that make the matrix $(\mathcal{I} - |\mathcal{L}|)^{-1}(|\mathcal{U}| + \Delta\mathcal{U})$ irreducible, one gets the assertion by letting $\Delta\mathcal{U}$ tend to zero since the eigenvalues are continuously dependent on the matrix elements.

REFERENCES

[1] G. Alefeld, Intervallrechnung über den komplexen Zahlen und einige Anwendungen. Ph.D. Thesis, Univ. Karlsruhe, 1968.

[2] J. Albrecht, Monotone Iterationsfolgen und ihre Verwendung zur Lösung linearer Gleichungssysteme. *Numer. Math.* **3**, 345–358 (1961).

[3] N. Apostolatos and U. Kulisch, Über die Konvergenz des Relaxationsverfahrens bei nichtnegativen und diagonaldominaten Matrizen. *Computing* **2**, 17–24 (1967).

[4] A. Berman and R. J. Plemmons, "Nonnegative Matrices in the Mathematical Sciences." Academic Press, New York, 1979.

[5] L. Collatz, "Funktionalanalysis und numerische Mathematik." Springer-Verlag, Berlin and New York, 1964.

[6] O. Mayer, Über intervallmäßige Iterationsverfahren bei linearen Gleichungssystemen und allgemeineren Intervallgleichungssystemen. *Z. Angew. Math. Mech.* **51**, 117–124 (1971).

[7] J. Schröder, Fehlerabschätzungen bei linearen Gleichungssystemen mit dem Brouwerschen Fixpunktsatz. *Arch. Rational Mech. Anal.* **3**, 28–44 (1959).

[8] R. Varga, "Matrix Iterative Analysis." Prentice-Hall, Englewood Cliffs, New Jersey, 1962.

Chapter 14 / OPTIMALITY OF THE SYMMETRIC SINGLE STEP METHOD WITH TAKING INTERSECTION AFTER EVERY COMPONENT

In this chapter it is assumed that all interval matrices are from $M_{nn}(R(\mathbb{C}))$ and all interval vectors are from $V_n(R(\mathbb{C}))$.

We now wish to investigate some modifications to the total step, the single step, and the symmetric single step method. If the total step method

$$x^{(k+1)} = \mathscr{A}x^{(k)} + b$$

is started with a vector $x^{(0)}$ for which $x^* \subseteq x^{(0)}$, then it follows from inclusion monotonicity that

$$x^* = \mathscr{A}x^* + b \subseteq \mathscr{A}x^{(0)} + b = x^{(1)}.$$

This means that $x^{(1)}$ contains the fixed point as well as $x^{(0)}$ and therefore also their intersection $x^{(0)} \cap x^{(1)}$. It is therefore natural to continue the iteration with this new containment. This results in the iteration procedure

$$x^{(k+1)} = \{\mathscr{A}x^{(k)} + b\} \cap x^{(k)}, \qquad k \geqslant 0,$$

which will be called the total step method with taking intersection after each step (TI).

If one carries out the same deliberations for the single step method, one obtains the iteration procedure

$$x^{(k+1)} = \{\mathscr{L}x^{(k+1)} + (\mathscr{D} + \mathscr{U})x^{(k)} + b\} \cap x^{(k)}, \qquad k \geqslant 0,$$

which will be called the single step method with taking intersection after each step (SI).

In the case of the single step method there is also the further possibility:

$$X_i^{(k+1)} = \left\{ \sum_{j=1}^{i-1} A_{ij} X_j^{(k+1)} + \sum_{j=i}^{n} A_{ij} X_j^{(k)} + B_i \right\} \cap X_i^{(k)}, \qquad 1 \leqslant i \leqslant n, \quad k \geqslant 0.$$

After the first component has been calculated, the intersection with the old approximation is formed giving a new approximation. This new approximation is now used to calculate the new approximation for the second component and so on. This modification is called the single step method with taking intersection after every component (SIC).

Again with the assumption that $x^* \subseteq x^{(0)}$ we finally consider the iteration procedure (for the case that all diagonal elements of $\mathscr{A}$ vanish)

$$\begin{cases} X_i^{(k+1/2)} = \left\{ \sum_{j=1}^{i-1} A_{ij} X_j^{(k+1/2)} + \sum_{j=i+1}^{n} A_{ij} X_j^{(k)} + B_i \right\} \cap X_i^{(k)}, & 1 \leqslant i \leqslant n, \\ X_i^{(k+1)} = \left\{ \sum_{j=1}^{i-1} A_{ij} X_j^{(k+1/2)} + \sum_{j=i+1}^{n} A_{ij} X_j^{(k+1)} + B_i \right\} \cap X_i^{(k+1/2)}, & 1 \leqslant i \leqslant n, \\ k \geqslant 0. & \end{cases}$$

We call this iteration procedure the symmetric single step method with taking intersection after every component (SSIC).

The method (SSIC) may be carried out in such a manner that – except for the first step – it uses the same number of interval multiplications as the method (SIC) in each step. In both cases $n^2 - n$ multiplications are required (assuming the diagonal elements vanish). See also Niethammer [11]. The following argument is used: Assume that the sums

$$\sum_{j=i+1}^{n} A_{ij} X_j^{(k)}, \qquad 1 \leqslant i \leqslant n-1,$$

are known for some $k > 0$. The calculation of $x^{(k+1/2)}$ according to (SSIC) then requires $\frac{1}{2}(n^2 - n)$ multiplications. If the $n - 1$ sums

$$\sum_{j=1}^{i-1} A_{ij} X_j^{(k+1/2)}, \qquad 2 \leqslant i \leqslant n,$$

occurring in this calculation are stored then the calculation of $x^{(k+1)}$ from $x^{(k+1/2)}$ requires another $\frac{1}{2}(n^2 - n)$ multiplications. In the same manner as for the iteration method (SIC), one therefore requires $n^2 - n$ multiplications for the calculation of an approximation $x^{(k+1)}$ from $x^{(k)}$. If the sums

$$\sum_{j=i+1}^{n} A_{ij} X_j^{(k+1)}$$

are stored when one calculates $x^{(k+1)}$ from $x^{(k+1/2)}$, then all the above considerations may be carried out with the index increased by one.

The following theorem states that the iteration methods (TI), (SI), (SIC), and (SSIC) all converge to the fixed point x^*.

Theorem 1: Let $\mathscr{A}$ be an interval matrix for which $\rho(|\mathscr{A}|) < 1$. If x^* is the fixed point of the equation $x = \mathscr{A}x + b$ and if $x^{(0)} \supseteq x^*$, then the iteration methods (TI), (SI), (SIC), and (SSIC) all converge to x^*.

Proof: We prove the theorem for the iteration method (TI). Since we take intersection we obtain a sequence $\{x^{(k)}\}_{k=0}^{\infty}$ of approximations that satisfy

$$x^{(0)} \supseteq x^{(1)} \supseteq \cdots \supseteq x^{(k)} \supseteq x^{(k+1)} \supseteq \cdots.$$

This sequence converges to a limiting value $\tilde{x}^*$ according to Corollary 10.8, and the limiting value furthermore satisfies $x^* \subseteq \tilde{x}^*$. The operation of taking intersections is also continuous (if the intersection is nonempty), which implies that when $k \to \infty$ we have

$$\tilde{x}^* = \{\mathscr{A}\tilde{x}^* + b\} \cap \tilde{x}^*,$$

from which it follows that

$$\mathscr{A}\tilde{x}^* + b \supseteq \tilde{x}^*.$$

We consider the total step method

$$y^{(k+1)} = \mathscr{A}y^{(k)} + b, \qquad k \geqslant 0,$$

with $y^{(0)} = \tilde{x}^*$. It follows that

$$y^{(1)} = \mathscr{A}y^{(0)} + b = \mathscr{A}\tilde{x}^* + b \supseteq \tilde{x}^* \supseteq x^*,$$

$$y^{(2)} = \mathscr{A}y^{(1)} + b \supseteq \mathscr{A}\tilde{x}^* + b \supseteq \tilde{x}^* \supseteq x^*,$$

and in general

$$y^{(k+1)} = \mathscr{A}y^{(k)} + b \supseteq \mathscr{A}\tilde{x}^* + b \supseteq \tilde{x}^* \supseteq x^*.$$

The sequence $\{y^{(k)}\}_{k=0}^{\infty}$ computed by the above iteration converges by Theorem 12.1 to x^*. From the last proven inclusion it follows that when $k \to \infty$ we get

$$x^* \supseteq \tilde{x}^* \supseteq x^*;$$

that is, $\tilde{x}^* = x^*$. The proofs are similar for the other methods. ■

Our aim is now to compare the methods (T), (S), (TI), (SI), (SIC), and (SSIC) with respect to their speeds of convergence when one starts the iteration with an interval vector containing the fixed point x^*.

The first theorem compares the methods (T) and (TI) as well as (S) and (SI) with each other.

Theorem 2: Let $\{x^{(k)}\}_{k=0}^{\infty}$ and $\{\tilde{x}^{(k)}\}_{k=0}^{\infty}$ be the sequences computed using thc iteration methods (T) and (TI) under the assumption that $x^{(0)} \supseteq \tilde{x}^{(0)} \supseteq x^*$. It then holds that

$$x^{(k)} \supseteq \tilde{x}^{(k)} \supseteq x^* \qquad \text{for all} \quad k \geqslant 0.$$

The same statement is valid for the sequences calculated using the iteration methods (S) and (SI).

Proof: We prove the theorem for the sequences computed using the iteration methods (T) and (TI). The relation $\tilde{x}^{(k)} \supseteq x^*$, $k \geqslant 0$, was already proved during the derivation of the iteration method (TI) under the assumption that $\tilde{x}^{(0)} \supseteq x^*$. Suppose

$$x^{(k)} \supseteq \tilde{x}^{(k)}$$

for some $k \geqslant 0$. This is certainly true for $k = 0$ from the assumption. Using inclusion monotonicity we get

$$x^{(k+1)} = \mathcal{A}x^{(k)} + b \supseteq \mathcal{A}\tilde{x}^{(k)} + b \supseteq \{\mathcal{A}\tilde{x}^{(k)} + b\} \cap \tilde{x}^{(k)} = \tilde{x}^{(k+1)}.$$

The proof for the sequences computed using the iteration methods (S) and (SI) may be carried out in a similar manner. ■

Theorem 3: Let $\{x^{(k)}\}_{k=0}^{\infty}$ and $\{\tilde{x}^{(k)}\}_{k=0}^{\infty}$ be the sequences computed using the iteration methods (TI) and (SIC) under the assumption that $x^{(0)} \supseteq \tilde{x}^{(0)} \supseteq x^*$. It then holds that

$$x^{(k)} \supseteq \tilde{x}^{(k)} \supseteq x^* \qquad \text{for all} \quad k \geqslant 0.$$

The same statement is valid for the sequences calculated using the iteration methods (SI) and (SIC).

Proof: We only have to prove the relation $x^{(k)} \supseteq \tilde{x}^{(k)}$. We carry out the proof for the sequences calculated using the iteration methods (TI) and (SIC). Assume

$$x^{(k)} \supseteq \tilde{x}^{(k)}$$

for some $k \geqslant 0$. This is certainly true for $k = 0$ from the assumption. It then holds with $x^{(k)} = (X_i^{(k)})$, $\tilde{x}^{(k)} = (\tilde{X}_i^{(k)})$, $\mathcal{A} = (A_{ij})$, $b = (B_i)$, that

$$X_1^{(k+1)} = \left(\sum_{j=1}^{n} A_{1j}X_j^{(k)} + B_1\right) \cap X_1^{(k)},$$

$$\tilde{X}_1^{(k+1)} = \left(\sum_{j=1}^{n} A_{1j}\tilde{X}_j^{(k)} + B_1\right) \cap \tilde{X}_1^{(k)}.$$

From $\tilde{X}_i^{(k)} \subseteq X_i^{(k)}$, $1 \leqslant i \leqslant n$, it follows from the inclusion monotonicity that

$$\sum_{j=1}^{n} A_{1j}\tilde{X}_j^{(k)} + B_1 \subseteq \sum_{j=1}^{n} A_{1j}X_j^{(k)} + B_1,$$

and therefore

$$\tilde{X}_1^{(k+1)} \subseteq X_1^{(k+1)}.$$

Since $\tilde{X}_1^{(k+1)} \subseteq \tilde{X}_1^{(k)} \subseteq X_1^{(k)}$ it follows that

$$A_{21}\tilde{X}_1^{(k+1)} + \sum_{j=2}^{n} A_{2j}\tilde{X}_j^{(k)} + B_2 \subseteq \sum_{j=1}^{n} A_{2j}X_j^{(k)} + B_2,$$

and because of

$$X_2^{(k+1)} = \left(\sum_{j=1}^{n} A_{2j}X_j^{(k)} + B_2\right) \cap X_2^{(k)},$$

$$\tilde{X}_2^{(k+1)} = \left(A_{21}\tilde{X}_1^{(k+1)} + \sum_{j=2}^{n} A_{2j}\tilde{X}_j^{(k)} + B_2\right) \cap \tilde{X}_2^{(k)},$$

it follows that

$$\tilde{X}_2^{(k+1)} \subseteq X_2^{(k+1)}.$$

In the same manner one shows $\tilde{X}_i^{(k+1)} \subseteq X_i^{(k+1)}$, $3 \leqslant i \leqslant n$; that is, $\tilde{x}^{(k+1)} \subseteq x^{(k+1)}$. The proof for the sequences calculated using the iteration methods (SI) and (SIC) may be carried out in a similar manner. ■

Theorem 4: Let $\{z^{(k)}\}_{k=0}^{\infty}$ and $\{x^{(k)}\}_{k=0}^{\infty}$ be the sequences computed using the iteration methods (SIC) and (SSIC) under the assumption that $z^{(0)} \supseteq x^{(0)} \supseteq x^*$. It then holds that

$$z^{(k)} \supseteq x^{(k)} \supseteq x^* \qquad \text{for all} \quad k \geqslant 0.$$

Proof: Assume $z^{(k)} \supseteq x^{(k)} \supseteq x^*$ for some $k \geqslant 0$. This is certainly true for $k = 0$ according to the assumptions. It then follows from the first part of (SSIC) and (SIC) using complete induction on the indices of the components that

$$z^{(k+1)} \supseteq x^{(k+1/2)} \supseteq x^*.$$

With the help of the second part of (SSIC) again using complete induction on the indices of the components we get

$$x^{(k+1/2)} \supseteq x^{(k+1)} \supseteq x^*.$$

Together the above inclusions give

$$z^{(k+1)} \supseteq x^{(k+1)} \supseteq x^*. \quad ■$$

With the statements proved above we can now give the optimal iteration method. Let (M) denote a method from the set

$$\{(\mathrm{T}), (\mathrm{S}), (\mathrm{TI}), (\mathrm{SI}), (\mathrm{SIC}), (\mathrm{SSIC})\}.$$

We allow as starting vectors all interval vectors $x^{(0)}$ satisfying $x^* \subseteq x^{(0)}$ where x^* is the fixed point of the equation $x = \mathscr{A}x + \ell$. We introduce a partial order on the above set of iteration methods by stating that $(M) \leqslant (\tilde{M})$ iff $x^{(k)} \subseteq \tilde{x}^{(k)}$, $k \geqslant 0$. Here $\{x^{(k)}\}_{k=0}^{\infty}$ and $\{\tilde{x}^{(k)}\}_{k=0}^{\infty}$ denote the sequences computed by the methods (M) and $(\tilde{M})$. From Theorem 2 we have

$$(\mathrm{TI}) \leqslant (\mathrm{T}) \qquad \text{and} \qquad (\mathrm{SI}) \leqslant (\mathrm{S}).$$

Correspondingly from Theorem 3 we have

$$(\mathrm{SIC}) \leqslant (\mathrm{TI}) \qquad \text{and} \qquad (\mathrm{SIC}) \leqslant (\mathrm{SI}).$$

Finally we have from Theorem 4 that

$$(\mathrm{SSIC}) \leqslant (M)$$

for all of the iteration methods (M) considered. We combine these results into

Theorem 5: Let $\mathscr{A}$ be an interval matrix for which $\rho(|\mathscr{A}|) < 1$ and let ℓ be an interval vector. If one starts the iteration methods (T), (S), (TI), (SI), (SIC), and (SSIC) with a starting vector $x^{(0)}$ satisfying $x^{(0)} \supseteq x^* = \mathscr{A}x^* + \ell$, then the method (SSIC) will always give the smallest (in the sense of inclusion) containing sequence for x^*. ■

The contents of Theorem 5 are made clear by the following diagram:

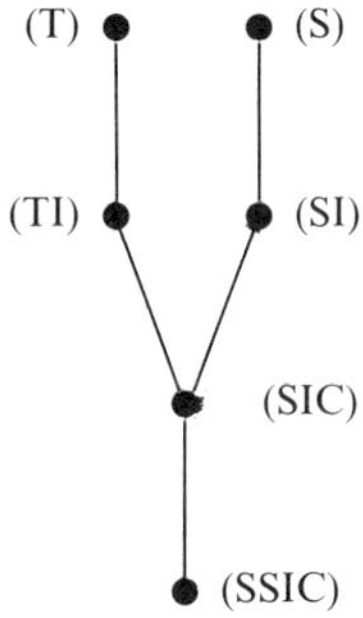

In order to illustrate Theorem 5 we have calculated a number of different examples on the UNIVAC 1108 computer (mantissa length eight decimal places) at the Computing Center of the University of Karlsruhe.

For each example we have given the interval vector $x^{(0)} = (X_i^{(0)})$ as well as the number of iterations k^* required for the iteration procedure to reach a stationary state. The examples show that the method (SIC) requires about 25% more iteration steps than the method (SSIC).

In the first two examples both the initial data and the fixed points are proper intervals. In this case we also display $x^{(k^*)}$. For the other examples we only give the largest width of the components of $x^{(k)}$; that is, $d^{(k)} = \max_{1 \leq i \leq n}\{d(X_i^{(k)})\}$. All the examples were transformed into the form $x = \mathcal{A}x + b$ in such a manner that the diagonal elements of $\mathcal{A}$ are zero.

I. Example: From Albrecht [1, Example 1, p. 356].

$$x^{(0)} = \begin{pmatrix} [0.9, & 1.2] \\ [0.4, & 0.7] \\ [0, & 0.2] \\ [-0.4, & -0.1] \end{pmatrix},$$

$$\text{(SIC)}: k^* = 37, \qquad \text{(SSIC)}: k^* = 31,$$

$$x^{(k^*)} = \begin{pmatrix} [1.032\,860\,1, & 1.059\,757\,9] \\ [0.550\,754\,40, & 0.574\,813\,98] \\ [0.099\,483\,623, & 0.122\,513\,79] \\ [-0.243\,545\,82, & -0.212\,698\,41] \end{pmatrix}.$$

II. Example: From Albrecht [1, Example 2, p. 357].

$$x^{(0)} = \begin{pmatrix} [0.8, & 1.0] \\ [0.65, & 0.85] \\ [0.55, & 0.7] \end{pmatrix}, \qquad x^{(k^*)} = \begin{pmatrix} [0.896\,368\,17, & 0.896\,479\,91] \\ [0.765\,057\,55, & 0.765\,202\,25] \\ [0.614\,247\,34, & 0.614\,521\,84] \end{pmatrix}$$

$$\text{(SIC)}: k^* = 18, \qquad \text{(SSIC)}: k^* = 15.$$

III. Example: From Schröder [14, Example 1, p. 332].

$$x^{(0)} = (X_i^{(0)}) \qquad \text{with} \quad X_i^{(0)} = [0, 1], \quad i = 1(1)8.$$

$$\text{(SIC)}: k^* = 27, \qquad \text{(SSIC)}: k^* = 22.$$

$d^{(k)}$	k: 0	5	10	15	20
(SIC)	1	4.0×10^{-2}	8.4×10^{-4}	1.8×10^{-5}	3.8×10^{-7}
(SSIC)	1	7.0×10^{-3}	5.4×10^{-5}	4.1×10^{-7}	3.7×10^{-9}

IV. Example: From Schröder [14, Example 2, p. 333].

$$x^{(0)} = (X_i^{(0)}) \qquad \text{with} \quad X_i^{(0)} = [0, 0.5], \quad 1 \leq i \leq 4.$$

$$\text{(SIC)}: k^* = 56, \qquad \text{(SSIC)}: k^* = 44.$$

$d^{(k)}$	k					
	0	5	10	20	30	40
(SIC)	5.0×10^{-1}	1.1×10^{-1}	2.9×10^{-2}	6.7×10^{-4}	2.7×10^{-5}	1.5×10^{-6}
(SSIC)	5.0×10^{-1}	3.7×10^{-2}	4.7×10^{-3}	7.4×10^{-5}	1.2×10^{-6}	1.8×10^{-8}

V. Example: From Schmidt [13, Example A, p. 85].

$$x^{(0)} = (X_i^{(0)}) \quad \text{with} \quad X_i^{(0)} = [-0.036016, 0.674056], \quad 1 \leqslant i \leqslant 3.$$

$$(\text{SIC}): k^* = 52, \qquad (\text{SSIC}): k^* = 42.$$

$d^{(k)}$	k				
	0	5	10	20	40
(SIC)	7.1×10^{-1}	1.2×10^{-1}	2.0×10^{-2}	5.3×10^{-4}	3.7×10^{-7}
(SSIC)	7.1×10^{-1}	5.5×10^{-2}	5.7×10^{-3}	6.2×10^{-5}	3.7×10^{-9}

VI. Example: As in Example V, but here we have

$$X_i^{(0)} = [0.059459, 0.643243], \; 1 \leqslant i \leqslant 3.$$

$$(\text{SIC}): k^* = 51, \qquad (\text{SSIC}): k^* = 41.$$

$d^{(k)}$	k				
	0	5	10	20	40
(SIC)	5.8×10^{-1}	1.0×10^{-1}	1.7×10^{-2}	4.3×10^{-4}	3.0×10^{-7}
(SSIC)	5.8×10^{-1}	4.5×10^{-2}	4.7×10^{-3}	5.1×10^{-5}	3.7×10^{-9}

The iteration using including sets is important when one carries out the computations on an electronic computer. If one starts the iteration in this case with a vector representable on the machine containing the fixed point of the equation $x^* = \mathscr{A}x^* + b$, then the new approximation as well as all further approximations have the property of again containing the fixed point. Since the calculations of the new approximation are contaminated by round-off errors we may in fact at some point "lose" the fixed point; that is, some newly computed interval will no longer contain the fixed point. If all the operations are executed using a machine interval arithmetic (see Chapter 4) then the including property will not be lost. If one uses a method that takes intersections, then one obtains a sequence

$$\tilde{x}^{(0)} \supseteq \tilde{x}^{(1)} \supseteq \cdots \supseteq \tilde{x}^{(k-1)} \supseteq \tilde{x}^{(k)} = \tilde{x}^{(k+1)} = \cdots.$$

The sequence of machine computed approximations, from a certain index k^* onward, will be identical. This follows from the fact that only a finite number of real numbers are representable on a digital computer.

We now show that it is not necessary to take intersections after a finite number of steps for the methods (TI), (SI), (SIC), and (SSIC). We formulate and prove a theorem to this effect for the iteration method (TI).

Theorem 6: Let $\mathscr{A}$ be an interval matrix for which $\rho(|\mathscr{A}|) < 1$. Furthermore let $x^{(0)} = (X_i^{(0)}) = ([i(X_i^{(0)}), s(X_i^{(0)})])$ be chosen such that for the fixed point $x^* = ([i(X_i^*), s(X_i^*)])$ of $x = \mathscr{A}x + b$ the relation $x^{(0)} \supset x^*$, defined by $i(X_i^{(0)}) < i(X_i^*) \leqslant s(X_i^*) < s(X_i^{(0)})$, $i = 1, 2, \ldots, n$, holds. Then there exists $\tilde{k} \geqslant 0$ so that for all $k \geqslant \tilde{k}$ for the iteration method

$$x^{(k+1)} = (\mathscr{A}x^{(k)} + b) \cap x^{(k)}, \qquad k \geqslant 0,$$

the equality

$$x^{(k+1)} = (\mathscr{A}x^{(k)} + b) \cap x^{(k)} = \mathscr{A}x^{(k)} + b,$$

holds which means that the inclusion

$$\mathscr{A}x^{(k)} + b \subseteq x^{(k)}$$

is valid.

Proof: We restrict ourselves to the case that $\mathscr{A}$, b, and $x^{(0)}$ have elements from $I(\mathbb{R})$ only. The case where the elements are from $R(\mathbb{C})$ may be proved in the same manner. We first show that the assumption $x^{(0)} \supset x^*$ implies that the relation

$$x^{(0)} \subseteq \mathscr{A}x^{(0)} + b$$

cannot be true. If this relation indeed was true, then it follows using the iteration method

$$z^{(k+1)} = \mathscr{A}z^{(k)} + b, \qquad k \geqslant 0,$$

with $z^{(0)} = x^{(0)}$, that

$$z^{(1)} = \mathscr{A}z^{(0)} + b = \mathscr{A}x^{(0)} + b \supseteq x^{(0)} \supset x^*,$$

$$z^{(2)} = \mathscr{A}z^{(1)} + b \supseteq \mathscr{A}x^{(0)} + b \supseteq x^{(0)} \supset x^*,$$

and in general

$$z^{(k+1)} = \mathscr{A}z^{(k)} + b \supseteq \mathscr{A}x^{(0)} + b \supseteq x^{(0)} \supset x^*, \qquad k \geqslant 0.$$

Since $\rho(|\mathscr{A}|) < 1$ we have $\lim_{k \to \infty} z^{(k)} = z^*$ with $z^* = \mathscr{A}z^* + b$. From the last relation it also follows that $z^* \supseteq x^{(0)} \supset x^*$. This contradicts the uniqueness of the fixed point of the equation $x = \mathscr{A}x + b$.

Now let

$$x^{(k)} = (X_i^{(k)}), \qquad \mathscr{A} = (A_{ij}), \qquad b = (B_i), \qquad y^{(k)} = (Y_i^{(k)}),$$

with

$$Y_i^{(k+1)} = \sum_{j=1}^{n} A_{ij}X_j^{(k)} + B_i, \qquad k \geqslant 0, \quad 1 \leqslant i \leqslant n.$$

From the above fact there must be at least one index i, $1 \leqslant i \leqslant n$, such that exactly one of the following two cases is valid

(a) $Y_i^{(1)} \subset X_i^{(0)}$; that is, $X_i^{(1)} = Y_i^{(1)} \cap X_i^{(0)} = Y_i^{(1)}$;
(b) $Y_i^{(1)} \not\subset X_i^{(0)}$ and $X_i^{(1)} = Y_i^{(1)} \cap X_i^{(0)} \subset X_i^{(0)}$.

Since $x^{(0)} \supseteq x^{(1)}$ it follows in the case of (a) from the inclusion monotonicity that

$$Y_i^{(2)} = \sum_{j=1}^{n} A_{ij}X_j^{(1)} + B_i \subseteq \sum_{j=1}^{n} A_{ij}X_j^{(0)} + B_i = Y_i^{(1)} = X_i^{(1)},$$
$$X_i^{(2)} = Y_i^{(2)} \cap X_i^{(1)} = Y_i^{(2)},$$

and in general since $x^{(k)} \supseteq x^{(k+1)}$ by complete induction that

$$Y_i^{(k+1)} = \sum_{j=1}^{n} A_{ij}X_j^{(k)} + B_i \subseteq \sum_{j=1}^{n} A_{ij}X_j^{(k-1)} + B_i = Y_i^{(k)} = X_i^{(k)},$$
$$X_i^{(k+1)} = Y_i^{(k+1)} \cap X_i^{(k)} = Y_i^{(k+1)}.$$

We now consider case (b). Setting $i(A) = a_1$, $s(A) = a_2$ for $A = [a_1, a_2] \in I(\mathbb{R})$ we can without restricting the generality assume that

$$i(Y_i^{(1)}) < i(X_i^{(0)}) \leqslant s(Y_i^{(1)}) < s(X_i^{(0)});$$

that is,

$$X_i^{(1)} = [i(X_i^{(0)}), s(Y_i^{(1)})].$$

(The other possible case is

$$i(X_i^{(0)}) < i(Y_i^{(1)}) \leqslant s(X_i^{(0)}) < s(Y_i^{(1)});$$

that is,

$$X_i^{(1)} = [i(Y_i^{(1)}), s(X_i^{(0)})].$$

This case may be similarly treated.) Since $x^{(0)} \supseteq x^{(1)}$ we have

$$Y_i^{(2)} = \sum_{j=1}^{n} A_{ij}X_j^{(1)} + B_i \subseteq \sum_{j=1}^{n} A_{ij}X_j^{(0)} + B_i = Y_i^{(1)};$$

that is,

$$i(Y_i^{(1)}) \leqslant i(Y_i^{(2)}), \qquad s(Y_i^{(2)}) \leqslant s(Y_i^{(1)}) = s(X_i^{(1)}),$$

as well as

$$X_i^{(2)} = [\max\{i(Y_i^{(2)}), i(X_i^{(0)})\}, s(Y_i^{(2)})].$$

Since $x^{(k)} \supseteq x^{(k+1)}$ one shows using complete induction that

$$i(Y_i^{(k)}) \leqslant i(Y_i^{(k+1)}), \qquad s(Y_i^{(k+1)}) \leqslant s(Y_i^{(k)}) = s(X_i^{(k)}), \qquad k \geqslant 1;$$

that is,

$$X_i^{(k+1)} = [\max\{i(Y_i^{(k+1)}), i(X_i^{(0)})\}, s(Y_i^{(k+1)})].$$

By the assumption we have $i(X_i^{(0)}) < i(X_i^*)$ and from Theorem 1 we have $\lim_{k\to\infty} i(X_i^{(k)}) = i(X_i^*)$. There is therefore a $k_0 \geqslant 1$ such that

$$\max\{i(Y_i^{(k_0+1)}), i(X_i^{(0)})\} = i(Y_i^{(k_0+1)})$$

or

$$X_i^{(k_0+1)} = [i(Y_i^{(k_0+1)}), s(Y_i^{(k_0+1)})];$$

that is,

$$X_i^{(k_0+1)} = Y_i^{(k_0+1)} \cap X_i^{(k_0)} = Y_i^{(k_0+1)}$$

holds. By complete induction one now shows that

$$X_i^{(k_0+\nu)} = Y_i^{(k_0+\nu)} \cap X_i^{(k_0+\nu-1)} = Y_i^{(k_0+\nu)}, \qquad \nu \geqslant 1,$$

since $x^{(k)} \supseteq x^{(k+1)}$. Because of $\lim_{k\to\infty} x^{(k)} = x^*$ and $x^{(0)} \supset x^*$ we must have for each i, $1 \leqslant i \leqslant n$, not satisfying (a) or (b) that after a finite number of further iteration steps either (a) or (b) must hold. Finally one obtains

$$x^{(k+1)} = (\mathscr{A} x^{(k)} + b) \cap x^{(k)} = \mathscr{A} x^{(k)} + b, \qquad k \geqslant \tilde{k} \geqslant 0. \quad ■$$

Remarks: Most of the results of this chapter are proved in the papers given in the list of references. Corresponding results are also given in papers dealing with relaxation methods (see, in particular, Alefeld [2] and Mayer [10]). Analogous results for so-called semilinear systems of equations are given in Alefeld [3]. Theorem 5 was first proved in Alefeld [4].

The theorems of this chapter may be carried over word for word to the corresponding iteration methods for determining a fixed point of the nonlinear equation

$$x = f(x),$$

where

$$f(x) = (f_i(x_1, \ldots, x_n; a_{i1}, \ldots, a_{im_i}))$$

is a $\mathscr{P}$ contraction. The essential property for the proof of these theorems is the inclusion monotonicity.

Analogous theorems for complex point systems of equations using circular arithmetic were given in Platzöder [12] modifying the process of taking intersections suitably.

REFERENCES

[1] J. Albrecht, Monotone Iterationsfolgen und ihre Verwendung zur Lösung linearer Gleichungssysteme. *Numer. Math.* **3**, 345–358 (1961).

[2] G. Alefeld, Intervallrechnung über den komplexen Zahlen und einige Anwendungen. Ph.D. Thesis, Univ. Karlsruhe (1968).

[3] G. Alefeld, Über die Existenz einer eindeutigen Lösung bei einer Klasse nichtlinearer Gleichungssysteme. *Apl. Mat.* **17**, 267–294 (1972).

[4] G. Alefeld, Das symmetrische Einzelschrittverfahren bei linearen Gleichungen mit Intervallen als Koeffizienten. *Computing* **18**, 329–340 (1977).

[5] O. Mayer, Über die in der Intervallrechnung auftretenden Räume und einige Anwendungen. Ph.D. Thesis, Univ. Karlsruhe (1968).

[6] O. Mayer, Algebraische und metrische Strukturen in der Intervallrechnung und einige Anwendungen. *Computing* **5**, 144–162 (1970).

[7] O. Mayer, Über die Bestimmung von Einschließungsmengen für die Lösungen linearer Gleichungssysteme mit fehlerbehafteten Koeffizienten. *Elektron. Datenverarbeitung* **12**, 164–167 (1970).

[8] O. Mayer, Über die intervallmäßige Durchführung einiger Iterationsverfahren. *Z. Angew. Math. Mech.* **50**, T 65–T 66 (1970).

[9] O. Mayer, Über eine Klasse komplexer Intervallgleichungssysteme mit iterationsfähiger Gestalt. *Computing* **6**, 104–106 (1970).

[10] O. Mayer, Über intervallmäßige Iterationsverfahren bei linearen Gleichungssystemen und allgemeineren Intervallgleichungssystemen. *Z. Angew. Math. Mech.* **51**, 117–124 (1971).

[11] W. Niethammer, Relaxation bei komplexen Matrizen. *Math. Z.* **86**, 34–40 (1964).

[12] L. Platzöder, Über optimale Iterationsverfahren bei komplexen Gleichungssystemen. Diplomarbeit, Inst. Angew. Math., Univ. Karlsruhe (1974).

[13] J. W. Schmidt, Ausgangsvektoren für monotone Iterationen bei linearen Gleichungssystemen. *Numer. Math.* **6**, 78–88 (1964).

[14] J. Schröder, Computing error bounds in solving linear systems. *Math. Comp.* **16**, 323–337 (1962).

Chapter 15 / ON THE FEASIBILITY OF THE GAUSSIAN ALGORITHM FOR SYSTEMS OF EQUATIONS WITH INTERVALS AS COEFFICIENTS

Let $\mathcal{A}$ be an interval matrix and $\mathfrak{b}$ an interval vector. We assume that the inverse $\mathcal{A}^{-1}$ exists for all $\mathcal{A} \in \mathcal{A}$. We wish to find the set

$$\mathfrak{L} = \{x \mid \mathcal{A}x = \mathfrak{b}, \mathcal{A} \in \mathcal{A}, \mathfrak{b} \in \mathfrak{b}\}.$$

This set may not in general be described in a simple manner. We therefore restrict ourselves to the inclusion of this set by an interval vector. An obvious way of determining such an interval vector is to employ the direct generalization of the Gaussian algorithm for linear equations to systems with interval coefficients. This means that we start with the coefficient tableau

$$\begin{matrix} A_{11} & \cdots & A_{1n} & B_1 \\ \vdots & & \vdots & \vdots \\ A_{n1} & \cdots & A_{nn} & B_n \end{matrix} .$$

Using the formulas

$$\begin{aligned} A'_{1j} &= A_{1j}, & 1 \leqslant j \leqslant n, \qquad B'_1 = B_1, \\ A'_{ij} &= A_{ij} - A_{1j}(A_{i1}/A_{11}), & 2 \leqslant i, j \leqslant n, \\ B'_i &= B_i - B_1(A_{i1}/A_{11}), & 2 \leqslant i \leqslant n, \\ A'_{i1} &= 0, & 2 \leqslant i \leqslant n, \end{aligned}$$

assuming $0 \notin A_{11}$, we compute a new coefficient tableau

$$\begin{array}{ccccc} A'_{11} & A'_{12} & \cdots & A'_{1n} & B'_1 \\ 0 & A'_{22} & \cdots & A'_{2n} & B'_2 \\ \vdots & \vdots & & \vdots & \vdots \\ 0 & A'_{n2} & \cdots & A'_{nn} & B'_n \end{array}.$$

We now show that

$$\{\underset{\cdot}{\mathscr{x}} \mid \underset{\cdot}{\mathscr{A}}\underset{\cdot}{\mathscr{x}} = \underset{\cdot}{\mathscr{b}}, \underset{\cdot}{\mathscr{A}} \in \mathscr{A}, \underset{\cdot}{\mathscr{b}} \in \mathscr{b}\} \subseteq \{\underset{\cdot}{\mathscr{y}} \mid \underset{\cdot}{\mathscr{A}}'\underset{\cdot}{\mathscr{y}} = \underset{\cdot}{\mathscr{b}}', \underset{\cdot}{\mathscr{A}}' \in \mathscr{A}', \underset{\cdot}{\mathscr{b}}' \in \mathscr{b}'\}$$

is valid. Assume $\underset{\cdot}{\mathscr{A}} \in \mathscr{A}$ and $\underset{\cdot}{\mathscr{b}} \in \mathscr{b}$ and consider the system of linear equations

$$\underset{\cdot}{\mathscr{A}}\underset{\cdot}{\mathscr{x}} = \underset{\cdot}{\mathscr{b}}.$$

We form the matrix $\underset{\cdot}{\mathscr{A}}' = (a'_{ij})$ and the vector $\underset{\cdot}{\mathscr{b}}' = (b'_i)$, where

$$a'_{1j} = a_{1j}, \qquad 1 \leqslant j \leqslant n, \qquad b'_1 = b_1,$$

and

$$\begin{aligned} a'_{ij} &= a_{ij} - a_{1j}(a_{i1}/a_{11}), & 2 \leqslant i, j \leqslant n, \\ b'_i &= b_i - b_1(a_{i1}/a_{11}), & 2 \leqslant i \leqslant n, \\ a'_{i1} &= 0, & 2 \leqslant i \leqslant n. \end{aligned}$$

It is well known from the theory of linear equations that the system $\underset{\cdot}{\mathscr{A}}'\underset{\cdot}{\mathscr{y}} = \underset{\cdot}{\mathscr{b}}'$ of equations has the same solution as $\underset{\cdot}{\mathscr{A}}\underset{\cdot}{\mathscr{x}} = \underset{\cdot}{\mathscr{b}}$. From the inclusion monotonicity it follows that $\underset{\cdot}{\mathscr{A}}' \in \mathscr{A}'$ and $\underset{\cdot}{\mathscr{b}}' \in \mathscr{b}'$, which proves the assertion.

If the above step is carried out $n - 1$ times, then the original coefficient tableau is changed to an upper triangular form as seen below:

$$\begin{array}{ccccc} \tilde{A}_{11} & \tilde{A}_{12} & \cdots & \tilde{A}_{1n} & \tilde{B}_1 \\ & \tilde{A}_{22} & & & \\ & & \ddots & \vdots & \vdots \\ & & & \tilde{A}_{nn} & \tilde{B}_n \end{array},$$

for which

$$\{\underset{\cdot}{\mathscr{x}} \mid \underset{\cdot}{\mathscr{A}}\underset{\cdot}{\mathscr{x}} = \underset{\cdot}{\mathscr{b}}, \underset{\cdot}{\mathscr{A}} \in \mathscr{A}, \underset{\cdot}{\mathscr{b}} \in \mathscr{b}\} \subseteq \{\underset{\cdot}{\tilde{\mathscr{x}}} \mid \underset{\cdot}{\tilde{\mathscr{A}}}\underset{\cdot}{\tilde{\mathscr{x}}} = \underset{\cdot}{\tilde{\mathscr{b}}}, \underset{\cdot}{\tilde{\mathscr{A}}} \in \tilde{\mathscr{A}}, \underset{\cdot}{\tilde{\mathscr{b}}} \in \tilde{\mathscr{b}}\}$$

is valid. Using the formulas

$$\begin{aligned} X_n &= \tilde{B}_n/\tilde{A}_{nn}, \\ X_i &= \left(\tilde{B}_i - \sum_{j=i+1}^{n} \tilde{A}_{ij}X_j\right)\Big/\tilde{A}_{ii}, \qquad 1 \leqslant i \leqslant n - 1, \end{aligned}$$

one then obtains an interval vector $\mathscr{x} = (X_i)$ satisfying

$$\{\underset{\cdot}{\mathscr{x}} \mid \underset{\cdot}{\mathscr{A}}\underset{\cdot}{\mathscr{x}} = \underset{\cdot}{\mathscr{b}}, \underset{\cdot}{\mathscr{A}} \in \mathscr{A}, \underset{\cdot}{\mathscr{b}} \in \mathscr{b}\} \subseteq \mathscr{x}.$$

If $\mathcal{A} = (a_{ij})$ is in particular a nonsingular point matrix, then the Gaussian algorithm is feasible for each right-hand interval vector. It might then be necessary to interchange rows during the elimination process. This is equivalent to multiplying the matrix $\mathcal{A}$ with a permutation matrix on the left prior to the elimination process.

We now define a map

$$g: M_{nn}(\mathbb{C}) \times V_n(\mathbb{C}) \to V_n(\mathbb{C})$$

for a nonsingular point matrix $\mathcal{A}$ and a point vector b. This map represents the application of the Gaussian algorithm to the linear system of equations

$$\mathcal{A}x = b$$

resulting in

$$x = g(\mathcal{A}, b).$$

The map g is unique. As usual, however, there are different expressions for g. For example, we have $\mathcal{A}^{-1}b = g(\mathcal{A}, b)$. Moreover, the expression for g that one obtains from the Gaussian algorithm also depends on the choice of the pivots.

The following properties are independent of the choice of pivots. The interval evaluation of g is denoted by $g(\mathcal{A}, b)$. The interval vector x obtained by executing the Gaussian algorithm described above is therefore given by $x = g(\mathcal{A}, b)$.

We have the following properties:

(1) $$\mathcal{A}, \mathcal{B} \in M_{nn}(I(\mathbb{C})), \qquad a, b \in V_n(I(\mathbb{C})),$$
$$\mathcal{A} \subseteq \mathcal{B}, \qquad a \subseteq b.$$

It then follows that

$$g(\mathcal{A}, a) \subseteq g(\mathcal{B}, b).$$

(2) $$\mathcal{A} \in M_{nn}(\mathbb{C}), \qquad b = u + v \in V_n(I(\mathbb{C})).$$

It then follows that

$$g(\mathcal{A}, b) = g(\mathcal{A}, u) + g(\mathcal{A}, v).$$

(3) $$\mathcal{A} \in M_{nn}(\mathbb{R}), \qquad b \in V_n(I(\mathbb{R})).$$

It then follows that

$$\mathcal{A}^{-1}b \subseteq g(\mathcal{A}, b).$$

(4) $\mathcal{A} \in M_{nn}(\mathbb{C})$, $\quad a, b \in V_n(I(\mathbb{C}))$, $\quad d(a) \leqslant \alpha\, d(b) \quad$ for some $\quad \alpha \geqslant 0$.

It then holds for the width that

$$d(g(\mathcal{A}, a)) \leqslant \alpha\, d(g(\mathcal{A}, b)).$$

In order to prove these properties, we remark that (1) follows immediately from inclusion monotonicity, and (2) follows using the relation $a(B + C) = aB + aC$ as well as the formulas for the Gaussian algorithm. The following fact is used to prove (3) (see also the discussion in Chapter 3):

If we have two rational expressions f_1 and f_2 for the function $f: \mathbb{R} \to \mathbb{R}$ with the property that f_1 contains the variable x exactly once and f_2 contains the variable m times, then we have for the interval evaluations of f_1 and f_2 that $f_1(X) \subseteq f_2(X)$. This also holds in the same manner for functions of several variables. Consider now the ith component, $1 \leqslant i \leqslant n$, of $\mathscr{A}^{-1}\ell$ and $\mathscr{g}(\mathscr{A}, \ell)$. For point vectors it holds that $(\mathscr{A}^{-1}\ell)_i = (\mathscr{g}(\mathscr{A}, \ell))_i$. From the formulas of the Gaussian algorithm it follows that the components of ℓ may occur several times in $(\mathscr{g}(\mathscr{A}, \ell))_i$. Since the components of ℓ only occur once in $(\mathscr{A}^{-1}\ell)_i$, (3) follows.

Finally (4) follows by using the formula for the Gaussian algorithm and the rules (2.10), (2.14), (6.12), and (6.16) as well as the assumption $d(a) \leqslant \alpha\, d(\ell)$.

The formulas for the Gaussian algorithm are only applicable to an interval matrix when we have $0 \notin A_{ii}$ in each of the $n - 1$ steps required to form upper triangular form. If this is not true and $0 \in A_{ii}$ for some i, $1 \leqslant i \leqslant n - 1$, then it might still be possible to rearrange the matrix by exchanging rows in such a manner that $0 \in A_{ii}$ is avoided. This is not always possible even if we assume that initially $\mathscr{A}^{-1}$ exists for all $\mathscr{A} \in \mathscr{A}$. The reason is as follows:

It is clear that one can always perform the first step of the Gaussian algorithm under the assumption that $\mathscr{A}^{-1}$ exists for all $\mathscr{A} \in \mathscr{A}$. If we were not able to perform this step because all members of the first column contained zero, then this would imply that the original matrix $\mathscr{A}$ contained a singular matrix $\mathscr{A}$ contradicting the assumptions. Let us now consider the $(n - 1) \times (n - 1)$ matrix $\mathscr{A}^{(1)} = (A_{ij}^{(1)})$ for which $A_{ij}^{(1)} = A'_{ij}$, $2 \leqslant i, j \leqslant n$. Let $\mathscr{U}$ be the $(n - 1) \times n$ interval matrix

$$\mathscr{U} = \begin{pmatrix} -A_{21}/A_{11} & 1 & & & 0 \\ -A_{31}/A_{11} & 0 & 1 & & \\ \vdots & & & \ddots & \\ -A_{n1}/A_{11} & 0 & \cdots & 0 & 1 \end{pmatrix},$$

and let $\mathscr{V}$ be the $n \times (n - 1)$ interval matrix

$$\mathscr{V} = \begin{pmatrix} A_{12} & \cdots & A_{1n} \\ \vdots & & \vdots \\ A_{n2} & \cdots & A_{nn} \end{pmatrix}.$$

We have $\mathscr{A}^{(1)} = \mathscr{U}\mathscr{V}$. Correspondingly let

$$\mathscr{R} = \begin{pmatrix} -a_{21}/a_{11} & 1 & & & 0 \\ -a_{31}/a_{11} & 0 & 1 & & \\ \vdots & & & \ddots & \\ -a_{n1}/a_{11} & 0 & \cdots & 0 & 1 \end{pmatrix},$$

$$\mathscr{S} = \begin{pmatrix} a_{12} & \cdots & a_{1n} \\ \vdots & & \vdots \\ a_{n2} & \cdots & a_{nn} \end{pmatrix}$$

for $a_{ij} \in A_{ij}$, $1 \leqslant i, j \leqslant n$. Then it follows that $\mathscr{Z} = \mathscr{R}\mathscr{S}$ is nonsingular since $\mathscr{A} \in \mathscr{A}$ is nonsingular. There is therefore at least a nonzero member of the first column of $\mathscr{Z}$. From inclusion monotonicity it follows that

$$\{\mathscr{Z} = \mathscr{R}\mathscr{S} \mid \mathscr{A} \in \mathscr{A}\} \subseteq \{\mathscr{A}^{(1)} \mid \mathscr{A}^{(1)} \in \mathscr{A}^{(1)} = \mathscr{U}\mathscr{V}\},$$

where the equality does not hold in general. This was shown by a simple example in Chapter 10. It is therefore not guaranteed that the first column of $\mathscr{A}^{(1)}$ will have at least one member not containing zero. The same considerations may be carried out for the remaining steps of the algorithm. We demonstrate this using a simple example as follows.

Example: The following interval matrix

$$\mathscr{A} = \begin{pmatrix} [1,5] + i[-1,1] & 1 \\ 25 & [-1,1] + i[-1,1] \end{pmatrix}$$

demonstrates the above. First we show that $\mathscr{A}$ does not contain a singular point matrix $\mathscr{A}$. Since

$$\det \mathscr{A} = a_{11}a_{22} - a_{21}a_{12} \in A_{11}A_{22} - A_{21}A_{12},$$

we get for $\mathscr{A} \in \mathscr{A}$ that

$$\det \mathscr{A} \in [-31, -19] + i[-6, 6].$$

Since the interval on the right-hand side does not contain zero it follows that $\mathscr{A}$ is nonsingular.

The elimination step of the Gaussian algorithm now generates the interval

$$A'_{22} = [-126, \tfrac{1}{26}] + i[-26, 26]$$

using Definition 5.3 and the remarks following this definition.

The Gaussian algorithm can therefore not be continued. Even if one decomposes the matrix $\mathscr{A}$ into its real and imaginary parts the 4×4 real system of interval equations is still not solvable using the Gaussian algorithm even though this starting matrix also contains no singular point matrices.

We shall now show that if $n \leqslant 2$ then the Gaussian algorithm can always be carried out when the coefficients are real intervals and all the point matrices $\mathscr{A} \in \mathscr{A}$ are nonsingular.

Theorem 1: Let $1 \leqslant n \leqslant 2$ and assume that the $n \times n$ interval matrix $\mathscr{A} = (A_{ij})$ does not contain a singular matrix $\mathscr{A}$. Then the Gaussian algorithm can be carried out.

Proof: For $n = 1$ the assumption is that $0 \notin A_{11}$, and $\mathscr{A} = (A_{11})$ therefore proves the theorem for this case. For $n = 2$ at least one of the intervals A_{11} and A_{21} must not contain zero. If this were not the case, then there would exist a singular matrix $\underset{\cdot}{\mathscr{A}} \in \mathscr{A}$ contradicting the assumption of the theorem. We assume $0 \notin A_{11}$ without loss of generality. If this is not the case, one may exchange the rows of $\mathscr{A}$. The Gaussian algorithm then gives

$$A'_{22} = A_{22} - (1/A_{11})A_{21}A_{12}.$$

We may consider A'_{22} as being the interval arithmetic evaluation (see Chapter 3) of the rational function a'_{22} of the four variables a_{11}, a_{21}, a_{12}, and a_{22}, defined by

$$a'_{22}(a_{11}, a_{12}, a_{21}, a_{22}) = a_{22} - (1/a_{11})a_{21}a_{12}.$$

From the assumption we have for all $\underset{\cdot}{\mathscr{A}} \in \mathscr{A}$ that

$$\det(\underset{\cdot}{\mathscr{A}}) = a_{11}a_{22} - a_{21}a_{12} \neq 0,$$

and furthermore

$$a'_{22}(a_{11}, a_{12}, a_{21}, a_{22}) = (1/a_{11})\det(\underset{\cdot}{\mathscr{A}}) \neq 0.$$

The interval evaluation will give the exact range when a_{11} is replaced by A_{11}, a_{12} by A_{12}, a_{21} by A_{21}, and a_{22} by A_{22} since each of the variables occurs only once in the expression for a'_{22} (see also the discussion prior to Theorem 3.2). We therefore have $0 \notin A'_{22}$, which means that the Gaussian algorithm can be carried out. ■

The above proof cannot be extended to the case $n \geqslant 3$. Even for the case $n = 2$ the theorem is false if the elements of the interval matrices are from $R(\mathbb{C})$ or $K(\mathbb{C})$.

We now wish to consider a particular class of interval matrices in which the Gaussian algorithm can always be carried out. In this class one may even dispense with the exchanging of rows or of columns. We limit ourselves in the following to systems of equations where the coefficient matrix as well as the constant vector on the right-hand side have elements from the set of real intervals or from the set of circular complex disks. Systems of equations where the elements are rectangular complex intervals can be reduced to the first case by decomposition into real and imaginary parts.

In order to unify the following proofs for real and circular complex intervals we note that a real interval of the form $A = [a_1, a_2]$ may also be written in the form

$$A = [a - r, a + r]$$

where

$$a = \tfrac{1}{2}(a_1 + a_2), \qquad r = \tfrac{1}{2}d(A) = \tfrac{1}{2}(a_2 - a_1).$$

Here a is the center of the interval and r the radius or half width of the interval. The notation

$$A = [a - r, a + r] =: \langle a, r\rangle$$

is therefore introduced in order to show the analogy to circular complex intervals. The arithmetic operations on real intervals may also be written in terms of center and width. Let $A = \langle a, r\rangle$, $B = \langle b, s\rangle$. The addition and subtraction of A and B may be written

$$A \pm B = \langle a \pm b, r + s\rangle.$$

This corresponds formally to the addition and subtraction of circular complex intervals. For the multiplication we only need the equation

$$[-r, r][-s, s] = \langle 0, r\rangle\langle 0, s\rangle = \langle 0, rs\rangle.$$

If we now note that the representation

$$\frac{1}{A} = \left[\frac{1}{a+r}, \frac{1}{a-r}\right] = \left[\frac{a}{a^2 - r^2} - \frac{r}{a^2 - r^2}, \frac{a}{a^2 - r^2} + \frac{r}{a^2 - r^2}\right]$$

is valid, if $0 \notin A = [a_1, a_2]$, then we get

$$\frac{1}{A} = \left\langle \frac{a}{a^2 - r^2}, \frac{r}{a^2 - r^2}\right\rangle.$$

This corresponds formally to the reciprocal of a circular complex interval. With the representation $A = \langle a, r\rangle$ then it also holds that for real intervals

$$|A| = \max\{|a_1|, |a_2|\} = |a| + r.$$

Furthermore we have

$$0 \notin A \Leftrightarrow |a| - r > 0.$$

Finally we have

$$A = \langle a, r\rangle \subseteq \langle 0, |A|\rangle = \langle 0, |a| + r\rangle.$$

We first prove

Lemma 2: Let $A = \langle a, r_1\rangle$, $B = \langle b, r_2\rangle$, $C = \langle c, r_3\rangle$, and $D = \langle d, r_4\rangle$ be real intervals or circular complex intervals and assume $0 \notin D$. Then for

$$Z = \langle z, r_5\rangle = A - (1/D)BC$$

the inequality

$$|a| - r_1 - |B|\,|C|/(|d| - r_4) \leqslant |z| - r_5$$

is valid.

Proof: From inclusion monotonicity it follows that

$$Z = \langle z, r_5 \rangle = A - BC\frac{1}{D} \subseteq A - \langle 0, |B| \rangle \langle 0, |C| \rangle \left\langle \frac{\bar{d}}{d\bar{d} - r_4^2}, \frac{r_4}{d\bar{d} - r_4^2} \right\rangle$$

$$= \langle a, r_1 \rangle - \left\langle 0, |B|\,|C| \frac{|d|}{|d\bar{d} - r_4^2|} + |B|\,|C| \frac{r_4}{d\bar{d} - r_4^2} \right\rangle$$

$$= \langle a, r_1 \rangle - \left\langle 0, |B|\,|C| \frac{1}{|d| - r_4} \right\rangle$$

$$= \left\langle a, r_1 + |B|\,|C| \frac{1}{|d| - r_4} \right\rangle =: \langle a, r_6 \rangle.$$

Since $Z \subseteq \langle a, r_6 \rangle$ it follows that

$$|a| - |z| \leqslant |a - z| \leqslant r_6 - r_5,$$

or

$$|z| - r_5 \geqslant |a| - r_1 - |B|\,|C|(1/(|d| - r_4)). \quad \blacksquare$$

In order to formulate the next statements we need the concept of an M matrix first introduced by Ostrowski [6]. We shall here use the equivalent definition from Varga [10]. A real matrix $\mathscr{B} = (b_{ij})$ is called an M matrix when the conditions

(1) $$b_{ij} \leqslant 0, \qquad i \neq j,$$

(2) $$\mathscr{B}^{-1} \geqslant \mathscr{O}$$

are satisfied. The partial order in (2) is defined componentwise. The condition (2) may be replaced by (see Fan [3])

(2′) There exists a real vector $u = (u_i)$ with $u_i > 0$, $1 \leqslant i \leqslant n$, such that $\mathscr{B}u > o$.

This fact as well as the fact that an M matrix has positive diagonal elements is used in the sequel.

Theorem 3: Let $\mathscr{A} = (A_{ij})$ with $A_{ij} = \langle a_{ij}, r_{ij} \rangle$, $1 \leqslant i, j \leqslant n$, be an interval matrix and let $\mathscr{B} = (b_{ij})$ be the real matrix defined by

$$b_{ij} = \begin{cases} |a_{ii}| - r_{ii}, & i = j \\ -|A_{ij}| & \text{otherwise.} \end{cases}$$

If $\mathscr{B}$ is an M matrix, then the Gaussian algorithm may be carried out for the interval matrix $\mathscr{A}$ without row or column exchanges.

Proof: Since $\mathscr{B}$ is an M matrix then by assumption there exists a vector $u = (u_i)$ with positive elements so that $\mathscr{B}u > o$; that is,

$$(|a_{ii}| - r_{ii})u_i > \sum_{j=1, j\neq i}^{n} |A_{ij}|u_j, \qquad 1 \leqslant i \leqslant n,$$

holds. Since $|a_{11}| - r_{11} > 0$ the condition $0 \notin A_{11}$ is satisfied and the formulas in the first part of this chapter are applicable. We now show that the conditions of the theorem are satisfied for the $(n-1) \times (n-1)$ interval matrix $\tilde{\mathscr{A}}' = (\tilde{A}'_{ij})$ with $\tilde{A}'_{ij} = A'_{ij} = \langle a'_{ij}, r'_{ij} \rangle$, $2 \leqslant i, j \leqslant n$. This would then constitute a proof for the theorem by complete induction.

For $i \geqslant 2$ it holds that

$$\sum_{j=2, j\neq i}^{n} |A'_{ij}|u_j = \sum_{j=2, j\neq i}^{n} \left| A_{ij} - A_{1j}\frac{A_{i1}}{A_{11}} \right| u_j$$
$$\leqslant \sum_{j=2, j\neq i}^{n} |A_{ij}|u_j + |A_{i1}| \left| \frac{1}{A_{11}} \right| \sum_{j=2, j\neq i}^{n} |A_{1j}|u_j.$$

The inequality

$$\sum_{j=2, j\neq i}^{n} |A_{1j}|u_j < (|a_{11}| - r_{11})u_1 - |A_{1i}|u_i,$$

valid from the assumptions of the theorem, may be used to estimate

$$\sum_{j=2, j\neq i}^{n} |A'_{ij}|u_j \leqslant \sum_{j=2, j\neq i}^{n} |A_{ij}|u_j + |A_{i1}| \frac{1}{|a_{11}| - r_{11}} \{(|a_{11}| - r_{11})u_1 - |A_{1i}|u_i\}$$
$$= \sum_{j=1, j\neq i}^{n} |A_{ij}|u_j - \frac{|A_{i1}||A_{1i}|}{|a_{11}| - r_{11}} u_i$$
$$< u_i \left(|a_{ii}| - r_{ii} - \frac{|A_{i1}||A_{1i}|}{|a_{11}| - r_{11}} \right)$$
$$\leqslant (|a'_{ii}| - r'_{ii})u_i,$$

where the last inequality holds because of Lemma 2. This finishes the proof. ■

An important class of interval matrices satisfying the conditions of Theorem 3 is given by the following

Definition 4: An interval matrix $\mathscr{A} = (A_{ij})$, whose components $A_{ij} = \langle a_{ij}, r_{ij} \rangle$ are either real intervals or circular complex intervals, is called a strictly diagonally dominant matrix if

$$|a_{ii}| - r_{ii} > \sum_{j=1, j\neq i}^{n} |A_{ij}|, \qquad 1 \leqslant i \leqslant n. \quad ■$$

This definition implies immediately that all the diagonal elements do not contain zero and that for every point matrix $\hat{\mathscr{A}} = (\hat{a}_{ij}) \in \mathscr{A}$ the relation

$$|\hat{a}_{ii}| > \sum_{j=1, j \neq i}^{n} |\hat{a}_{ij}|, \qquad 1 \leqslant i \leqslant n,$$

is valid. Each point matrix $\hat{\mathscr{A}} \in \mathscr{A}$ is therefore a strictly diagonally dominant matrix in the usual sense and therefore nonsingular (see, for example, Varga [10]).

The conditions of Theorem 3 are satisfied for a strictly diagonally dominant interval matrix if the vector $\mathscr{u} = (u_i)$ is chosen such that $u_i = 1$, $1 \leqslant i \leqslant n$. We therefore have

Corollary 5: Let $\mathscr{A}$ be a strictly diagonally dominant interval matrix. Then the Gaussian algorithm can be carried out for the interval matrix $\mathscr{A}$ without row or column interchanges. ■

The condition of strict diagonal dominance for the feasibility of the Gaussian algorithm may be relaxed if the given interval matrix has the form

$$\mathscr{A} = \begin{pmatrix} A_1 & C_1 & & & 0 \\ B_2 & A_2 & C_2 & & \\ & \ddots & \ddots & \ddots & \\ 0 & & & B_n & A_n \end{pmatrix};$$

that is, the matrix is a tridiagonal interval matrix. We furthermore assume that $C_i \neq 0$, $1 \leqslant i \leqslant n-1$, and $B_i \neq 0$, $2 \leqslant i \leqslant n$. Otherwise the problem is decomposed into smaller partial problems for which these conditions are fulfilled.

Theorem 6: Let $\mathscr{A}$ be a tridiagonal interval matrix such that

$$\begin{aligned} A_i &= \langle a_i, r_i \rangle, && 1 \leqslant i \leqslant n, \\ B_i &= \langle b_i, s_i \rangle \neq 0, && 2 \leqslant i \leqslant n, \\ C_i &= \langle c_i, t_i \rangle \neq 0, && 1 \leqslant i \leqslant n-1. \end{aligned}$$

Furthermore assume

$$\begin{aligned} |a_1| - r_1 &> |C_1|, \\ |a_i| - r_i &\geqslant |B_i| + |C_i|, \qquad 2 \leqslant i \leqslant n-1, \\ |a_n| - r_n &> |B_n|. \end{aligned}$$

The Gaussian algorithm may then be carried out without row or column interchanges.

Remark: The tridiagonal matrix $\mathscr{A}$ only satisfies the conditions of Definition 4 in the first and last row.

Proof of Theorem 6: The first step in the Gaussian algorithm consists in generating a tridiagonal matrix $\mathscr{A}'$ for which

$$
\begin{aligned}
&A_1' = A_1, \qquad C_1' = C_1, && \\
&B_2' = 0, \qquad B_i' = B_i, && 3 \leqslant i \leqslant n, \\
&A_2' = A_2 - C_1 B_2 (1/A_1), \qquad A_i' = A_i, && 3 \leqslant i \leqslant n, \\
&C_i' = C_i, && 2 \leqslant i \leqslant n-1.
\end{aligned}
$$

We show that the matrix $\mathscr{A}'$ satisfies the strict row sum criterion in the second as well as in the first row, that is, that

$$|a_2'| - r_2' > |C_2| = |C_2'|$$

holds. We have

$$
\begin{aligned}
A_2' &= A_2 - C_1(B_2/A_1) \subseteq A_2 - |1/A_1|\langle 0, |C_1|\rangle\langle 0, |B_2|\rangle \\
&= \langle a_2, r_2 + |C_1|\,|B_2|/(|a_1| - r_1)\rangle;
\end{aligned}
$$

that is,

$$|a_2'| - r_2' \geqslant |a_2| - \left(r_2 + \frac{|C_1|\,|B_2|}{|a_1| - r_1}\right).$$

Since

$$|a_1| - r_1 > |C_1| > 0, \qquad |B_2| > 0,$$

and

$$-|B_2| - r_2 + |a_2| \geqslant |C_2|,$$

it follows that

$$
\begin{aligned}
|a_2'| - r_2' &\geqslant |a_2| - \left(r_2 + \frac{|C_1|\,|B_2|}{|a_1| - r_1}\right) \\
&> -|B_2| - r_2 + |a_2| \geqslant |C_2| = |C_2'|.
\end{aligned}
$$

After $n-1$ steps of this type one arrives at a matrix $\hat{\mathscr{A}}$ that has nonzero elements only on the main diagonal and the super diagonal and where all the main diagonal elements do not contain zero. The third assumption $|a_n| - r_n > |B_n|$ is used in the $(n-1)$th step. ■

We remark that the proof of Theorem 6 may be shortened somewhat. The assumptions of Theorem 6 imply that the matrix $\mathscr{B}$ defined in Theorem 3 is irreducibly diagonally dominant (see Varga [10], Definition 1.7) and therefore an *M* matrix. This means that the statement is simply a special case of Theorem 3.

We now consider a system in a form suitable for iteration,

$$x = \mathscr{C}x + c,$$

with a real interval matrix $\mathscr{C} = (C_{ij})$, $c_{ij} = \langle c_{ij}, r_{ij}\rangle$, $1 \leqslant i,\ j \leqslant n$, and a real interval vector c.

The iteration method

$$x^{(k+1)} = \mathscr{C}x^{(k)} + c, \qquad k = 0, 1, 2, \ldots,$$

converges according to Theorem 12.1 for each interval vector $x^{(0)}$ to the unique fixed point x^* of the equation $x = \mathscr{C}x + x$ iff the spectral radius of the real matrix $|\mathscr{C}| = (|C_{ij}|)$ is smaller than one. We wish to show that the assumptions of Theorem 3 are always fulfilled for the matrix $\mathscr{A} = \mathscr{I} - \mathscr{C} = (A_{ij})$ where $\mathscr{I}$ is the unit matrix.

We have

$$A_{ij} = \begin{cases} \langle 1 - c_{ii}, r_{ii}\rangle, & i = j \\ -\ C_{ij} & \text{otherwise.} \end{cases}$$

From $\rho(|\mathscr{C}|) < 1$ it follows that $|C_{ii}| = |c_{ii}| + r_{ii} < 1$, $1 \leqslant i \leqslant n$. The matrix $\mathscr{B} = (b_{ij})$ defined in Theorem 3 has the elements

$$b_{ij} = \begin{cases} |1 - c_{ii}| - r_{ii}, & i = j \\ -\ |C_{ij}| & \text{otherwise,} \end{cases}$$

when $\mathscr{A} = \mathscr{I} - \mathscr{C}$.

We now consider the real matrix $\mathscr{B}_1 = \mathscr{I} - |\mathscr{C}|$. Since $\rho(|\mathscr{C}|) < 1$ the inverse $\mathscr{B}_1^{-1}$ exists from Theorem 3.8 in Varga [10], and we have $\mathscr{B}_1^{-1} \geqslant \mathscr{O}$. Consider furthermore the decomposition $\mathscr{B}_1 = \mathscr{M}_1 - \mathscr{N}_1$ of $\mathscr{B}_1$ where

$$\mathscr{M}_1 = \operatorname{diag}(1 - |C_{ii}|), \qquad \mathscr{N}_1 = -(\mathscr{B}_1 - \mathscr{M}_1) = \mathscr{M}_1 - \mathscr{B}_1.$$

We have $\mathscr{M}_1^{-1} \geqslant \mathscr{O}$, $\mathscr{N}_1 \geqslant \mathscr{O}$. From this it follows that $\rho(\mathscr{M}_1^{-1}\mathscr{N}_1) < 1$ according to Theorem 3.13 in Varga [10].

Finally consider the decomposition $\mathscr{B} = \mathscr{M} - \mathscr{N}$ of $\mathscr{B}$ with

$$\mathscr{M} = \operatorname{diag}(|1 - c_{ii}| - r_{ii}), \qquad \mathscr{N} = -(\mathscr{B} - \mathscr{M}) = \mathscr{M} - \mathscr{B}.$$

We have

$$|1 - c_{ii}| - r_{ii} \geqslant 1 - |c_{ii}| - r_{ii} = 1 - |C_{ii}| > 0, \qquad 1 \leqslant i \leqslant n,$$

and therefore

$$\mathscr{M} \geqslant \mathscr{M}_1; \qquad \text{that is,} \quad \mathscr{M}_1^{-1} \geqslant \mathscr{M}^{-1}.$$

Since $\mathscr{N} = \mathscr{N}_1$ it follows that $\mathscr{M}^{-1}\mathscr{N} \leqslant \mathscr{M}_1^{-1}\mathscr{N}_1$.

The theorem of Perron–Frobenius ([10, Theorem 2.7]) now implies that $\rho(\mathscr{M}^{-1}\mathscr{N}) \leqslant \rho(\mathscr{M}_1^{-1}\mathscr{N}_1)$; that is, $\rho(\mathscr{M}^{-1}\mathscr{N}) < 1$. Using Theorem 3.10 of [10] we finally obtain the result that $\mathscr{B}$ is an M matrix.

Remarks: The statements about the feasibility of the Gaussian algorithm (Theorem 3) may be found in Alefeld [1]. The proof of Theorem 3 is a direct generalization of the corresponding proof for the same statements for point matrices. (See Wilkinson [11].) Let $\mathscr{A}^{\mathrm{T}}$ be the transposed interval matrix of $\mathscr{A}$. Corresponding to Corollary 5 we then have: If $\mathscr{A}^{\mathrm{T}}$ is strictly diagonally dominant (in the sense of Definition 4), then the Gaussian algorithm may be carried out on the interval matrix $\mathscr{A}$ without row interchanges. The proof follows as a special case of Theorem 3 since the matrix $\mathscr{B}$ defined there is again an M matrix. The analogous statement is valid for a tridiagonal matrix $\mathscr{A}$ if the assumptions of Theorem 6 are satisfied for $\mathscr{A}^{\mathrm{T}}$.

In Chapter 19 we give some applications where the statements of this section are used to solve systems of nonlinear point equations.

The question of the feasibility of the Gaussian algorithm has so far not been investigated in a satisfactory manner. In [8], Reichmann showed that it is possible to carry out the algorithm for a particular class of equations.

The 2×2 example of a complex interval matrix where the Gaussian algorithm fails, was given by Platzöder [7]. A class of 3×3 interval matrices which contain no singular point matrices and for which the Gaussian algorithm fails for every choice of pivots was given by Reichmann [9]. An explicit description of the set

$$\mathfrak{L} = \{x \mid \mathscr{A}x = b, \mathscr{A} \in \mathscr{A}, b \in \mathscr{b}\}$$

defined at the beginning of this section was recently given in Hartfiel [4].

REFERENCES

[1] G. Alefeld, Über die Durchführbarkeit des Gaußschen Algorithmus bei Gleichungen mit Intervallen als Koeffizienten. *Computing Suppl.* **1**, 15–19 (1977).

[2] J. W. Barth and E. Nuding, Optimale Lösung von Intervallgleichungssystemen. *Computing* **12**, 117–125 (1974).

[3] K. Fan, Topological proof for certain theorems on matrices with nonnegative elements. *Monatsh. Math.* **62**, 219–237 (1958).

[4] D. J. Hartfiel, Concerning the solution $Ax = b$ when $P \leqslant A \leqslant Q$ and $p \leqslant b \leqslant q$. *Numer. Math.* **35**, 355–359 (1980).

[5] M. Hebgen, Eine scaling-invariante Pivotsuche für Intervallmatrizen. *Computing* **12**, 99–106 (1974).

[6] A. M. Ostrowski, Über die Determinanten mit überwiegender Hauptdiagonale. *Comment. Math. Helv.* **10**, 69–96 (1937).

[7] L. Platzöder, Unpublished manuscript (1979).

[8] K. Reichmann, Ein hinreichendes Kriterium für die Durchführbarkeit des Intervall-Gauß-Algorithmus bei Intervall-Hessenberg-Matrizen ohne Pivotsuche. *Z. Angew. Math. Mech.* **59**, 373–379 (1979).

[9] K. Reichmann, Abbruch beim Intervall-Gauß-Algorithmus. *Computing* **22**, 355–361 (1979).

[10] R. Varga, "Matrix Iterative Analysis." Prentice-Hall, Englewood Cliffs, New Jersey, 1962.

[11] J. H. Wilkinson, Error analysis of direct methods of matrix inversion. *J. Assoc. Comput. Mach.* **8**, 281–330 (1960).

Chapter 16 / **HANSEN'S METHOD**

Systems of interval equations that are not strongly diagonally dominant (Definition 15.4) can be solved using a transformation suggested by Hansen. This transformation tries to transform the system of interval equations into a strongly diagonally dominant system. Let $\mathscr{A} = (A_{ij})$ where $A_{ij} = \langle a_{ij}, r_{ij} \rangle$ be the given interval matrix with elements from $I(\mathbb{R})$ or $K(\mathbb{C})$. Assume furthermore that the inverse of all point matrices $\underset{\cdot}{\mathscr{A}} \in A$ exists. The point matrix $m(\mathscr{A}) := (a_{ij})$ is then inverted and the interval matrix

$$\tilde{\mathscr{A}} = m(\mathscr{A})^{-1}\mathscr{A}$$

as well as the interval vector

$$\tilde{\mathscr{b}} = m(\mathscr{A})^{-1}\mathscr{b}$$

is formed using $m(\mathscr{A})^{-1}$. We then have

$$\{\underset{\cdot}{x} \mid \underset{\cdot}{\mathscr{A}}\underset{\cdot}{x} = \underset{\cdot}{b}, \underset{\cdot}{\mathscr{A}} \in \mathscr{A}, \underset{\cdot}{b} \in \mathscr{b}\} \subseteq \{\underset{\cdot}{y} \mid \underset{\cdot}{\tilde{\mathscr{A}}}\underset{\cdot}{y} = \underset{\cdot}{\tilde{b}}, \underset{\cdot}{\tilde{\mathscr{A}}} \in \tilde{\mathscr{A}}, \underset{\cdot}{\tilde{b}} \in \tilde{\mathscr{b}}\}.$$

To show this, suppose $\underset{\cdot}{x}$ is a member of the left-hand set; that is,

$$\underset{\cdot}{\mathscr{A}}\underset{\cdot}{x} = \underset{\cdot}{b} \qquad \text{with} \quad \underset{\cdot}{\mathscr{A}} \in \mathscr{A}, \quad \underset{\cdot}{b} \in \mathscr{b}.$$

Then

$$m(\mathscr{A})^{-1}\underset{\cdot}{\mathscr{A}}\underset{\cdot}{x} = m(\mathscr{A})^{-1}\underset{\cdot}{b}$$

and because of

$$m(\mathscr{A})^{-1}\underset{\cdot}{\mathscr{A}} \in \tilde{\mathscr{A}}, \qquad m(\mathscr{A})^{-1}\underset{\cdot}{b} \in \tilde{\mathscr{b}}$$

the claim follows.

The idea behind this transformation is that if the widths of the elements of $\mathscr{A}$ are not too large then the matrix $\tilde{\mathscr{A}}$ will be strongly diagonally dominant. The Gaussian algorithm may then be applied. In the limit we have $d(\mathscr{A}) = \mathscr{O}$, that is, $\tilde{\mathscr{A}} = \mathscr{I}$, and this shows that $\tilde{\mathscr{A}}$ is certainly strongly diagonally dominant in this case. If the widths of the components of $\mathscr{A}$ are small, then the matrix $\tilde{\mathscr{A}}$ does not differ a great deal from $\mathscr{I}$.

Clearly the question of strong diagonal dominance for the matrix $\tilde{\mathscr{A}} = m(\mathscr{A})^{-1}\mathscr{A}$ does not only depend on the widths of the components of $\mathscr{A}$. In fact, if one represents the components of the interval matrix (these may be real or circular complex intervals) as

$$A_{ij} = \langle a_{ij}, r_{ij} \rangle, \qquad 1 \leqslant i, j \leqslant n,$$

then one obtains with $\mathscr{D} = (D_{ij})$, $D_{ij} = \langle 0, r_{ij} \rangle$, $1 \leqslant i, j \leqslant n$, that

$$\begin{aligned} \tilde{\mathscr{A}} = m(\mathscr{A})^{-1}\mathscr{A} &= m(\mathscr{A})^{-1}(m(\mathscr{A}) + \mathscr{D}) \\ &= \mathscr{I} + m(\mathscr{A})^{-1}\mathscr{D} = \mathscr{I} + |m(\mathscr{A})^{-1}|\mathscr{D} \\ &= \mathscr{I} + \mathscr{H}, \end{aligned}$$

where

$$\mathscr{H} = |m(\mathscr{A})^{-1}|\mathscr{D}.$$

Since

$$\begin{aligned} \| |\mathscr{H}| \| \leqslant \| |m(\mathscr{A})^{-1}| \| \cdot \| |\mathscr{D}| \| &= \tfrac{1}{2} \| |m(\mathscr{A})^{-1}| \| \, \|d(\mathscr{A})\| \\ &= \tfrac{1}{2} \| |m(\mathscr{A})^{-1}| \| \cdot \|m(\mathscr{A})\| \cdot \|d(\mathscr{A})\| / \|m(\mathscr{A})\|, \end{aligned}$$

it follows that for a given point matrix $m(\mathscr{A})$ the matrix $\mathscr{A}$ is more likely strongly diagonally dominant the smaller the quantity

$$\tilde{\kappa} = \| |m(\mathscr{A})^{-1}| \| \, \|m(\mathscr{A})\|.$$

If $m(\mathscr{A})$ is a real point matrix and if one uses a monotonic matrix norm, then one has

$$\tilde{\kappa} = \|m(\mathscr{A})^{-1}\| \, \|m(\mathscr{A})\|$$

where $\tilde{\kappa}$ is the well-known condition number of the matrix $m(\mathscr{A})$. The feasibility of Hansen's method for real interval matrices depends therefore not only on the magnitude of $d(\mathscr{A})$ but also critically on the condition number of $m(\mathscr{A})$.

In the last chapter we described a method for transforming a set of linear interval equations to upper triangular form. Other methods from the theory of real linear equation systems may also be extended to linear interval equation systems. We mention the Gauss–Jordan method where the given matrix is transformed to a diagonal matrix. The calculation of the solution of the original system then requires n additional divisions. A detailed description of this version of the Gaussian algorithm is found in Isaacson and Keller [3]. If one uses this method for a system of linear interval equations, then one may show as in Theorem 15.3 that the method is always feasible if the given interval matrix is strongly diagonally dominant.

Let a real interval matrix $\mathscr{A} = (A_{ij})$ be given and assume that the inverse $\underset{\cdot}{\mathscr{A}}^{-1}$ exists for all $\underset{\cdot}{\mathscr{A}} \in \mathscr{A}$ and let $\mathscr{b} = (B_i)$ be a real interval vector.

Then we denote by

$$\ell = (L_1, L_2, \ldots, L_n)^{\mathrm{T}}$$

the real interval vector whose components L_i, $1 \leqslant i \leqslant n$, are obtained from the set

$$\mathfrak{L} = \{\underset{\cdot}{x} \mid \underset{\cdot}{\mathscr{A}}\underset{\cdot}{x} = \underset{\cdot}{b}, \underset{\cdot}{\mathscr{A}} \in \mathscr{A}, \underset{\cdot}{b} \in \mathscr{b}\}$$

by projection of $\mathfrak{L}$ on the ith coordinate axis; that is,

$$L_i = L_i(\mathscr{A}, \mathscr{b}) = \{l_i \mid (l_1, \ldots, l_i, \ldots, l_n)^{\mathrm{T}} \in \mathfrak{L}\}, \qquad 1 \leqslant i \leqslant n.$$

ℓ is the interval vector of the smallest width among all the interval vectors that enclose the set $\mathfrak{L}$.

We now wish to investigate how the interval vectors computed using Hansen's method approximate the vector ℓ. For the "solution" of the transformed system of equations we then use the Gauss–Jordan algorithm. It will be shown that the difference of the width of this vector and the width of the vector ℓ tends to zero with the square of the widths of the elements of $\mathscr{A}$ and $\mathscr{b}$. This shows that Hansen's method computes a vector that is reasonably close to the vector ℓ provided the widths of the original data are not too large.

Again we represent a real interval $A = [a_1, a_2]$ by the midpoint $a = \frac{1}{2}(a_1 + a_2)$ and the half-width $r = \frac{1}{2}d(A) = \frac{1}{2}(a_2 - a_1)$ and we write this

$$A = \langle a, r \rangle.$$

Lemma 1: Let $\underset{\cdot}{\mathscr{A}}\underset{\cdot}{x} = \underset{\cdot}{b}$, where $\underset{\cdot}{\mathscr{A}} = (a_{ij})$ is a real nonsingular matrix having the inverse $\underset{\cdot}{\mathscr{A}}^{-1} = \underset{\cdot}{\mathscr{B}} = (b_{ij})$, and let $\underset{\cdot}{x} = (x_j)$, $\underset{\cdot}{b} = (b_i)$ be point vectors. Furthermore, let $\mathscr{A} = (A_{ij})$ be an interval matrix for which $A_{ij} = \langle a_{ij}, r_{ij} \rangle$, $1 \leqslant i, j \leqslant n$, and $\mathscr{b} = (B_i)$ a real interval vector with $B_i = \langle b_i, r_i \rangle$, $1 \leqslant i \leqslant n$. Then the relation

$$\tfrac{1}{2}d(L_k) = \sum_{i=1}^{n} \sum_{j=1}^{n} |b_{ki}x_j r_{ij}| + \sum_{i=1}^{n} |b_{ki}r_i| + O(d^2)$$

holds for the kth component L_k of the interval vector $\ell = (L_i)$ defined above. Here

$$d = \max\left\{ \max_{1 \leqslant i, j \leqslant n} \{r_{ij}\}, \max_{1 \leqslant i \leqslant n} \{r_i\} \right\}.$$

We denote by $O(d^2)$ a real function f of d for which

$$|f/d^2| \leqslant \gamma \qquad \text{for} \quad d \leqslant d_0,$$

where $\gamma \geqslant 0$, $d_0 > 0$ are constants.

Proof: We know from Cramer's rule that L_k is the image of an $(n^2 + n)$-dimensional hypercube under the mapping

$$x_k = x_k(\mathcal{A}, b).$$

From the mean-value theorem we get

$$x_k(\hat{\mathcal{A}}, \hat{b}) = x_k(\mathcal{A}, b) + \sum_{i=1}^{n} \sum_{j=1}^{n} \frac{\partial x_k}{\partial a_{ij}} (\hat{a}_{ij} - a_{ij})$$

$$+ \sum_{i=1}^{n} \frac{\partial x_k}{\partial b_i} (\hat{b}_i - b_i) + \tfrac{1}{2} x_k''(u + t(v - u))(v - u)(v - u).$$

The Hessian of the map x_k is denoted by x_k'' and u, v are vectors from $V_{n^2+n}(\mathbb{R})$ whose components take on the values of the elements of the matrix $\mathcal{A}$ as well as the vector b, respectively, $\hat{\mathcal{A}}$ and $\hat{b}$. Furthermore, $t \in (0, 1)$.

If one now differentiates the n equations

$$\sum_{j=1}^{n} a_{ij} x_j = b_i, \qquad 1 \leqslant i \leqslant n,$$

with respect to a_{ij} then one obtains the equation

$$\mathcal{A} \frac{\partial}{\partial a_{ij}} x = - x_j e^i,$$

using

$$\frac{\partial}{\partial a_{ij}} x = \left(\frac{\partial x_1}{\partial a_{ij}}, \frac{\partial x_2}{\partial a_{ij}}, \ldots, \frac{\partial x_n}{\partial a_{ij}} \right)^{\mathrm{T}},$$

where e^i is the ith unit vector. From this one obtains

$$\frac{\partial}{\partial a_{ij}} x = - x_j \mathcal{A}^{-1} e^i \qquad \text{or} \qquad \frac{\partial x_k}{\partial a_{ij}} = - b_{ki} x_j.$$

From the equation

$$x = \mathcal{B} b$$

one gets

$$\frac{\partial x_k}{\partial b_i} = b_{ki}.$$

If one now uses these derivatives in the mean-value theorem one obtains

$$x_k(\hat{\mathscr{A}},\hat{\ell}) \in x_k(\mathscr{A},\ell) + \sum_{i=1}^{n}\sum_{j=1}^{n} |b_{ki}x_j|\langle 0, r_{ij}\rangle + \sum_{i=1}^{n} |b_{ki}|\langle 0, r_i\rangle + \cdots,$$

since

$$\hat{\mathscr{A}} \in \mathscr{A}, \qquad \hat{\ell} \in \ell;$$

that is,

$$\tfrac{1}{2}d(L_k) = \sum_{i=1}^{n}\sum_{j=1}^{n} |b_{ki}x_j r_{ij}| + \sum_{i=1}^{n} |b_{ki}r_i| + O(d^2). \quad \blacksquare$$

Lemma 2: Let $\mathscr{A} = (a_{ij})$ be a real nonsingular matrix for which $\mathscr{A}^{-1} = \mathscr{B} = (b_{ij})$ and let $\ell = (b_i)$ be a real vector. Furthermore let $\mathscr{A} = (A_{ij})$ be a real interval matrix with $A_{ij} = \langle a_{ij}, r_{ij}\rangle$, $1 \leqslant i, j \leqslant n$, and $\ell = (B_i)$ a real interval vector with $B_i = \langle b_i, r_i\rangle$, $1 \leqslant i \leqslant n$. The matrix $\tilde{\mathscr{A}} = (\tilde{A}_{ij})$ is defined to be the interval matrix $\mathscr{A}^{-1}\mathscr{A} = \mathscr{B}\mathscr{A}$, and $\tilde{\ell} = (\tilde{B}_i)$ is defined to be the interval vector $\mathscr{A}^{-1}\ell = \mathscr{B}\ell$. It will be assumed that all the point matrices from $\tilde{\mathscr{A}}$ are nonsingular. The interval vector

$$\tilde{\ell} = (\tilde{L}_1, \tilde{L}_2, \ldots, \tilde{L}_n)^{\mathrm{T}},$$

where

$$\tilde{L}_i = L_i(\tilde{\mathscr{A}}, \tilde{\ell}) = \{\tilde{l}_i \mid (\tilde{l}_1, \ldots, \tilde{l}_i, \ldots, \tilde{l}_n)^{\mathrm{T}} \in \tilde{\mathfrak{L}}\}, \qquad 1 \leqslant i \leqslant n,$$

$$\tilde{\mathfrak{L}} = \{\tilde{x} \mid \tilde{\mathscr{A}}\tilde{x} = \tilde{\ell}, \tilde{\mathscr{A}} \in \tilde{\mathscr{A}}, \tilde{\ell} \in \tilde{\ell}\},$$

then satisfies the relation

$$d(\tilde{L}_k) = d(L_k) + O(d^2), \qquad 1 \leqslant k \leqslant n.$$

Proof: Let $\mathscr{A}x = \ell$. It then follows from Lemma 1 that

$$\tfrac{1}{2}d(L_k) = \sum_{i=1}^{n}\sum_{j=1}^{n} |b_{ki}x_j r_{ij}| + \sum_{i=1}^{n} |b_{ki}r_i| + O(d^2).$$

The interval matrix $\tilde{\mathscr{A}} = (\tilde{A}_{ij})$ has the elements

$$\tilde{A}_{ij} = \left\langle \delta_{ij}, \sum_{m=1}^{n} |b_{im}r_{mj}| \right\rangle, \qquad 1 \leqslant i, j \leqslant n,$$

where δ_{ij} is the Kronecker delta. The interval vector $\tilde{\ell} = (\tilde{B}_i)$ has the elements

$$\tilde{B}_i = \left\langle x_i, \sum_{m=1}^{n} |b_{im}r_m| \right\rangle, \qquad 1 \leqslant i \leqslant n.$$

Again using Lemma 1 we obtain

$$\begin{aligned}
\tfrac{1}{2}d(\tilde{L}_k) &= \sum_{i=1}^{n}\sum_{j=1}^{n}\left|\delta_{ki}x_j\left(\sum_{m=1}^{n}|b_{im}r_{mj}|\right)\right| + \sum_{i=1}^{n}\left|\delta_{ki}\left(\sum_{m=1}^{n}|b_{im}r_m|\right)\right| + O(\tilde{d}^2)\\
&= \sum_{j=1}^{n}\left|x_j\left(\sum_{m=1}^{n}|b_{km}r_{mj}|\right)\right| + \sum_{m=1}^{n}|b_{km}r_m| + O(\tilde{d}^2)\\
&= \sum_{i=1}^{n}\sum_{j=1}^{n}|b_{ki}x_jr_{ij}| + \sum_{i=1}^{n}|b_{ki}r_i| + O(\tilde{d}^2),
\end{aligned}$$

where

$$\tilde{d} = \max\left\{\max_{1\leqslant i,j\leqslant n}\{\tilde{r}_{ij}\},\ \max_{1\leqslant i\leqslant n}\{\tilde{r}_i\}\right\}.$$

From the representation of the elements of $\tilde{\mathscr{A}}$ and $\tilde{\ell}$ we clearly have that

$$\tilde{d} = O(d).$$

A comparison with the above expression for $\frac{1}{2}d(L_k)$ gives

$$d(\tilde{L}_k) = d(L_k) + O(d^2),$$

which proves the lemma. ■

The Gauss–Jordan algorithm now generates a diagonal matrix from the interval matrix $\tilde{\mathscr{A}}$ in a finite number of steps. Starting from the interval matrix $\tilde{\mathscr{A}}$ a new matrix is generated in each step having at least one more off-diagonal zero. We assume that an interval matrix $\mathscr{H} = (H_{ij})$ and an interval vector $\hbar = (H_i)$ are given for which

$$\begin{aligned}
H_{ij} &= \langle h_{ij}, e_{ij}\rangle, \qquad 1\leqslant i,j\leqslant n,\\
H_i &= \langle h_i, e_i\rangle, \qquad 1\leqslant i\leqslant n,
\end{aligned}$$

where $h_{ij} = \delta_{ij}$, $h_i = x_i$ and

$$\max\left\{\max_{1\leqslant i,j\leqslant n}\{e_{ij}\},\ \max_{1\leqslant i\leqslant n}\{e_i\}\right\} = O(d).$$

Furthermore, with $\mathscr{A} = (A_{ij})$, $\ell = (b_i)$ where

$$\begin{aligned}
A_{ij} &= \langle a_{ij}, r_{ij}\rangle, \qquad 1\leqslant i,j\leqslant n,\\
B_i &= \langle b_i, r_i\rangle, \qquad 1\leqslant i\leqslant n,
\end{aligned}$$

we define

$$d = \max\left\{\max_{1\leqslant i,j\leqslant n}\{r_{ij}\},\ \max_{1\leqslant i\leqslant n}\{r_i\}\right\}.$$

With these assumptions we may now perform an induction proof for the algorithm. The assumptions are clearly valid for the matrix $\mathscr{H} := \tilde{\mathscr{A}} = \underset{.}{\mathscr{A}}^{-1}\mathscr{A}$ and the interval vector $\mathscr{h} := \tilde{\mathscr{b}} = \underset{.}{\mathscr{A}}^{-1}\mathscr{b}$ as shown in Lemma 2. Furthermore, from Lemma 2 it follows that

$$d(\tilde{L}_k) = d(L_k(\mathscr{H}, \mathscr{h})) = d(L_k) + O(d^2), \qquad 1 \leqslant k \leqslant n.$$

In order to generate a zero for the pair (r, s) of indices $(r \neq s)$, a matrix $\mathscr{H}'$ and a vector $\mathscr{h}'$ are generated from the matrix $\mathscr{H}$ and the vector $\mathscr{h}$ according to the following scheme

$$\begin{aligned} H'_{ij} &= H_{ij}, & i &\neq r, \\ H'_{rj} &= H_{rj} - H_{sj}H_{rs}/H_{ss}, & j &\neq s, \\ H'_{rs} &= 0, \\ H'_{i} &= H_{i}, & i &\neq r, \\ H'_{r} &= H_{r} - H_{s}H_{rs}/H_{ss}. \end{aligned}$$

Since $h_{ij} = \delta_{ij}$ it follows that

$$\begin{aligned} H'_{rj} &= \langle h_{rj}, e_{rj}\rangle - \langle h_{sj}, e_{sj}\rangle\langle h_{rs}, e_{rs}\rangle/\langle h_{ss}, e_{ss}\rangle \\ &= \langle \delta_{rj}, e_{rj}\rangle - \langle 0, e_{sj}\rangle\langle 0, e_{rs}\rangle/\langle h_{ss}, e_{ss}\rangle \\ &= \langle \delta_{rj}, e_{rj}\rangle - \langle 0, e_{sj}e_{rs}\rangle(1/(1 - e_{ss})) \\ &= \langle \delta_{rj}, e_{rj} + [e_{sj}e_{rs}/(1 - e_{ss})]\rangle \end{aligned}$$

and

$$\begin{aligned} H'_{r} &= \langle x_r, e_r\rangle - \langle x_s, e_s\rangle\langle 0, e_{rs}\rangle/\langle h_{ss}, e_{ss}\rangle \\ &= \langle x_r, e_r + ((|x_s|e_{rs} + e_{rs}e_s)/(1 - e_{ss}))\rangle. \end{aligned}$$

The representation

$$H'_{ij} = \langle \delta_{ij}, e'_{ij}\rangle, \qquad H'_i = \langle x_i, e'_i\rangle$$

is therefore also valid for the matrix $\mathscr{H}' = (H'_{ij})$ and the vector $\mathscr{h}' = (H'_i)$.

It follows from Lemma 1 that

$$\begin{aligned} \tfrac{1}{2} d(L_k(\mathscr{H}, \mathscr{h})) &= \sum_{i=1}^{n} \sum_{j=1}^{n} |\delta_{ki} x_j e_{ij}| + \sum_{i=1}^{n} |\delta_{ki} e_i| + O(d^2) \\ &= \sum_{j=1}^{n} |x_j e_{kj}| + e_k + O(d^2). \end{aligned}$$

Correspondingly we get

$$\tfrac{1}{2} d(L_k(\mathscr{H}', \mathscr{h}')) = \sum_{j=1}^{n} |x_j e'_{kj}| + e'_k + O(d^2).$$

For $k \neq r$ we have $e'_{kj} = e_{kj}$ and $e'_k = e_k$. From the hypothesis

$$d(L_k(\mathscr{H}, \mathscr{h})) = d(L_k) + O(d^2),$$

we therefore get

$$d(L_k(\mathscr{H}', \mathscr{h}')) = d(L_k) + O(d^2), \qquad k \neq r.$$

For $k = r$ we have

$$\sum_{j=1}^{n} |x_k e'_{kj}| = \sum_{j=1, j\neq s}^{n} |x_j e_{kj}| + O(d^2)$$

and

$$e'_k = e_k + |x_s e_{ks}| + O(d^2),$$

and therefore

$$\tfrac{1}{2} d(L_k(\mathscr{H}', \mathscr{h}')) = \sum_{j=1}^{n} |x_j e_{kj}| + e_k + O(d^2);$$

that is,

$$d(L_k(\mathscr{H}', \mathscr{h}')) = d(L_k) + O(d^2) \qquad \text{for} \quad k = r.$$

This completes the induction proof.

The solution of the diagonal system of equations does not increase the widths of the elements so that the relation

$$d(L_k(\hat{\mathscr{H}}, \hat{\mathscr{h}})) = d(L_k(\mathscr{A}, \mathscr{b})) + O(d^2)$$

is valid for the final diagonal matrix $\hat{\mathscr{H}}$ with the corresponding interval vector $\hat{\mathscr{h}}$.

Remarks: The "solution" of a system of equations with interval coefficients using the methods described in this chapter was suggested in Hansen [1, 2]. Lemmas 1 and 2 and their implications for Hansen's method are due to Miller [4].

REFERENCES

[1] E. Hansen, Interval arithmetic in matrix computations. Part I. *SIAM J. Numer. Anal.* **2**, 308–320 (1965).

[2] E. Hansen and R. Smith, Interval arithmetic in matrix computations. Part II. *SIAM J. Numer. Anal.* **4**, 1–9 (1967).

[3] E. Isaacson and H. B. Keller, "Analysis of Numerical Methods." Wiley, New York, 1966.

[4] W. Miller, On an interval-arithmetic matrix method. *BIT* **12**, 213–219 (1972).

Chapter 17 / THE PROCEDURE OF KUPERMANN AND HANSEN

Let $\mathcal{A}$ be an interval matrix for which $\underset{\cdot}{\mathcal{A}}$ is nonsingular for all $\underset{\cdot}{\mathcal{A}} \in \mathcal{A}$ and let

$$\mathfrak{L} = \{\underset{\cdot}{x} \mid \underset{\cdot}{\mathcal{A}}\underset{\cdot}{x} = \underset{\cdot}{b}, \underset{\cdot}{\mathcal{A}} \in \mathcal{A}, \underset{\cdot}{b} \in b\}$$

be the solution set for a given interval vector b. Simple examples already show that Hansen's method as described in the last chapter only computes a superset of the vector ℓ with the components

$$L_k = \{l_k \mid (l_1, \ldots, l_k, \ldots, l_n)^{\mathrm{T}} \in \mathfrak{L}\}.$$

Kupermann [5] described a noninterval procedure that gives an improved inclusion of the set $\mathfrak{L}$ in certain cases. This procedure was then generalized by Hansen to an interval method.

Let us consider the set of linear equations

$$\underset{\cdot}{\mathcal{A}}\underset{\cdot}{x} = \underset{\cdot}{b}, \qquad \underset{\cdot}{x} = (x_i),$$

where $\underset{\cdot}{\mathcal{A}}$ is a nonsingular point matrix and $\underset{\cdot}{b}$ a real vector. If the partial derivative satisfies

$$\frac{\partial x_k}{\partial a_{ij}} \geqslant 0,$$

then we have that x_k is a monotonically nondecreasing function of a_{ij}. If now a_{ij} is allowed to vary in a real interval

$$A_{ij} = [a_{ij}^1, a_{ij}^2],$$

then it follows that x_k, considered as a function of a_{ij} ranging over the interval $A_{ij} = [a^1_{ij}, a^2_{ij}]$, takes on its smallest value for $a_{ij} = a^1_{ij}$ and its largest value for $a_{ij} = a^2_{ij}$. The same is also true for the dependence of the component x_k on b_i.

Now let $\mathscr{A} = (A_{ij})$ be a given real interval matrix and $\mathscr{b} = (B_i)$ be a given interval vector. In order to enclose the intervals

$$L_k = [l^1_k, l^2_k], \qquad 1 \leqslant k \leqslant n,$$

we proceed as follows.

Starting with the real interval matrix $\mathscr{A} = (A_{ij})$, $A_{ij} = [a^1_{ij}, a^2_{ij}]$, $1 \leqslant i, j \leqslant n$, and the real interval vector $\mathscr{b} = (B_i)$, $B_i = [b^1_i, b^2_i]$, $1 \leqslant i \leqslant n$, we form the interval matrices

$$\tilde{\mathscr{A}} = (\tilde{A}_{ij}) \qquad \text{and} \qquad \hat{\mathscr{A}} = (\hat{A}_{ij})$$

as well as the interval vectors

$$\tilde{\mathscr{b}} = (\tilde{B}_i) \qquad \text{and} \qquad \hat{\mathscr{b}} = (\hat{B}_i)$$

according to the following rules:

$$\tilde{A}_{ij} = \begin{cases} [a^1_{ij}, a^1_{ij}] & \text{if } \partial x_k/\partial a_{ij} \geqslant 0 \quad \text{for all } A \in \mathscr{A} \text{ and } b \in \mathscr{b} \\ [a^2_{ij}, a^2_{ij}] & \text{if } \partial x_k/\partial a_{ij} \leqslant 0 \quad \text{for all } A \in \mathscr{A} \text{ and } b \in \mathscr{b} \\ A_{ij} & \text{otherwise;} \end{cases}$$

$$\hat{A}_{ij} = \begin{cases} [a^2_{ij}, a^2_{ij}] & \text{if } \partial x_k/\partial a_{ij} \geqslant 0 \quad \text{for all } A \in \mathscr{A} \text{ and } b \in \mathscr{b} \\ [a^1_{ij}, a^1_{ij}] & \text{if } \partial x_k/\partial a_{ij} \leqslant 0 \quad \text{for all } A \in \mathscr{A} \text{ and } b \in \mathscr{b} \\ A_{ij} & \text{otherwise;} \end{cases}$$

$$\tilde{B}_i = \begin{cases} [b^1_i, b^1_i] & \text{if } \partial x_k/\partial b_i \geqslant 0 \quad \text{for all } A \in \mathscr{A} \text{ and } b \in \mathscr{b} \\ [b^2_i, b^2_i] & \text{if } \partial x_k/\partial b_i \leqslant 0 \quad \text{for all } A \in \mathscr{A} \text{ and } b \in \mathscr{b} \\ B_i & \text{otherwise;} \end{cases}$$

$$\hat{B}_i = \begin{cases} [b^2_i, b^2_i] & \text{if } \partial x_k/\partial b_i \geqslant 0 \quad \text{for all } A \in \mathscr{A} \text{ and } b \in \mathscr{b} \\ [b^1_i, b^1_i] & \text{if } \partial x_k/\partial b_i \leqslant 0 \quad \text{for all } A \in \mathscr{A} \text{ and } b \in \mathscr{b} \\ B_i & \text{otherwise.} \end{cases}$$

Using, for example, Hansen's method from the last chapter we now compute including intervals

$$X_k = [x^1_k, x^2_k] \qquad \text{and} \qquad Y_k = [y^1_k, y^2_k]$$

for the intervals

$$L_k(\tilde{\mathscr{A}}, \tilde{\mathscr{b}}) = [\tilde{l}^1_k, \tilde{l}^2_k] \qquad \text{and} \qquad L_k(\hat{\mathscr{A}}, \hat{\mathscr{b}}) = [\hat{l}^1_k, \hat{l}^2_k].$$

From the above discussion we get

$$l_k^1 \geqslant \tilde{l}_k^1 \qquad \text{and} \qquad l_k^2 \leqslant \hat{l}_k^2$$

for $L_k(\mathscr{A}, \ell) = [l_k^1, l_k^2]$. From this it follows that

$$[\tilde{l}_k^1, \hat{l}_k^2] \supseteq L_k(\mathscr{A}, \ell)$$

holds. Therefore we also have

$$[x_k^1, y_k^2] \supseteq [\tilde{l}_k^1, \hat{l}_k^2] \supseteq L_k(\mathscr{A}, \ell).$$

In order to construct the interval matrices $\tilde{\mathscr{A}}$, $\hat{\mathscr{A}}$ and the interval vectors $\tilde{\ell}$, $\hat{\ell}$ for a fixed k one needs the partial derivatives $\partial x_k/\partial a_{ij}$ and $\partial x_k/\partial b_i$. These were already developed in the last chapter and we have in fact

$$\frac{\partial x_k}{\partial a_{ij}} = -\bar{a}_{ki} x_j, \qquad \frac{\partial x_k}{\partial b_i} = \bar{a}_{ki},$$

using $\mathscr{A}^{-1} = (\bar{a}_{ij})$.

In order to determine the distribution of signs for these derivatives we calculate—for example, by using Hansen's method—an interval vector $\bar{\ell} = (\bar{L}_i)$ containing the set $\mathfrak{L}$ as well as an interval matrix $\bar{\mathscr{A}} = (\bar{A}_{ij})$ containing the inverses of all $\mathscr{A} \in \mathscr{A}$. If now the lower bound of $\bar{A}_{ki}\bar{L}_i$ is nonnegative for a fixed k, then it follows that $\partial x_k/\partial a_{ij} \leqslant 0$ for all $\mathscr{A} \in \mathscr{A}$ and $\ell \in \ell$. In the same manner, if the upper bound of $\bar{A}_{ki}\bar{L}_i$ is nonpositive then $\partial x_k/\partial a_{ij} \geqslant 0$. A corresponding statement is valid for $\partial x_k/\partial b_i$. If $0 \in \bar{A}_{ki}\bar{L}_i$ (resp., $0 \in \bar{A}_{ki}$), then we can still have $\partial x_k/\partial a_{ij} \geqslant 0$ or $\leqslant 0$ (resp., $\partial x_k/\partial b_i \geqslant 0$ or $\leqslant 0$), because $\bar{\ell}$ is only a superset of $\mathfrak{L}$ and $\bar{\mathscr{A}}$ is only an including superset of the inverses of $\mathscr{A} \in \mathscr{A}$. During the construction of matrices $\tilde{\mathscr{A}}$, $\hat{\mathscr{A}}$ and the vectors $\tilde{\ell}$, $\hat{\ell}$ the elements of $\mathscr{A}$ and ℓ are in this case replaced as if $\partial x_k/\partial a_{ij}$ (resp., $\partial x_k/\partial b_i$) are changing signs since the computed quantities do not allow any other decisions.

This method gives in general much better inclusions than Hansen's method of Chapter 16. The disadvantage is the large amount of computation that has to be done. In addition to the computation of an interval matrix $\bar{\mathscr{A}}$ containing the inverses of all $\mathscr{A} \in \mathscr{A}$ we have to solve two systems of interval equations for each component of $L_k(\mathscr{A}, \ell)$ in general.

Remarks: The method of Kupermann and Hansen may be carried over to the iteration methods of Chapter 12. A complete discussion of this case including numerical results and an ALGOL 60 program is found in Alefeld and Herzberger [1].

We also mention that the formulas for the generation of the $2n$ system of equations were given in Braess [2] without explicit use of interval arithmetic.

Furthermore, it should be remarked here that one could seek a method based on solving the $2n$ systems of equations using the Gaussian algorithm. This was done in Wißkirchen [7]. The inclusions are, however, always better using Hansen's method.

REFERENCES

[1] G. Alefeld and J. Herzberger, Über die Verbesserung von Schranken für die Lösung bei linearen Gleichungssystemen. *Angew. Informat.* **14**, 107–112 (1971).

[2] D. Braess, Die Berechnung der Fehlergrenzen bei linearen Gleichungssystemen mit fehlerhaften Koeffizienten. *Arch. Rational. Mech. Anal.* **19**, 74–80 (1965).

[3] E. Hansen, "Topics in Interval Analysis." Oxford Univ. Press (Clarendon), London and New York, 1969.

[4] E. Hansen, On linear algebraic equations with interval coefficients. *In* "Topics in Interval Analysis." Oxford Univ. Press (Clarendon), London and New York, 1969.

[5] I. B. Kupermann, Approximate Linear Algebraic Equations and Rounding Error Estimation. Ph.D. Thesis. Dept. of Applied Math., Univ. of Witwatersrand, Johannesburg (1967).

[6] I. B. Kupermann, "Approximate Linear Algebraic Equations," p. 2. Van Nostrand Reinhold, New York, 1971.

[7] P. Wißkirchen, Ein Steuerungsprinzip und dessen Anwendung auf den Gaußschen Algorithmus. *Gesellsch. Math. Datenverarbeit.* Nr. 20, Bonn (1969).

Chapter 18 / ITERATION METHODS FOR THE INCLUSION OF THE INVERSE MATRIX AND FOR TRIANGULAR DECOMPOSITIONS

Let a nonsingular $n \times n$ matrix $\mathcal{A} \in M_{nn}(\mathbb{R})$ and an interval matrix $\mathcal{X}^{(0)} \in M_{nn}(I(\mathbb{R}))$ be given such that $\mathcal{X}^{(0)}$ includes the inverse of $\mathcal{A}$, that is $\mathcal{A}^{-1} \in \mathcal{X}^{(0)}$. We shall consider here iteration procedures that improve the inclusion matrix $\mathcal{X}^{(0)}$ iteratively.

In the treatment of these methods we employ a map m that maps the set of $n \times n$ interval matrices into the set of $n \times n$ real point matrices. This map maps each interval matrix into the point matrix having as elements the midpoints of the interval matrix elements. That is, using the notation

$$i(X) = x_1, \qquad s(X) = x_2 \qquad \text{for} \quad X = [x_1, x_2] \in I(\mathbb{R}),$$

we define

(1) $$m: M_{nn}(I(\mathbb{R})) \ni \mathcal{X} \to m(\mathcal{X}) = \tfrac{1}{2}(i(X_{ij}) + s(X_{ij})) \in M_{nn}(\mathbb{R}).$$

The midpoint mapping of interval matrices is clearly a continuous map. It has the following properties:

(2) $$m(\mathcal{X} \pm \mathcal{Y}) = m(\mathcal{X}) \pm m(\mathcal{Y}), \qquad \mathcal{X}, \mathcal{Y} \in M_{nn}(I(\mathbb{R})),$$

(3) $$m(\mathcal{B}\mathcal{X}) = \mathcal{B}m(\mathcal{X}), \quad m(\mathcal{X}\mathcal{B}) = m(\mathcal{X})\mathcal{B}, \quad \mathcal{B} \in M_{nn}(\mathbb{R}), \quad \mathcal{X} \in M_{nn}(I(\mathbb{R})),$$

(4) $$m(\mathcal{B}) = \mathcal{B}, \qquad \mathcal{B} \in M_{nn}(\mathbb{R}).$$

A short proof for (3) is given by

$$m(\mathscr{B}\mathscr{X}) = m\left(\sum_{k=1}^{n} b_{lk}X_{kj}\right) = \left(\sum_{k=1}^{n} m(b_{lk}[i(X_{kj}), s(X_{kj})])\right)$$
$$= \left(\sum_{k=1}^{n} b_{lk}\tfrac{1}{2}(i(X_{kj}) + s(X_{kj}))\right) = \left(\sum_{k=1}^{n} b_{lk}m(X_{kj})\right)$$
$$= \mathscr{B}m(\mathscr{X}).$$

Statements (2) and (4) may be proved in a similar manner.

We now wish to formulate the first method for calculating a sequence of inclusions for the inverse $\mathscr{A}^{-1}$. Let $r > 1$ be a fixed natural number, and let

$$\mathscr{B}^0 = \mathscr{I} \qquad (\mathscr{I} = \text{unit matrix})$$

for an arbitrary point matrix $\mathscr{B}$.

Let us consider the procedure

$$\mathscr{X}^{(k+1)} = m(\mathscr{X}^{(k)}) \sum_{\nu=0}^{r-2} (\mathscr{I} - \mathscr{A}m(\mathscr{X}^{(k)}))^{\nu} + \mathscr{X}^{(k)}(\mathscr{I} - \mathscr{A}m(\mathscr{X}^{(k)}))^{r-1}, \qquad k \geqslant 0, \tag{5}$$

for calculating the iterates. For the case $r = 2$ we get

$$\mathscr{X}^{(k+1)} = m(\mathscr{X}^{(k)}) + \mathscr{X}^{(k)}(\mathscr{I} - \mathscr{A}m(\mathscr{X}^{(k)})), \qquad k \geqslant 0,$$

which may be considered to be the interval version of the method of Schulz for calculating the inverse matrix [15].

The properties of the iteration method (5) are collected in the following theorem.

Theorem 1: Let $\mathscr{A}$ be a nonsingular $n \times n$ matrix and $\mathscr{X}^{(0)}$ an $n \times n$ interval matrix such that $\mathscr{A}^{-1} \in \mathscr{X}^{(0)}$. A sequence $\{\mathscr{X}^{(k)}\}_{k=0}^{\infty}$ of interval matrices is calculated according to (5). Then

(6) each iterate $\mathscr{X}^{(k)}$, $k \geqslant 0$, contains $\mathscr{A}^{-1}$;

(7) the sequence $\{\mathscr{X}^{(k)}\}_{k=0}^{\infty}$ converges to $\mathscr{A}^{-1}$ iff the spectral radius $\rho(\mathscr{I} - \mathscr{A}m(\mathscr{X}^{(0)}))$ is smaller than 1;

(8) using a matrix norm $\|\cdot\|$ the sequence $\{d(\mathscr{X}^{(k)})\}_{k=0}^{\infty}$ satisfies

$$\|d(\mathscr{X}^{(k+1)})\| \leqslant \gamma \|d(\mathscr{X}^{(k)})\|^r, \qquad \gamma \geqslant 0;$$

that is, the R order of the method (5) satisfies $O_R((5), \mathscr{A}^{-1}) \geqslant r$ (see Appendix A, Theorem 2).

Proof: Of (6): The relation

$$m(\mathscr{X}^{(k)}) \sum_{\nu=0}^{r-2} (\mathscr{I} - \mathscr{A}m(\mathscr{X}^{(k)}))^{\nu} = \mathscr{A}^{-1} - \mathscr{A}^{-1}(\mathscr{I} - \mathscr{A}m(\mathscr{X}^{(k)}))^{r-1}$$

may be easily verified for an arbitrary $m(\mathscr{X}^{(k)}) \in M_{nn}(\mathbb{R})$. (6) is valid for $k = 0$ according to the assumptions. Now suppose that $\mathscr{A}^{-1} \in \mathscr{X}^{(k)}$. Using the above relation as well as (10.10′) we get

$$\begin{aligned}\mathscr{A}^{-1} &= m(\mathscr{X}^{(k)}) \sum_{\nu=0}^{r-2} (\mathscr{I} - \mathscr{A}m(\mathscr{X}^{(k)}))^{\nu} + \mathscr{A}^{-1}(\mathscr{I} - \mathscr{A}m(\mathscr{X}^{(k)}))^{r-1} \\ &\in m(\mathscr{X}^{(k)}) \sum_{\nu=0}^{r-2} (\mathscr{I} - \mathscr{A}m(\mathscr{X}^{(k)}))^{\nu} + \mathscr{X}^{(k)}(\mathscr{I} - \mathscr{A}m(\mathscr{X}^{(k)}))^{r-1} \\ &= \mathscr{X}^{(k+1)}.\end{aligned}$$

This proves (6) by complete induction.

Of (7): Using the rules (2)–(4) for the midpoint mapping in the iteration procedure (5) we get the following recursive formula for the sequence $\{m(\mathscr{X}^{(k)})\}_{k=0}^{\infty}$:

$$m(\mathscr{X}^{(k+1)}) = m(\mathscr{X}^{(k)}) \sum_{\nu=0}^{r-1} (\mathscr{I} - \mathscr{A}m(\mathscr{X}^{(k)}))^{\nu}.$$

This is a generalization of the iteration procedure of Schulz (see Schulz [15] and Altmann [8]). Multiplying both sides of this equation by $\mathscr{A}$ one obtains

$$\begin{aligned}\mathscr{A}m(\mathscr{X}^{(k+1)}) &= (\mathscr{I} - (\mathscr{I} - \mathscr{A}m(\mathscr{X}^{(k)}))) \sum_{\nu=0}^{r-1} (\mathscr{I} - \mathscr{A}m(\mathscr{X}^{(k)}))^{\nu} \\ &= \mathscr{I} - (\mathscr{I} - \mathscr{A}m(\mathscr{X}^{(k)}))^{r}\end{aligned}$$

or

$$\mathscr{I} - \mathscr{A}m(\mathscr{X}^{(k+1)}) = (\mathscr{I} - \mathscr{A}m(\mathscr{X}^{(k)}))^{r} = (\mathscr{I} - \mathscr{A}m(\mathscr{X}^{(0)}))^{r^{k+1}}.$$

From this it follows that

$$\begin{aligned}\lim_{k\to\infty} m(\mathscr{X}^{(k)}) = \mathscr{A}^{-1} &\Leftrightarrow \lim_{k\to\infty} (\mathscr{I} - \mathscr{A}m(\mathscr{X}^{(0)}))^{k} = \mathscr{O} \\ &\Leftrightarrow \rho(\mathscr{I} - \mathscr{A}m(\mathscr{X}^{(0)})) < 1.\end{aligned}$$

We shall now show that the sequence $\{\mathscr{X}^{(k)}\}_{k=0}^{\infty}$ converges to $\mathscr{A}^{-1}$ iff the sequence of midpoint matrices $\{m(\mathscr{X}^{(k)})\}_{k=0}^{\infty}$ converges to $\mathscr{A}^{-1}$. This follows from the consideration of the sequence $\{d(\mathscr{X}^{(k)})\}_{k=0}^{\infty}$ of widths that satisfy the recursion

$$d(\mathscr{X}^{(k+1)}) = d(\mathscr{X}^{(k)})|(\mathscr{I} - \mathscr{A}m(\mathscr{X}^{(k)}))^{r-1}|$$

by (10.12) and (10.19). If we now have $\lim_{k\to\infty} m(\mathscr{X}^{(k)}) = \mathscr{A}^{-1}$, then the last relation implies that $\lim_{k\to\infty} d(\mathscr{X}^{(k)}) = \mathscr{O}$. Conversely using the continuity of m and (4) it follows trivially that $\lim_{k\to\infty} \mathscr{X}^{(k)} = \mathscr{A}^{-1}$ implies $\lim_{k\to\infty} m(\mathscr{X}^{(k)}) = \mathscr{A}^{-1}$. Since, however, it was shown above that the condition

$$\rho(\mathscr{I} - \mathscr{A}m(\mathscr{X}^{(0)})) < 1$$

was necessary and sufficient for the convergence of $\{m(\mathscr{X}^{(k)})\}_{k=0}^{\infty}$ it follows that (7) holds true.

Of (8): We have

$$d(\mathscr{X}^{(k+1)}) = d(\mathscr{X}^{(k)})|(\mathscr{I} - \mathscr{A}m(\mathscr{X}^{(k)}))^{r-1}| = d(\mathscr{X}^{(k)})|(\mathscr{A}\mathscr{A}^{-1} - \mathscr{A}m(\mathscr{X}^{(k)}))^{r-1}|$$
$$\leqslant d(\mathscr{X}^{(k)})(|\mathscr{A}|\,|\mathscr{A}^{-1} - m(\mathscr{X}^{(k)})|)^{r-1} \leqslant d(\mathscr{X}^{(k)})2^{-(r-1)}(|\mathscr{A}|d(\mathscr{X}^{(k)}))^{r-1}.$$

We first use a monotonic and multiplicative matrix norm $\|\cdot\|'$ which implies that

$$\|d(\mathscr{X}^{(k+1)})\|' \leqslant 2^{-(r-1)}\|\mathscr{A}\|'^{r-1}\|d(\mathscr{X}^{(k)})\|'^{r}$$

using the above relation. The inequality

$$\|\mathscr{B}\|\gamma_1 \leqslant \|\mathscr{B}\|' \leqslant \gamma_2\|\mathscr{B}\|, \qquad \gamma_1 > 0, \quad \gamma_2 > 0,$$

is valid for every matrix norm $\|\cdot\|$. From this inequality it follows that

$$\|d(\mathscr{X}^{(k+1)})\|\gamma_1 \leqslant 2^{-(r-1)}\gamma_2^{r-1}\|\mathscr{A}\|^{r-1}\gamma_2^{r}\|d(\mathscr{X}^{(k)})\|^{r}.$$

This proves (8). ■

From the proof we can see that the convergence statement is also valid even if $\mathscr{X}^{(0)}$ is an arbitrary interval matrix not necessarily containing $\mathscr{A}^{-1}$. In this case, however, the iterates do not necessarily contain $\mathscr{A}^{-1}$. We note that the criterion (7) depended only on the midpoint matrix $m(\mathscr{X}^{(0)})$ of the given inclusion matrix $\mathscr{X}^{(0)}$. The width $d(\mathscr{X}^{(0)})$ may on the other hand be arbitrary. This means that if one has a suitable approximation $m(\mathscr{X}^{(0)})$ for $\mathscr{A}^{-1}$ that satisfies $\rho(\mathscr{I} - \mathscr{A}m(\mathscr{X}^{(0)})) < 1$, then using certain norm estimates one is always able to produce an interval matrix $\mathscr{X}^{(0)}$ such that $\mathscr{A}^{-1} \in \mathscr{X}^{(0)}$. The sequence of iterates generated according to (5) then converges to $\mathscr{A}^{-1}$ by Theorem 1.

Since the sequence of iterates from (5) always contains $\mathscr{A}^{-1}$ according to (6) it seems natural to form the intersection of the new iterate with the previous iterate and then to continue the iteration with this new potentially improved iterate. This leads to the iteration procedure

$$(9)\quad \begin{cases} \mathscr{Y}^{(k+1)} = m(\mathscr{X}^{(k)}) \sum_{v=0}^{r-2} (\mathscr{I} - \mathscr{A}m(\mathscr{X}^{(k)}))^{v} + \mathscr{X}^{(k)}(\mathscr{I} - \mathscr{A}m(\mathscr{X}^{(k)}))^{r-1}, \\ \mathscr{X}^{(k+1)} = \mathscr{Y}^{(k+1)} \cap \mathscr{X}^{(k)}, \qquad k \geqslant 0. \end{cases}$$

Using this iteration procedure one obtains a monotonic sequence

$$\mathscr{X}^{(0)} \supseteq \mathscr{X}^{(1)} \supseteq \mathscr{X}^{(2)} \supseteq \cdots$$

of inclusions for $\mathscr{A}^{-1}$. The following numerical example does show, however, that the convergence criterion (7) is not sufficient for convergence in general.

We chose $r = 2$ and let

$$\mathscr{A} = \begin{pmatrix} 0.4 & 0.6 \\ -0.6 & 0.4 \end{pmatrix}, \qquad \mathscr{X}^{(0)} = \begin{pmatrix} [-2,4] & [-3,3] \\ [-3,3] & [-2,4] \end{pmatrix},$$

which implies that $m(\mathscr{X}^{(0)}) = \mathscr{I}$. We get

$$\mathscr{I} - \mathscr{A}m(\mathscr{X}^{(0)}) = \begin{pmatrix} 0.6 & -0.6 \\ 0.6 & 0.6 \end{pmatrix}$$

and therefore

$$\rho(\mathscr{I} - \mathscr{A}m(\mathscr{X}^{(0)})) < 1.$$

The procedure (5) therefore converges to $\mathscr{A}^{-1}$ using this interval matrix. Using (9) we calculate

$$\mathscr{Y}^{(1)} = m(\mathscr{X}^{(0)}) + \mathscr{X}^{(0)}(\mathscr{I} - \mathscr{A}m(\mathscr{X}^{(0)})) = \begin{pmatrix} [-2,5.2] & [-4.2,3] \\ [-3,4.2] & [-2,5.2] \end{pmatrix},$$

which implies that $\mathscr{X}^{(1)} = \mathscr{X}^{(0)}$. The sequence of iterates computed according to (9) therefore does not converge to $\mathscr{A}^{-1}$ in contrast to the sequence computed according to (5). A convergence statement for the iteration (9) is contained in the following theorem.

Theorem 2: Let $\mathscr{A}$ be a nonsingular $n \times n$ matrix and $\mathscr{X}^{(0)}$ an $n \times n$ interval matrix for which $\mathscr{A}^{-1} \in \mathscr{X}^{(0)}$. Then

(6′) each iterate $\mathscr{X}^{(k)}$, $k \geqslant 0$, contains $\mathscr{A}^{-1}$;

(10) if the inequality $\rho(|\mathscr{I} - \mathscr{A}\mathscr{X}|) < 1$ is satisfied for all $\mathscr{X} \in \mathscr{X}^{(0)}$, then the sequence $\{\mathscr{X}^{(k)}\}_{k=0}^{\infty}$ converges toward $\mathscr{A}^{-1}$;

(8′) the sequence $\{d(\mathscr{X}^{(k)})\}_{k=0}$ may be bounded as follows using a matrix norm $\|\cdot\|$:

$$\|d(\mathscr{X}^{(k+1)})\| \leqslant \gamma' \|d(\mathscr{X}^{(k)})\|^r, \qquad \gamma' \geqslant 0,$$

that is the R order of the iteration procedure (9) satisfies $O_R((9), \mathscr{A}^{-1}) \geqslant r$ (see Appendix A, Theorem 2).

Proof: Of (6′): As in the proof of (6) one first shows that $\mathscr{A}^{-1} \in \mathscr{Y}^{(k+1)}$, from which it follows immediately that $\mathscr{A}^{-1} \in \mathscr{X}^{(k+1)}$ since $\mathscr{A}^{-1} \in \mathscr{X}^{(k)}$.

Of (10): The sequence

$$\mathscr{X}^{(0)} \supseteq \mathscr{X}^{(1)} \supseteq \mathscr{X}^{(2)} \supseteq \cdots$$

of iterates always converges to an interval matrix $\mathscr{X}$ according to Corollary 10.8. We now show that under the assumptions of the theorem we must necessarily have $d(\mathscr{X}) = \mathscr{O}$. We therefore define

$$\mathscr{Y} = m(\mathscr{X}) \sum_{\nu=0}^{r-2} (\mathscr{I} - \mathscr{A}m(\mathscr{X}))^{\nu} + \mathscr{X}(\mathscr{I} - \mathscr{A}m(\mathscr{X}))^{r-1}$$

and obtain $\mathscr{X} = (X_{ij} \cap Y_{ij}) \subseteq \mathscr{Y}$ from (9). Using (10.11) we obtain $d(\mathscr{X}) \leqslant d(\mathscr{Y})$. For $d(\mathscr{X})$ we get from (9) that

$$d(\mathscr{X})|\mathscr{I} - \mathscr{A}m(\mathscr{X})|^{r-1} \geqslant d(\mathscr{X})|(\mathscr{I} - \mathscr{A}m(\mathscr{X}))^{r-1}| = d(\mathscr{Y}) \geqslant d(\mathscr{X}),$$

which implies that

$$d(\mathscr{X})(\mathscr{I} - |\mathscr{I} - \mathscr{A}m(\mathscr{X})|^{r-1}) \leqslant \mathscr{O}$$

The assumption $\rho(|\mathscr{I} - \mathscr{A}m(\mathscr{X})|) < 1$ implies the existence of $(\mathscr{I} - |\mathscr{I} - \mathscr{A}m(\mathscr{X})|^{r-1})^{-1}$. This inverse is also nonnegative (compare with Varga [17, p. 83]). From this it follows that $d(\mathscr{X}) \leqslant \mathscr{O}$; that is, necessarily $d(\mathscr{X}) = \mathscr{O}$. Because of (6′) we therefore have $\mathscr{X} = \mathscr{A}^{-1}$.

Of (8′): As in the proof of (8) one first shows that the inequality

$$\|d(\mathscr{Y}^{(k+1)})\|' \leqslant \gamma \|d(\mathscr{X}^{(k)})\|'^{r}$$

holds for a monotonic and multiplicative matrix norm $\|\cdot\|'$. From this it follows that the inequality

$$\|d(\mathscr{X}^{(k+1)})\|' \leqslant \|d(\mathscr{Y}^{(k+1)})\|' \leqslant \gamma \|d(\mathscr{X}^{(k)})\|'^{r}$$

is valid since $\mathscr{X}^{(k+1)} \subseteq \mathscr{Y}^{(k+1)}$ as well as using (10.11) and the monotonicity of the norm $\|\cdot\|'$. Analogous to the proof of (8) we use the norm equivalence theorem to prove the final statement. ■

The convergence criterion (10) depends on the width of the inclusion matrix $\mathscr{X}^{(0)}$ for $\mathscr{A}^{-1}$ as opposed to the criterion (7). Formulas may easily be given for this dependence. If, for example, an interval matrix $\mathscr{X}^{(0)}$ satisfies the inequality $\|\mathscr{I} - \mathscr{A}m(\mathscr{X}^{(0)})\| < 1$, for a monotonic and multiplicative norm $\|\cdot\|$, then we have that

$$\|d(\mathscr{X}^{(0)})\| < 2(1 - \|\mathscr{I} - \mathscr{A}m(\mathscr{X}^{(0)})\|)/\|\mathscr{A}\| \tag{11}$$

is a sufficient criterion for the statement that $\|\mathscr{I} - \mathscr{A}\mathscr{X}\| < 1$ for all $\mathscr{X} \in \mathscr{X}^{(0)}$. We now briefly consider how to determine a suitable interval matrix $\mathscr{X}^{(0)}$. For this we assume that $\mathscr{A}$ may be represented as

$$\mathscr{A} = \mathscr{I} - \mathscr{B} \qquad \text{with} \quad \|\mathscr{B}\| < 1.$$

With $m(\mathscr{X}^{(0)}) := \mathscr{I}$ we then have

$$\|\mathscr{I} - \mathscr{A}m(\mathscr{X}^{(0)})\| = \|\mathscr{B}\| < 1,$$

and therefore we have the convergence of (5) for every interval matrix $\mathscr{X}^{(0)}$ for which $m(\mathscr{X}^{(0)}) = \mathscr{I}$ according to the criterion (7). In order to guarantee $\mathscr{A}^{-1} \in \mathscr{X}^{(0)}$ we consider the equation

$$\mathscr{A}\mathscr{X} = (\mathscr{I} - \mathscr{B})\mathscr{X} = \mathscr{I}$$

or

$$\mathscr{X} = \mathscr{B}\mathscr{X} + \mathscr{I}.$$

From this it follows using a multiplicative matrix norm $\|\cdot\|$ that

$$\|\mathscr{X}\| \leqslant a := 1/(1 - \|\mathscr{B}\|).$$

If one now uses either the row-sum or the column-sum norm, then one has

$$-a \leqslant x_{ij} \leqslant a, \qquad 1 \leqslant i, j \leqslant n,$$

for all the elements of $\mathscr{X} = (x_{ij})$. For the matrix $\mathscr{X}^{(0)}$ with

$$X_{ij}^{(0)} = \begin{cases} [-a, a] & \text{for} \quad i \neq j \\ [-a, 2 + a] & \text{for} \quad i = j, \end{cases}$$

we then have $\mathscr{A}^{-1} \in \mathscr{X}^{(0)}$ as well as $m(\mathscr{X}^{(0)}) = \mathscr{I}$. According to Theorem 1 the iteration (5) therefore converges to $\mathscr{A}^{-1}$. If (11) is now satisfied after a certain iteration step, then the iteration may be continued as the procedure (9). A detailed description of such a combined procedure may be found in Alefeld and Herzberger [4].

In the practical execution of the algorithm (5) one evaluates the expression in a Horner scheme fashion.

This results in

$$(5') \quad \mathscr{X}^{(k+1)} = (\cdots(\mathscr{X}^{(k)}\mathscr{T}^{(k)} + m(\mathscr{X}^{(k)}))\mathscr{T}^{(k)} + m(\mathscr{X}^{(k)}))\mathscr{T}^{(k)}\cdots)\mathscr{T}^{(k)} + m(\mathscr{X}^{(k)})$$

with

$$\mathscr{T}^{(k)} = \mathscr{I} - \mathscr{A}m(\mathscr{X}^{(k)}).$$

Since the matrix multiplication is no longer associative if at least one interval matrix is involved (see also Chapter 10) we have in general that

$$\mathscr{X}^{(k)}(\mathscr{I} - \mathscr{A}m(\mathscr{X}^{(k)}))^{r-1} \neq (\cdots(\mathscr{X}^{(k)}(\mathscr{I} - \mathscr{A}m(\mathscr{X}^{(k)})))\cdots)(\mathscr{I} - \mathscr{A}m(\mathscr{X}^{(k)})).$$

Starting from the same initial interval matrix the sequences generated according to the iteration procedures (5) and (5′) are in general different. Theorem 1 is still valid, however, for the iteration (5) (see Alefeld and Herzberger [4]). We now consider the computational effort required in each iteration step according to (5′).

If $\mathscr{A}$ is an $n \times n$ matrix, then (5′) requires

$$rn^3 \text{ multiplications} \qquad \text{and} \qquad rn^3 - n^2 + n \text{ additions}$$

per step. In practical calculations terms in (5′) that do not contain any intervals explicitly, for example, $\mathscr{T}^{(k)}$, must also be calculated using interval arithmetic in order to guarantee the inclusion of $\mathscr{A}^{-1}$. If we neglect lower powers of n, we see that the computational cost of the algorithm (5′) is proportional to r.

We now wish to estimate the number of iteration steps $\bar{k}$ that for a given $\mathscr{X}^{(0)}$ is required to produce a $\|d(\mathscr{X}^{(\bar{k})})\|$ that is smaller than a given prescribed tolerance. Analogously to the proof of Theorem 1 we get the following relation for (5′)

$$d(\mathscr{X}^{(k+1)}) = d(\mathscr{X}^{(k)})|(\mathscr{I} - \mathscr{A}m(\mathscr{X}^{(k)}))|^{r-1}$$
$$\leqslant d(\mathscr{X}^{(k)})|(\mathscr{I} - \mathscr{A}m(\mathscr{X}^{(0)}))|^{r^k(r-1)};$$

that is,

$$d(\mathscr{X}^{(k+1)}) \leqslant d(\mathscr{X}^{(0)}) \prod_{v=0}^{k} |(\mathscr{I} - \mathscr{A}m(\mathscr{X}^{(0)}))|^{r^v(r-1)}.$$

If we again use a monotonic and multiplicative matrix norm $\|\cdot\|$ and if we furthermore assume that $\|\mathscr{I} - \mathscr{A}m(\mathscr{X}^{(0)})\| < 1$, then it follows that

$$\|d(\mathscr{X}^{(k+1)})\| \leqslant \|d(\mathscr{X}^{(0)})\| \left(\prod_{v=0}^{k} \|\mathscr{I} - \mathscr{A}m(\mathscr{X}^{(0)})\|^{r^v} \right)^{r-1}$$
$$= \|d(\mathscr{X}^{(0)})\| \, \|\mathscr{I} - \mathscr{A}m(\mathscr{X}^{(0)})\|^{r^{k+1}-1}.$$

This expression allows us to estimate $\bar{k}$ under the above conditions.

Starting with the relation

$$\|d(\mathscr{X}^{(\bar{k})})\| \leqslant \|d(\mathscr{X}^{(0)})\| \, \|\mathscr{I} - \mathscr{A}m(\mathscr{X}^{(0)})\|^{r^{\bar{k}}-1}$$

we shall determine that iteration method in dependence on r that requires the least computational effort in order to achieve the prescribed accuracy for $\|d(\mathscr{X}^{(k)})\|$. According to our previous considerations this computational effort may be set proportional to r. Let therefore $r^{(1)} > 1$ and $r^{(2)} > 1$ be given and assume $r^{(1)} \neq r^{(2)}$. After $p^{(1)}$ (resp., $p^{(2)}$) steps in the iteration (5′) with $r = r^{(1)}$ (resp., $r = r^{(2)}$) and starting with the same $\mathscr{X}^{(0)}$ we have expended the same computational effort. That is, we have

$$r^{(1)}p^{(1)} = r^{(2)}p^{(2)}.$$

The accuracy achieved in both methods may be estimated by $(r^{(1)})^{p^{(1)}} - 1$ (resp., $(r^{(2)})^{p^{(2)}} - 1$). We require for an "optimal" iteration procedure with $r = r^{(1)}$ that for all other values $r^{(2)}$ and all steps $p^{(1)}$, $p^{(2)}$ we have

$$(r^{(1)})^{p^{(1)}} > (r^{(2)})^{p^{(2)}}$$

or because $p^{(2)} = r^{(1)}p^{(1)}/r^{(2)}$ that

$$(r^{(1)})^{1/r^{(1)}} > (r^{(2)})^{1/r^{(2)}}$$

is always valid. Since the function $x^{1/x}$ has a maximum at $x = 3$ for integer x we have in this sense an optimal version of the iteration (5′) for this value of x.

We remark additionally that method (5) may also be carried out for complex matrices using the arithmetic in $R(\mathbb{C})$. The same statements are valid as those in Theorem 1. We refer in this connection to the more general considerations found in Alefeld and Herzberger [5] as well as the paper [B281] in the Bibliography.

We shall now consider methods for the monotonic inclusion of the inverse matrix that have quite similar properties to the method (9). These methods were considered by Mönch [13]. The basic calculation uses no interval arithmetic. It calculates the upper and lower bounds (matrices) by a separate formula. This method is only applicable, however, whenever $\mathscr{A}^{-1} \geqslant \mathscr{O}$.

Let therefore $\mathscr{A}$ be a nonsingular matrix and let r be a natural number for which $r \geqslant 2$. We then consider the iteration procedure

$$(12)\quad \begin{cases} \mathscr{X}^{(k+1)} = \mathscr{X}^{(k)} + (\mathscr{I} - \mathscr{X}^{(k)}\mathscr{A}) \sum_{\nu=0}^{r-2} (\mathscr{I} - \mathscr{X}^{(k)}\mathscr{A})^{\nu}\mathscr{X}^{(k)}, \\ \mathscr{Y}^{(k+1)} = \mathscr{Y}^{(k)} + (\mathscr{I} - \mathscr{Y}^{(k)}\mathscr{A}) \sum_{\nu=0}^{r-2} (\mathscr{I} - \mathscr{X}^{(k)}\mathscr{A})^{\nu}\mathscr{X}^{(k)}, \qquad k \geqslant 0, \end{cases}$$

with $\mathscr{X}^{(0)}$, $\mathscr{Y}^{(0)}$ prescribed.

When we execute this procedure we obtain two sequences of point matrices $\{\mathscr{X}^{(k)}\}_{k=0}^{\infty}$, $\{\mathscr{Y}^{(k)}\}_{k=0}^{\infty}$ for which the following theorem holds.

Theorem 3: Let $\mathscr{A}$ be a nonsingular $n \times n$ matrix for which $\mathscr{A}^{-1} \geqslant \mathscr{O}$. Furthermore, let $\mathscr{X}^{(0)}$, $\mathscr{Y}^{(0)}$ be two $n \times n$ matrices for which

$$\mathscr{X}^{(0)} \geqslant \mathscr{O} \qquad \text{and} \qquad \mathscr{X}^{(0)}\mathscr{A} \leqslant \mathscr{I} \leqslant \mathscr{Y}^{(0)}\mathscr{A}$$

hold. The sequences $\{\mathscr{X}^{(k)}\}_{k=0}^{\infty}$ and $\{\mathscr{Y}^{(k)}\}_{k=0}^{\infty}$ are calculated according to (12). Then the following statements are valid:

$$(13)\quad \mathscr{O} \leqslant \mathscr{X}^{(0)} \leqslant \cdots \leqslant \mathscr{X}^{(k)} \leqslant \mathscr{X}^{(k+1)} \leqslant \cdots \leqslant \mathscr{A}^{-1} \leqslant \cdots \leqslant \mathscr{Y}^{(k+1)} \leqslant \mathscr{Y}^{(k)} \leqslant \cdots \leqslant \mathscr{Y}^{(0)}.$$

(7′) Both sequences $\{\mathscr{X}^{(k)}\}_{k=0}^{\infty}$, $\{\mathscr{Y}^{(k)}\}_{k=0}^{\infty}$ converge to $\mathscr{A}^{-1}$ iff the spectral radius $\rho(\mathscr{I} - \mathscr{X}^{(0)}\mathscr{A})$ is less than 1.

(14) If the procedure converges then the quantities

$$d^{(k)} = \|\mathscr{Y}^{(k)} - \mathscr{X}^{(k)}\|$$

satisfy the relation

$$d^{(k+1)} \leqslant \gamma(d^{(k)})^{r}, \qquad \gamma \geqslant 0.$$

If one therefore interprets (12) as an iteration method for calculating the interval matrices $([x_{ij}^{(k)}, y_{ij}^{(k)}])$ then it holds that

$$O_R((12), \mathscr{A}^{-1}) \geqslant r.$$

(See Appendix A, Theorem 2.)

Proof: Of (13): We prove by complete induction that the relations

$$\mathcal{O} \leqslant \mathcal{X}^{(0)} \leqslant \cdots \leqslant \mathcal{X}^{(k-1)} \leqslant \mathcal{X}^{(k)},$$

$$\mathcal{Y}^{(k)} \leqslant \mathcal{Y}^{(k-1)} \leqslant \cdots \leqslant \mathcal{Y}^{(0)},$$

$$\mathcal{X}^{(k)}\mathcal{A} \leqslant \mathcal{I} \leqslant \mathcal{Y}^{(k)}\mathcal{A}$$

are valid for $k \geqslant 0$.

These inequalities are trivially true for $k = 0$ from the assumptions. From

$$\mathcal{I} - \mathcal{Y}^{(k)}\mathcal{A} \leqslant \mathcal{O} \leqslant \mathcal{I} - \mathcal{X}^{(k)}\mathcal{A}, \qquad \mathcal{X}^{(k)} \geqslant \mathcal{O}$$

it follows that

$$(\mathcal{I} - \mathcal{Y}^{(k)}\mathcal{A}) \sum_{\nu=0}^{r-2} (\mathcal{I} - \mathcal{X}^{(k)}\mathcal{A})^{\nu}\mathcal{X}^{(k)}$$

$$\leqslant \mathcal{O} \leqslant (\mathcal{I} - \mathcal{X}^{(k)}\mathcal{A}) \sum_{\nu=0}^{r-2} (\mathcal{I} - \mathcal{X}^{(k)}\mathcal{A})^{\nu}\mathcal{X}^{(k)};$$

that is,

$$\mathcal{X}^{(k)} \leqslant \mathcal{X}^{(k+1)}, \qquad \mathcal{Y}^{(k+1)} \leqslant \mathcal{Y}^{(k)}.$$

From

$$\mathcal{I} - \mathcal{X}^{(k+1)}\mathcal{A} = \mathcal{I} - \mathcal{X}^{(k)}\mathcal{A} - (\mathcal{I} - \mathcal{X}^{(k)}\mathcal{A}) \sum_{\nu=0}^{r-2} (\mathcal{I} - \mathcal{X}^{(k)}\mathcal{A})^{\nu}\mathcal{X}^{(k)}\mathcal{A}$$

$$= (\mathcal{I} - \mathcal{X}^{(k)}\mathcal{A})^{r} \geqslant \mathcal{O}$$

and

$$\mathcal{Y}^{(k+1)}\mathcal{A} - \mathcal{I} = \mathcal{Y}^{(k)}\mathcal{A} - \mathcal{I} - (\mathcal{Y}^{(k)}\mathcal{A} - \mathcal{I}) \sum_{\nu=0}^{r-2} (\mathcal{I} - \mathcal{X}^{(k)}\mathcal{A})^{\nu}\mathcal{X}^{(k)}\mathcal{A}$$

$$= (\mathcal{Y}^{(k)}\mathcal{A} - \mathcal{I})(\mathcal{I} - \mathcal{X}^{(k)}\mathcal{A})^{r-1} \geqslant \mathcal{O},$$

it follows that

$$\mathcal{X}^{(k+1)}\mathcal{A} \leqslant \mathcal{I} \leqslant \mathcal{Y}^{(k+1)}\mathcal{A}.$$

Since $\mathcal{A}^{-1} \geqslant \mathcal{O}$ one therefore obtains

$$\mathcal{X}^{(k+1)} \leqslant \mathcal{A}^{-1} \leqslant \mathcal{Y}^{(k+1)}.$$

Of (7′): Using the relations

$$\mathcal{I} - \mathcal{X}^{(k+1)}\mathcal{A} = (\mathcal{I} - \mathcal{X}^{(k)}\mathcal{A})^{r},$$

$$\mathcal{Y}^{(k+1)}\mathcal{A} - \mathcal{I} = (\mathcal{Y}^{(k)}\mathcal{A} - \mathcal{I})(\mathcal{I} - \mathcal{X}^{(k)}\mathcal{A})^{r-1}$$

from the proof of (13) one is able to show using complete induction that

$$\mathscr{I} - \mathscr{X}^{(k)}\mathscr{A} = (\mathscr{I} - \mathscr{X}^{(0)}\mathscr{A})^{r^k},$$

$$\mathscr{Y}^{(k)}\mathscr{A} - \mathscr{I} = (\mathscr{Y}^{(0)}\mathscr{A} - \mathscr{I})(\mathscr{I} - \mathscr{X}^{(0)}\mathscr{A})^{r^k - 1}$$

from which the statement follows.

Of (14): Again using the relations used in the proof of (13) we get

$$\|\mathscr{A}^{-1} - \mathscr{X}^{(k+1)}\| \leqslant \|\mathscr{A}\|^{r-1}\|\mathscr{A}^{-1} - \mathscr{X}^{(k)}\|^r,$$

$$\|\mathscr{Y}^{(k+1)} - \mathscr{A}^{-1}\| \leqslant \|\mathscr{A}\|^{r-1}\|\mathscr{Y}^{(k)} - \mathscr{A}^{-1}\|\,\|\mathscr{A}^{-1} - \mathscr{X}^{(k)}\|^{r-1}.$$

Using a monotonic matrix norm this may then be estimated further as

$$\begin{aligned}\|\mathscr{Y}^{(k+1)} - \mathscr{X}^{(k+1)}\| &\leqslant \|\mathscr{A}^{-1} - \mathscr{X}^{(k+1)}\| + \|\mathscr{Y}^{(k+1)} - \mathscr{A}^{-1}\| \\ &\leqslant 2\|\mathscr{A}\|^{r-1}\|\mathscr{Y}^{(k)} - \mathscr{X}^{(k)}\|^r;\end{aligned}$$

that is, $d^{(k+1)} \leqslant \gamma(d^{(k)})^r$, $\gamma = 2\|\mathscr{A}\|^{r-1}$. From Theorem 2, Appendix A, the relation then follows. ■

The method (12) uses no interval operations. In spite of this it generates a monotonic sequence of bounds for $\mathscr{A}^{-1}$ in a manner similar to method (9). A necessary and sufficient criterion for convergence has also been given similar to the one proven for (5). The applicability of (12) is limited, however, by the requirement that $\mathscr{A}^{-1} \geqslant \mathscr{O}$.

For certain classes of matrices one may give quite general starting values $\mathscr{X}^{(0)}, \mathscr{Y}^{(0)}$ for which we always have convergence in (12). If the matrix $\mathscr{A} = (a_{ij})$ satisfies the assumptions

$$a_{ii} > 0, \qquad 1 \leqslant i \leqslant n,$$

$$a_{ij} \leqslant 0, \qquad 1 \leqslant i, j \leqslant n, \quad i \neq j,$$

$$\sum_{i=1}^{n} a_{ij} > 0, \qquad 1 \leqslant j \leqslant n,$$

then $\mathscr{A}$ is an M matrix (see, for example, Varga [17, p. 85ff]). The starting matrices $\mathscr{X}^{(0)} = (x_{ij})$ and $\mathscr{Y}^{(0)} = (y_{ij})$ where

$$x_{ij} = \begin{cases} 1/a_{ii} & \text{for} \quad i = j \\ 0 & \text{otherwise} \end{cases} \qquad \text{and} \qquad y_{ij} = 1/\textstyle\sum_{\nu=1}^{n} a_{\nu i}, \quad 1 \leqslant i, j \leqslant n,$$

satisfy the conditions of Theorem 3.

We now consider a nonsingular real $n \times n$ matrix $\mathscr{A} = (a_{ij})$ whose rows have been interchanged to allow the factorization

$$\mathscr{A} = (\mathscr{I} + \mathscr{L}^*)\mathscr{U}^* \tag{15}$$

where $\mathscr{I}$ is the identity matrix, $\mathscr{L}^*$ a strictly lower triangular matrix, and $\mathscr{U}^*$ an upper triangular matrix. This factorization may be computed, as is well known, with the aid of the Gaussian algorithm (for example, see Stoer and Bulirsch [16]). We wish to develop an iteration procedure that improves bounds for the elements of $\mathscr{L}^*$ and $\mathscr{U}^*$ iteratively. Such bounds may be computed when one executes the Gaussian algorithm for a point matrix $\mathscr{A}$ that includes all the rounding errors by using a machine interval arithmetic that rounds outward.

It is therefore assumed that $\mathscr{L}^{(0)}$ is a given strictly lower triangular interval matrix and $\mathscr{U}^{(0)}$ is a given upper triangular interval matrix for which

$$\mathscr{L}^* \in \mathscr{L}^{(0)}, \qquad \mathscr{U}^* \in \mathscr{U}^{(0)}. \tag{16}$$

In order to develop the method it is first assumed that $\mathscr{L}$ and $\mathscr{U}$ are arbitrary but fixed strictly lower (resp., upper) triangular matrices. From (15) we then have

$$\mathscr{R} = (r_{ik}) := \mathscr{A} - (\mathscr{I} + \mathscr{L})\mathscr{U} = (\mathscr{I} + \mathscr{L}^*)(\mathscr{U}^* - \mathscr{U}) + (\mathscr{L}^* - \mathscr{L})\mathscr{U}.$$

We consider the matrices $\mathscr{U}^* - \mathscr{U}$ and $\mathscr{L}^* - \mathscr{L}$ to be unknowns and the factor $\mathscr{I} + \mathscr{L}^*$ in front of $\mathscr{U}^* - \mathscr{U}$ as known and then solve this equation by solving first for a row of $\mathscr{U}^* - \mathscr{U}$ and then for a column of $\mathscr{L}^* - \mathscr{L}$ in turn. Under the assumption that $u_{ii} = 0$ for $1 \leqslant i \leqslant n$ one then obtains

$$\left\{\begin{aligned} u_{ik}^* - u_{ik} &= r_{ik} - \sum_{j=1}^{i-1} l_{ij}^*(u_{jk}^* - u_{jk}) - \sum_{j=1}^{i-1} (l_{ij}^* - l_{ij})u_{jk}, && i \leqslant k \leqslant n, \\ l_{ki}^* - l_{ki} &= \frac{1}{u_{ii}}\left\{r_{ki} - \sum_{j=1}^{i} l_{kj}^*(u_{ji}^* - u_{ji}) - \sum_{j=1}^{i-1} (l_{kj}^* - l_{kj})u_{ji}\right\}, && i < k \leqslant n, \end{aligned}\right.$$

$$1 \leqslant i \leqslant n.$$

Letting

$$\left.\begin{aligned} r_{ik} &= a_{ik} - \sum_{j=1}^{i-1} l_{ij}u_{jk} - u_{ik}, && i \leqslant k \leqslant n \\ r_{ki} &= a_{ki} - \sum_{j=1}^{i} l_{kj}u_{ji}, && i+1 \leqslant k \leqslant n \end{aligned}\right\} \quad 1 \leqslant i \leqslant n,$$

one obtains the relations:

$$\begin{cases} u_{ik}^* = a_{ik} - \sum_{j=1}^{i-1} l_{ij}^* u_{jk} - \sum_{j=1}^{i-1} \mathbf{l}_{ij}^*(u_{jk}^* - u_{jk}), & i \leqslant k \leqslant n, \\ l_{ki}^* = \dfrac{1}{u_{ii}}\left\{a_{ki} - \sum_{j=1}^{i-1} l_{kj}^* u_{ji} - \sum_{j=1}^{i} \mathbf{l}_{kj}^*(u_{ji}^* - u_{ji})\right\}, & i < k < n, \\ 1 \leqslant i \leqslant n. \end{cases}$$

The italic quantities that are boldfaced are from the factor $\mathcal{I} + \mathcal{L}^*$ that was assumed to be known. If one uses the assumption that $\mathcal{L}^* \in \mathcal{L}^{(0)}$ for these quantities and if one choses $\mathcal{U} := \mathcal{U}^{(0)} \in \mathcal{U}^{(0)}$ with $u_{ii}^{(0)} \neq 0$, $1 \leqslant i \leqslant n$, then one gets the following including intervals for the elements of $\mathcal{L}^*$ and $\mathcal{U}^*$:

$$\begin{cases} u_{ik}^* \in U_{ik}^{(1)} = \left\{a_{ik} - \sum_{j=1}^{i-1} L_{ij}^{(1)} u_{jk}^{(0)} - \sum_{j=1}^{i-1} L_{ij}^{(0)}(U_{jk}^{(1)} - u_{jk}^{(0)})\right\} \cap U_{ik}^{(0)}, \\ \qquad i \leqslant k \leqslant n, \\ l_{ki}^* \in L_{ki}^{(1)} = \left\{\dfrac{1}{u_{ii}^{(0)}}\left(a_{ki} - \sum_{j=1}^{i-1} L_{kj}^{(1)} u_{ji}^{(0)} - \sum_{j=1}^{i} L_{kj}^{(0)}(U_{ji}^{(1)} - u_{ji}^{(0)})\right)\right\} \cap L_{ki}^{(0)}, \\ \qquad i < k \leqslant n, \\ 1 \leqslant i \leqslant n. \end{cases}$$

The systematic repetition of this procedure leads to the following iteration method:

$$(17)\quad \begin{cases} U_{ik}^{(m+1)} = \left\{a_{ik} - \sum_{j=1}^{i-1} L_{ij}^{(m+1)} u_{jk}^{(m)} - \sum_{j=1}^{i-1} L_{ij}^{(m)}(U_{jk}^{(m+1)} - u_{jk}^{(m)})\right\} \cap U_{ik}^{(m)}, \\ \qquad i \leqslant k \leqslant n, \\ L_{ki}^{(m+1)} = \left\{\dfrac{1}{u_{ii}^{(m)}}\left(a_{ki} - \sum_{j=1}^{i-1} L_{kj}^{(m+1)} u_{ji}^{(m)} - \sum_{j=1}^{i} L_{kj}^{(m)}(U_{ji}^{(m+1)} - u_{ji}^{(m)})\right)\right\} \cap L_{ki}^{(m)}, \\ \qquad i < k \leqslant n, \\ 1 \leqslant i \leqslant n, \\ m \geqslant 0. \end{cases}$$

In the same manner as in the development of the first step of the method, one shows in general that the following statements are valid.

Theorem 4: Let the matrix $\mathcal{A}$ have the factorization $\mathcal{A} = (\mathcal{I} + \mathcal{L}^*)\mathcal{U}^*$. Let $\mathcal{L}^* \in \mathcal{L}^{(0)}$, $\mathcal{U}^* \in \mathcal{U}^{(0)}$. If $\mathcal{L}^{(m)} \in \mathcal{L}^{(m)}$, $\mathcal{U}^{(m)} \in \mathcal{U}^{(m)}$, it follows that

$$(18)\qquad \mathcal{L}^* \in \mathcal{L}^{(m+1)}, \qquad \mathcal{U}^* \in \mathcal{U}^{(m+1)}$$

for all $m \geqslant 0$. ■

We note that (17) may be carried out with starting intervals of arbitrary but finite width provided the assumption (16) is satisfied. The only division occurs in the calculation of $L_{ki}^{(m+1)}$. Since $0 \neq u_{ii}^* \in U_{ii}^{(m)}$ one may always chose $u_{ii}^{(m)} \in U_{ii}^{(m)}$ where $u_{ii}^{(m)} \neq 0$.

We now show that method (17) gives the exact triangular decomposition in a finite number of steps if no rounding errors occur.

Theorem 5: Method (17) computes the exact factorization of the $n \times n$ matrix $\mathscr{A}$ in at most $2n - 1$ steps under the assumption that the conditions of the previous theorem hold.

Proof: For $m = 0$ we get from (17) for $i = 1$ that

$$U_{1k}^{(1)} = a_{1k} \cap U_{1k}^{(0)} = a_{1k} = u_{1k}^*, \qquad 1 \leqslant k \leqslant n.$$

This means that for arbitrary starting matrices having the inclusion property (16), the matrix $\mathscr{U}^{(1)}$ has the exact values of the triangular decomposition in the first row. $\mathscr{U}^{(2)}$ has in the same manner as $\mathscr{U}^{(1)}$ the exact values in the first row. For the first column of $\mathscr{L}^{(2)}$ one therefore gets from (17) with $m = 1$, $i = 1$ that

$$L_{k1}^{(2)} = \left(\frac{a_{k1}}{u_{11}}\right) \cap L_{k1}^{(1)} = \frac{a_{k1}}{a_{11}} = l_{k1}^*, \qquad 2 \leqslant k \leqslant n.$$

Therefore, after the second iteration step both the first column of $\mathscr{L}^{(2)}$ as well as the first row of $\mathscr{U}^{(2)}$ are exact. We now show that if the first i rows of $\mathscr{U}^{(m)}$ and the first i columns of $\mathscr{L}^{(m)}$ are exact then we have that $\mathscr{U}^{(m+1)}$ has at least $i + 1$ exact rows (and $\mathscr{L}^{(m+1)}$ at least i exact columns). This holds true for $i = 0$ by the above argument. For $i > 0$ it will be shown using complete induction.

We first note that $\mathscr{L}^{(m+1)}$ and $\mathscr{U}^{(m+1)}$ have the same elements as $\mathscr{L}^{(m)}$ and $\mathscr{U}^{(m)}$ in the first i rows and columns. That is, these elements are still exact. This follows immediately from (17). From (17) one also obtains for the $(i + 1)$th row of $\mathscr{U}^{(m+1)}$ that

$$U_{i+1,k}^{(m+1)} = \left\{a_{i+1,k} - \sum_{j=1}^{i-1} l_{i+1,j}^* u_{jk}^*\right\} \cap U_{i+1,k}^{(m)} = u_{i+1,k}^*, \qquad i + 1 \leqslant k \leqslant n.$$

In order to complete the proof we have to show that if the first i rows of $\mathscr{U}^{(m)}$ and the first $i - 1$ columns of $\mathscr{L}^{(m)}$ are exact for some $m \geqslant 0$ then it follows that $\mathscr{L}^{(m+1)}$ has at least i exact columns (and $\mathscr{U}^{(m+1)}$ has at least i exact rows).

This was proved above for $i = 1$. For $i > 1$ we note from (17) that $\mathscr{U}^{(m+1)}$ has at least the same number of exact rows as $\mathscr{U}^{(m)}$ and $\mathscr{L}^{(m+1)}$ has at least the same number of exact columns as $\mathscr{L}^{(m)}$. For the ith column of $\mathscr{L}^{(m+1)}$ it then follows from (17) that

$$L_{ki}^{(m+1)} = \left\{\frac{1}{u_{ii}^*}\left(a_{ki} - \sum_{j=1}^{i-1} l_{kj}^* u_{ji}^*\right)\right\} \cap L_{ki}^{(m)} = l_{ki}^*, \qquad i + 1 \leqslant k \leqslant n.$$

After at most $2n - 1$ steps one then obtains the exact solution. ■

The next theorem shows that the method has the so-called "quadratic" convergence property, well known from the Newton–Raphson method.

Theorem 6: Let $d^{(m)} := \max_{1 \leqslant i, j \leqslant n}\{\max\{d(L_{ij}^{(m)}), d(U_{ij}^{(m)})\}\}$. Then it holds for (17) that

$$d^{(m+1)} \leqslant \alpha(d^{(m)})^2$$

with a nonnegative real number α, which is independent of m: The widths of the intervals are approximately squared in each step.

Proof (*by mathematical induction*): We get from (17):

$$(17')\begin{cases} d(U_{ik}^{(m+1)}) \leqslant \sum_{j=1}^{i-1} d(L_{ij}^{(m+1)})|u_{jk}^{(m)}| + \sum_{j=1}^{i-1} d(L_{ij}^{(m)})|U_{jk}^{(m+1)} - u_{jk}^{(m)}| \\ \qquad + \sum_{j=1}^{i-1} |L_{ij}^{(m)}|\, d(U_{jk}^{(m+1)}), \qquad i \leqslant k \leqslant n, \\ d(L_{ki}^{(m+1)}) \leqslant \dfrac{1}{|u_{ii}^{(m)}|}\left\{ \sum_{j=1}^{i-1} d(L_{kj}^{(m+1)})|u_{ji}^{(m)}| \right. \\ \qquad \left. + \sum_{j=1}^{i} d(L_{kj}^{(m)})|U_{ji}^{(m+1)} - u_{ji}^{(m)}| + \sum_{j=1}^{i} |L_{kj}^{(m)}|\, d(U_{ji}^{(m+1)}) \right\}, \\ \qquad\qquad i < k \leqslant n, \end{cases}$$

$$1 \leqslant i \leqslant n.$$

Now define for $0 \notin U_{ii}^{(0)}$, $1 \leqslant i \leqslant n$,

$$(17'')\begin{cases} \alpha_{ik} = \begin{cases} \sum_{j=1}^{i-1} (\beta_{ij}|U_{jk}^{(0)}| + 1 + |L_{ij}^{(0)}|\alpha_{jk}), & i \leqslant k \leqslant n \\ 0 & \text{otherwise}, \end{cases} \\ \beta_{ki} = \begin{cases} \left|\dfrac{1}{U_{ii}^{(0)}}\right| \left\{ \sum_{j=1}^{i-1} \beta_{kj}|U_{ji}^{(0)}| + \sum_{j=1}^{i} (1 + |L_{kj}^{(0)}|\alpha_{ji}) \right\}, & i < k \leqslant n \\ 0 & \text{otherwise}, \end{cases} \end{cases}$$

and finally

$$\alpha = \max_{1 \leqslant i, k \leqslant n} \{\max\{\alpha_{ik}, \beta_{ki}\}\}.$$

Using definition (17″) we immediately get from (17′) for $i = 1$ that

$$d(U_{1k}^{(m+1)}) \leqslant \alpha_{1k}(d^{(m)})^2, \qquad 1 \leqslant k \leqslant n,$$

and

$$d(L_{k1}^{(m+1)}) \leqslant \beta_{k1}(d^{(m)})^2, \qquad 1 < k \leqslant n.$$

Assume now that for the first $i \neq 1$ rows and columns

$$(17''') \qquad \begin{cases} d(U_{lk}^{(m+1)}) \leqslant \alpha_{lk}(d^{(m)})^2, & l \leqslant k \leqslant n, \\ d(L_{kl}^{(m+1)}) \leqslant \beta_{kl}(d^{(m)})^2, & l \leqslant k \leqslant n, \end{cases}$$

hold ($1 \leqslant l \leqslant i-1$). This is certainly true for $l = 1$. Then we get from (17′) and (17″)

$$\begin{aligned} d(U_{ik}^{(m+1)}) &\leqslant \sum_{j=1}^{i-1} \beta_{ij}|U_{jk}^{(0)}|(d^{(m)})^2 + \sum_{j=1}^{i-1} (d^{(m)})^2 + \sum_{j=1}^{i-1} |L_{ij}^{(0)}|\alpha_{jk}(d^{(m)})^2 \\ &= \alpha_{ik}(d^{(m)})^2, \qquad i \leqslant k \leqslant n. \end{aligned}$$

Similarly,

$$\begin{aligned} d(L_{ki}^{(m+1)}) &\leqslant \left|\frac{1}{U_{ii}^{(0)}}\right| \left\{ \sum_{j=1}^{i-1} \beta_{kj}|U_{ji}^{(0)}|(d^{(m)})^2 + \sum_{j=1}^{i} (d^{(m)})^2 + \sum_{j=1}^{i} |L_{kj}^{(0)}|\alpha_{ji}(d^{(m)})^2 \right\} \\ &= \beta_{ki}(d^{(m)})^2, \qquad i < k \leqslant n. \end{aligned}$$

From these relations the assertion

$$d^{(m+1)} \leqslant \alpha(d^{(m)})^2$$

follows. ■

If one performs (17) on a digital computer using a machine interval arithmetic, then in contrast to the content of Theorem 5 one will usually not have computed the exact triangular decomposition of $\mathscr{A}$ after a finite number of steps. We shall now discuss the final accuracy that one can reach. For this discussion we make the same assumption that led us to formulas (4.15a) and (4.15b) and therefore also to (4.22) and (4.23). (4.22) and (4.23) were used in order to prove (4.24) and (4.25). These two formulas are now applied to (17). This gives for the width of computed elements

$$\begin{cases} d(\bar{U}_{ik}^{(m+1)}) \leqslant d(U_{ik}^{(m+1)}) + 2\varepsilon \displaystyle\sum_{j=1}^{2i-2} |\bar{S}_j| \\ \qquad + 2\varepsilon(3 + 3\varepsilon + \varepsilon^2) \displaystyle\sum_{j=1}^{i-1} (|(L_{ij}^{(m)}|\,|\bar{U}_{jk}^{(m+1)} - u_{jk}^{(m)}| + |\bar{L}_{ij}^{(m+1)}|\,|u_{jk}^{(m)}|), \\ \qquad\qquad i \leqslant k \leqslant n, \\ d(\bar{L}_{ki}^{(m+1)}) \leqslant d(L_{ki}^{(m+1)}) + \left|\dfrac{1}{u_{ii}^{(m)}}\right| \left\{ 2\varepsilon \displaystyle\sum_{j=1}^{2i-1} |\bar{T}_j| + 2\varepsilon_{2i}|\bar{T}_{2i}| \right. \\ \qquad + 2\varepsilon(3 + 3\varepsilon + \varepsilon^2)\left(\displaystyle\sum_{j=1}^{i-1} |\bar{L}_{kj}^{(m+1)}|\,|u_{ji}^{(m)}| \right. \\ \qquad \left.\left. + \displaystyle\sum_{j=1}^{i} |L_{kj}^{(m)}|\,|\bar{U}_{ji}^{(m+1)} - u_{ji}^{(m)}| \right)\right\}, \qquad i < k \leqslant n, \end{cases}$$

$1 \leqslant i \leqslant n.$

In these inequalities $\bar{S}_j$ and $\bar{T}_j$ denote the actually computed intermediate results. These inequalities may be interpreted in the following manner: Assume that the elements of the matrix $\mathscr{L}^*$ are all not greater than 1 in absolute value. This assumption does hold at least approximately if the rows of the matrix $\mathscr{A}$ are ordered in such a way that in the process of performing Gaussian elimination with column pivoting numerically no rows have to be exchanged. If the widths of the intervals $L_{ij}^{(m)}$ are not too large, then $|L_{ij}^{(m)}|$ and, because of forming intersection in (17), also $|L_{ij}^{(m+1)}|$ are not much greater than 1. Under the same assumptions and with the same reasoning it follows that $|\bar{U}_{jk}^{(m+1)}| - |u_{jk}^{(m)}|$ and $|\bar{U}_{ji}^{(m+1)}| - |u_{ji}^{(m)}|$ are small. For small widths of the elements in the mth step, we therefore conclude that the difference between $d(\bar{U}_{ik}^{(m+1)})$ and $d(U_{ik}^{(m+1)})$ is essentially dependent on the size of $|u_{jk}^{(m)}|$ and on the size of the intermediate results. The same is true for the difference of $d(\bar{L}_{ki}^{(m+1)})$ and $d(L_{ki}^{(m+1)})$ with the additional property that small values of $|u_{ii}^{(m)}|$ can make the difference even worse.

Since the first terms in both of the above inequalities are by Theorem 6 approximately the squares of the corresponding terms of the preceding step, one will obtain small width if the following conditions hold:

(a) the elements of $\mathscr{U}^*$ are in absolute value not much larger than one;
(b) the diagonal elements of $\mathscr{U}^*$ are in absolute value not much smaller than one;
(c) the elements of $\mathscr{L}^*$ are in absolute value not greater than one;
(d) the computed intermediate results in (17) are in absolute value not too large.

REFERENCES

[1] J. Albrecht, Bemerkungen zum Iterationsverfahren von SCHULZ zur Matrixinversion. *Z. Angew. Math. Mech.* **41**, 262–263 (1961).

[2] G. Alefeld and J. Herzberger, Über die Berechnung der inversen Matrix mit Hilfe der Intervallrechnung. *Elektron. Rech.* **12**, 259–261 (1970).

[3] G. Alefeld and J. Herzberger, Verfahren höherer Ordnung zur iterativen Einschließung der inversen Matrix. *Z. Angew. Math. Phys.* **21**, 819–824 (1970).

[4] G. Alefeld and J. Herzberger, Matrizeninvertierung mit Fehlererfassung. *Elektron. Datenverarbeitung* **12**, 410–416 (1970).

[5] G. Alefeld and J. Herzberger, Zur Invertierung linearer und beschränkter Operatoren. *Math. Z.* **120**, 309–317 (1971).

[6] G. Alefeld and J. Herzberger, Einschließungsverfahren zur Berechnung des inversen Operators. *Z. Angew. Math. Mech.* **52**, T 197–T 198 (1972).

[7] G. Alefeld and J. G. Rokne, On the iterative improvement of approximate triangular factorizations. *Beiträge Numer. Math.* (to be published).

[8] M. Altmann, An optimum cubically convergent iterative method of inverting a linear bounded operator in Hilbert space. *Pacific J. Math.* **10**, 1107–1113 (1960).

[9] W. Dück, Fehlerabschätzung für das Iterationsverfahren von SCHULZ zur Berechnung der Inversen einer Matrix. *Z. Angew. Math. Mech.* **40**, 192–194 (1960).

[10] R. T. Gregory and D. L. Karney, "A Collection of Matrices for Testing Computational Algorithms." Wiley (Interscience), New York, 1969.

[11] P. Forster, Bemerkungen zum Iterationsverfahren von Schulz zur Bestimmung der Inversen einer Matrix. *Numer. Math.* **12**, 211–214 (1968).

[12] F. Krückeberg, Inversion von Matrizen mit Fehlererfassung. *Z. Angew. Math. Mech.* **46**, T 69 (1966).

[13] W. Mönch, Monotone Einschließung von positiven Inversen. *Z. Angew. Math. Mech.* **53**, 207–208 (1973).

[14] W. V. Petryshyn, On generalized inverses and on the convergence of $(I - \beta K)^n$ with application to iterative methods. *J. Math. Anal. Appl.* **18**, 417–439 (1967).

[15] G. Schulz, Iterative Berechnung der reziproken Matrix. *Z. Angew. Math. Mech.* **13**, 57–59 (1933).

[16] J. Stoer and R. Bulirsch, "Introduction to Numerical Analysis." Springer-Verlag, Berlin and New York, 1980.

[17] R. S. Varga, "Matrix Iterative Analysis." Prentice-Hall, Englewood Cliffs, New Jersey, 1962.

Chapter 19 / NEWTON-LIKE METHODS FOR NONLINEAR SYSTEMS OF EQUATIONS

We consider methods for the iterative inclusion of a solution of a nonlinear system of equations. We are given a set of functions $f_i(\underset{\cdot}{x})$, $1 \leqslant i \leqslant n$, of the vector variable $\underset{\cdot}{x} = (x_1, \ldots, x_n)^T$ that we combine into a vector function $\underset{\cdot}{f}(\underset{\cdot}{x}) = (f_1(\underset{\cdot}{x}), \ldots, f_n(\underset{\cdot}{x}))^T$. Let the Frechét derivative $\underset{\cdot}{f}'(\underset{\cdot}{x})$ of $\underset{\cdot}{f}(\underset{\cdot}{x})$ exist in $\dot{\mathfrak{B}} \subseteq V_n(\mathbb{R})$ and assume that $x^{(0)} = (X_1^{(0)}, \ldots, X_n^{(0)})^T \stackrel{\cdot}{\subseteq} \dot{\mathfrak{B}}$. Furthermore we assume that the Frechét derivative has an interval arithmetic evaluation over $x^{(0)}$. Let now $\underset{\cdot}{y} = (y_1, \ldots, y_n)^T \in x^{(0)}$ be a vector satisfying the equation

$$(1) \qquad \underset{\cdot}{f}(\underset{\cdot}{y}) = \underset{\cdot}{o}.$$

From the above assumptions we may approximate $f_i(\underset{\cdot}{x})$ linearly at the point $\underset{\cdot}{y} \in x^{(0)}$ using the Taylor series approximation

$$(2) \qquad f_i(\underset{\cdot}{x}) = f_i(\underset{\cdot}{y}) + \sum_{r=1}^{n} \frac{\partial}{\partial x_r} f_i(\underset{\cdot}{y} + \theta_i(\underset{\cdot}{x} - \underset{\cdot}{y}))(x_r - y_r),$$

$$0 < \theta_i < 1, \qquad 1 \leqslant i \leqslant n, \qquad \underset{\cdot}{x} \in x^{(0)}.$$

From (1) it follows that

$$\underset{\cdot}{f}(\underset{\cdot}{x}) = \underset{\cdot}{\mathcal{J}}(\underset{\cdot}{x})(\underset{\cdot}{x} - \underset{\cdot}{y})$$

is satisfied by $\underset{\cdot}{y}$, where the matrix $\underset{\cdot}{J}(\underset{\cdot}{x})$ is defined by

$$(3) \qquad \underset{\cdot}{J}(\underset{\cdot}{x}) := \left(\frac{\partial}{\partial x_r} f_i(\underset{\cdot}{z}) \bigg|_{\underset{\cdot}{z} = \underset{\cdot}{y} + \theta_i(\underset{\cdot}{x} - \underset{\cdot}{y})} \right).$$

By the matrix $\underset{\cdot}{J}(\underset{\cdot}{x})$ we understand

$$\underset{\cdot}{J}(\underset{\cdot}{x}, \underset{\cdot}{y}, \theta_1, \ldots, \theta_n)$$

since it is dependent on all the quantities occurring in the definition. Since the values θ_i lie in the open interval $(0, 1)$ we have that

$$\underset{\cdot}{y} + \theta_i(\underset{\cdot}{x} - \underset{\cdot}{y}) \in x^{(0)}.$$

Let the interval matrix $f'(x^{(0)})$ denote the interval evaluation of the Frechét derivative of $\underset{\cdot}{f}(\underset{\cdot}{x})$ over the interval $x^{(0)}$. It therefore follows that the set of linear equations

$$(4) \qquad \underset{\cdot}{f}(\underset{\cdot}{x}) = \underset{\cdot}{F} \cdot (\underset{\cdot}{x} - \underset{\cdot}{z}) \qquad \text{with} \quad \underset{\cdot}{F} \in f'(x^{(0)})$$

contains the vector $\underset{\cdot}{y}$ as one of the solutions $\underset{\cdot}{z}$ since $\underset{\cdot}{J}(\underset{\cdot}{x}) \in f'(x^{(0)})$. The task is now to calculate an including vector $x^{(1)}$ for the solution set of (4) and then use this interval vector as a potentially improved inclusion for the solution vector $\underset{\cdot}{y}$. Prior to tackling this task we also consider a further possibility for constructing a set of equations along the lines of (4) as given in Hansen [6]. For this one applies a linear approximation to $\underset{\cdot}{f}(\underset{\cdot}{x})$ at the point $\underset{\cdot}{y}$ that is slightly different from (2).

Let us consider the function $f_i(x_1, \ldots, x_n)$ only in dependence on x_1, and then let it be expanded at the point y_1 as

$$f_i(x_1, \ldots, x_n) = f_i(y_1, x_2, \ldots, x_n)$$
$$+ (x_1 - y_1) \frac{\partial}{\partial x_1} f_i(y_1 + \theta_{i1}(x_1 - y_1), x_2, \ldots, x_n),$$

with $\theta_{i1} \in (0, 1)$.

We then expand $f_i(y_1, x_2, \ldots, x_n)$ as a function of x_2 at the point y_2 and get

$$f_i(y_1, x_2, \ldots, x_n) = f_i(y_1, y_2, x_3, \ldots, x_n)$$
$$+ (x_2 - y_2) \frac{\partial}{\partial x_2} f_i(y_1, y_2 + \theta_{i2}(x_2 - y_2), x_3, \ldots, x_n),$$

with $\theta_{i2} \in (0, 1)$.

We now expand $f_i(y_1, y_2, x_3, \ldots, x_n)$ again as a function only of x_3 at the point y_3 and so on until $f_i(y_1, y_2, \ldots, y_{n-1}, x_n)$ is expanded at the point y_n. If we now insert the expressions formed in this manner into each other, then we get

$$f_i(x_1, \dots, x_n) = f_i(y_1, \dots, y_n)$$
$$+ \sum_{j=1}^{n} (x_j - y_j) \frac{\partial}{\partial x_j} f_i(y_1, \dots, y_{j-1}, y_j + \theta_{ij}(x_j - y_j), x_{j+1}, \dots, x_n)$$

with $\theta_{ij} \in (0, 1)$, $1 \leqslant i \leqslant n$.

From this we have the relation

$$\mathfrak{f}(\mathfrak{x}) = \tilde{\mathfrak{J}}(\mathfrak{x})(\mathfrak{x} - \mathfrak{y}),$$

where the matrix $\tilde{\mathfrak{J}}(\mathfrak{x})$ is defined by

$$(5) \qquad \tilde{\mathfrak{J}}(\mathfrak{x}) = \left(\frac{\partial}{\partial x_j} f_i(y_1, \dots, y_{j-1}, z_j, x_{j+1}, \dots, x_n) \bigg|_{z_j = y_j + \theta_{ij}(x_j - y_j)} \right).$$

We may take $\mathfrak{y}$ to be a solution for the set of linear equations

$$(6) \qquad \mathfrak{f}(\mathfrak{x}) = \tilde{\mathfrak{F}}(\mathfrak{x} - \mathfrak{z}) \qquad \text{with} \quad \tilde{\mathfrak{F}} \in \tilde{\mathfrak{J}}(\mathfrak{x}^{(0)})$$

since both $y_j \in X_j^{(0)}$ and $y_j + \theta_{ij}(x_j - y_j) \in X_j^{(0)}$ for $1 \leqslant j \leqslant n$. The interval matrix $\tilde{\mathfrak{J}}(\mathfrak{x}^{(0)})$ is here defined by

$$\tilde{\mathfrak{J}}(\mathfrak{x}^{(0)}) = \left(\frac{\partial}{\partial x_j} f_i(X_1^{(0)}, \dots, X_{j-1}^{(0)}, X_j^{(0)}, x_{j+1}, \dots, x_n) \right)$$

and it clearly follows that for the interval matrix $\mathfrak{f}'(\mathfrak{x}^{(0)})$ we have

$$\tilde{\mathfrak{J}}(\mathfrak{x}^{(0)}) \subseteq \mathfrak{f}'(\mathfrak{x}^{(0)}).$$

We therefore have in (6) a set containing no more systems of linear equations than the set in (4). Both in the case $n = 1$, as well as in the dependence of the partial derivatives $\partial f_i / \partial x_j$ only on x_j, we have that the interval matrices $\mathfrak{f}'(x^{(0)})$ and $\tilde{\mathfrak{J}}(\mathfrak{x}^{(0)})$ are identical.

We now return to the original task of improving the inclusion vector $\mathfrak{x}^{(0)}$ for the solution vector $\mathfrak{y}$. Following the above development we calculate an interval vector that contains the set of solutions of (4), respectively, (6), and therefore also contains the vector $\mathfrak{y}$. We first consider (4). Now let $\mathfrak{y}$ be a simple zero of $\mathfrak{f}(\mathfrak{x})$ in $\mathfrak{x}^{(0)}$ and assume furthermore that all matrices $\mathfrak{F} \in \mathfrak{f}'(\mathfrak{x}^{(0)})$ are nonsingular. Let $\mathfrak{V}$ be an interval matrix containing all the inverses $\mathfrak{F}^{-1}$ of $\mathfrak{F} \in \mathfrak{f}'(\mathfrak{x}^{(0)})$. We have already demonstrated the practicality of computing such an interval matrix $\mathfrak{V}$ in Chapters 12–17. We now compute the interval vector

$$\mathfrak{y}^{(1)} = m(\mathfrak{x}^{(0)}) - \mathfrak{V} \cdot \mathfrak{f}(m(\mathfrak{x}^{(0)})),$$

using $\mathfrak{V}$, where $m(\mathfrak{x}^{(0)})$ is determined according to (18.1). Through this we therefore obtain an inclusion for the set of solutions of (4). We used the relation

$\mathcal{J}(m(x^{(0)})) \in f'(x^{(0)})$. Equation (6) may be used in a similar manner. Since $y \in x^{(0)}$ we can form the interval vector

$$x^{(1)} = \{m(x^{(0)}) - \mathcal{V} \cdot f(m(x^{(0)}))\} \cap x^{(0)} \subseteq x^{(0)}$$

thereby obtaining a new inclusion for y. A repeated execution of this calculating procedure leads to the iteration method

$$x^{(k+1)} = \{m(x^{(k)}) - \mathcal{V} \cdot f(m(x^{(k)}))\} \cap x^{(k)}, \qquad k \geq 0, \tag{7}$$

where $m(x^{(k)})$ is as in (18.1).

This iteration generates a monotonic sequence

$$x^{(0)} \supseteq x^{(1)} \supseteq x^{(2)} \supseteq \cdots$$

of interval vectors for which we prove

Theorem 1: Let $x^{(0)}$ be an interval vector, and let $y \in x^{(0)}$ be a zero of the function $f(x)$. Furthermore let $\mathcal{V}$ be an interval matrix that contains the inverses $\mathcal{J}(x)^{-1}$ (compare with (3)) for $x \in x^{(0)}$. The sequence $\{x^{(k)}\}_{k=0}^{\infty}$ of interval vectors computed according to (7) then satisfies:

(8) each interval vector $x^{(k)}$, $k \geq 0$, contains the zero y;

(9) if each matrix $V \in \mathcal{V}$ is nonsingular, then it follows that $\lim_{k\to\infty} x^{(k)} = y$.

Proof: Of (8): According to the assumptions we have that $y \in x^{(0)}$. Since $m(x^{(0)}) \in x^{(0)}$ we also have from the conditions of the theorem that $\mathcal{J}(m(x^{(0)}))^{-1} \in \mathcal{V}$. From (10.10′) it therefore follows that

$$y = m(x^{(0)}) - \mathcal{J}(m(x^{(0)}))^{-1} f(m(x^{(0)})) \in m(x^{(0)}) - \mathcal{V} \cdot f(m(x^{(0)}))$$

and therefore

$$y \in \{m(x^{(0)}) - \mathcal{V} \cdot f(m(x^{(0)}))\} \cap x^{(0)} = x^{(1)}.$$

The claim (8) may in this manner be proved by complete induction.

Of (9): The monotonically including sequence of iterates $x^{(0)} \supseteq x^{(1)} \supseteq x^{(2)} \cdots$ converges according to Corollary 10.8 towards an interval vector x. We show that we must have $d(x) = o$ from which it follows that the relation $\lim_{k\to\infty} x^{(k)} = y$ holds from (8). We therefore assume that $d(x) \neq o$. From the continuity of iteration (7) it follows that x satisfies the equation

$$x = \{m(x) - \mathcal{V} \cdot f(m(x))\} \cap x.$$

We consider $m(x) \in x$ and we are able to conclude that $m(x) \neq y$ since $d(x) \neq o$. We have, however, that

$$m(x) \in m(x) - \mathcal{V} \cdot f(m(x)).$$

Furthermore it follows from (10.1) that if an interval matrix $\mathscr{A}$ is multiplied by a real vector x, then

$$\mathscr{A}x = \{Ax \mid A \in \mathscr{A}\}$$

is satisfied. This implies that the representation

$$m(x) = m(x) - V \cdot f(m(x))$$

is valid for some matrix $V \in \mathscr{V}$. We therefore obtain

$$o = V \cdot f(m(x)),$$

which provides a contradiction since $m(x) \neq y$ and since V is non-singular. ■

The sufficient convergence criterion (9) for the iteration method (7) used the weaker assumption

$$\{J(x)^{-1} \mid x \in x^{(0)}\} \subseteq \mathscr{V}$$

for $\mathscr{V}$ than the assumption used in the development of the method. Originally we required the condition

$$\{F^{-1} \mid F \in f'(x^{(0)})\} \subseteq \mathscr{V},$$

which is easier to check.

The proof of (9) may be modified in such a manner that for the sequence $\{x^{(k)}\}_{k=0}^{\infty}$ it is shown that if $m(x^{(k)}) \neq y$, then

$$m(x^{(k)}) \notin x^{(k+1)}, \qquad k \geqslant 0.$$

This means that for the method (7) at least one component of the width $d(x^{(k)})$ is more than halved in the $(k+1)$th step. The method (7) may be improved by determining a new interval matrix $\mathscr{V}^{(k)}$ in each step. From the development of (7) it is clear that for the relation (8) one only needs

$$\{J(x)^{-1} \mid x \in x^{(k)}\} \subseteq \mathscr{V}^{(k)},$$

respectively,

$$\{F^{-1} \mid F \in f'(x^{(k)})\} \subseteq \mathscr{V}^{(k)}.$$

That is, one determines an including set for the set of equations

$$f(m(x^{(k)})) = F \cdot (m(x^{(k)}) - z) \qquad \text{with} \quad F \in f'(x^{(k)})$$

in the $(k+1)$th step. We shall now show a possibility for a suitable determination of the sequence $\{\mathscr{V}^{(k)}\}_{k=0}^{\infty}$. First, we shall determine an including set for

$$\{F^{-1} \mid F \in f'(x^{(k)})\}$$

iteratively using the methods described in Chapters 12 and 14. For this we use the shorthand notation $\mathscr{F}^{(k)}$ for the interval matrix $f'(x^{(k)})$ and $\tilde{\mathscr{F}}^{(k)}$ for the interval matrix $\tilde{\mathscr{J}}(x^{(k)})$.

We start with an interval matrix $\mathscr{W}^{(0)}$ that has been determined, for example, by the use of the usual norm estimating techniques such that

$$\{\mathscr{F}^{-1} \mid \mathscr{F} \in \mathscr{F}^{(k)}\} \subseteq \mathscr{W}^{(0)}$$

is satisfied. We then carry out the iteration procedure

$$(10) \quad \mathscr{W}^{(v+1)} = \{(\mathscr{I} - (m(\mathscr{F}^{(k)}))^{-1}\mathscr{F}^{(k)})\mathscr{W}^{(v)} + (m(\mathscr{F}^{(k)}))^{-1}\} \cap \mathscr{W}^{(v)}, \quad v \geq 0.$$

The sequence

$$\mathscr{W}^{(0)} \supseteq \mathscr{W}^{(1)} \supseteq \mathscr{W}^{(2)} \supseteq \cdots$$

will always converge towards an interval matrix $\mathscr{W} \subseteq \mathscr{W}^{(0)}$ that is uniquely determined when the spectral radius satisfies

$$\rho(|\mathscr{I} - (m(\mathscr{F}^{(k)}))^{-1}\mathscr{F}^{(k)}|) < 1$$

(compare with Chapters 12 and 14). The matrix $\mathscr{I} - (m(\mathscr{F}^{(k)}))^{-1}\mathscr{F}^{(k)}$ in (10) is symmetric around the midpoint $\mathscr{O}$. This means that the procedure (10) can be decomposed into two equal, real iteration procedures for the matrix of upper (resp., lower) bounds of $\mathscr{W}$. In order to calculate $\mathscr{W}$ one therefore only needs to solve one real system of equations.

We now define

$$(11) \qquad \mathscr{V}^{(k)} := \mathscr{W} \supseteq \{\mathscr{F}^{-1} \mid \mathscr{F} \in \mathscr{F}^{(k)}\}.$$

Using this matrix $\mathscr{V}^{(k)}$ we carry out the iteration step

$$x^{(k+1)} = \{m(x^{(k)}) - \mathscr{V}^{(k)} f(m(x^{(k)}))\} \cap x^{(k)}.$$

The new matrix $\mathscr{F}^{(k+1)} = f'(x^{(k+1)})$ is calculated and then (10) is started. Since $x^{(k+1)} \subseteq x^{(k)}$ it clearly follows that $\mathscr{F}^{(k+1)} \subseteq \mathscr{F}^{(k)}$ and $\{\mathscr{F}^{-1} \mid \mathscr{F} \in \mathscr{F}^{(k+1)}\} \subseteq \mathscr{V}^{(k)}$. The matrix $\mathscr{V}^{(k)}$ may therefore be used as a new starting element for (10). We therefore obtain the following iteration procedure, in which $\mathscr{V}^{(0)}$ is a superset of $\{\mathscr{F}^{-1} \mid \mathscr{F} \in \mathscr{F}^{(0)}\}$:

(12) $x^{(k+1)} = \{m(x^{(k)}) - \mathscr{V}^{(k)} f(m(x^{(k)}))\} \cap x^{(k)}, \quad k \geq 0,$ where $\mathscr{V}^{(k)}$ with property (11) is the matrix calculated from (10), when $\mathscr{V}^{(0)}$ is used as the starting matrix in (10) for $k = 1$.

We prove the following statement for the method (12).

Theorem 2: Let $x^{(0)}$ be an interval vector and let $y \in x^{(0)}$ be a zero of the function $f(x)$. Let furthermore $\mathscr{V}^{(0)}$ be an interval matrix containing the inverses $\mathscr{F}^{-1}$ for $\mathscr{F} \in \mathscr{F}^{(0)} = f'(x^{(0)})$. It is moreover assumed that the Frechét derivative $f'(x)$ for $x^{(0)}$ satisfies the conditions of Theorem 3.5 for each element. It then holds for the sequence of interval vectors $\{x^{(k)}\}_{k=0}^{\infty}$ calculated using the iteration (12) that

(8) each iterate $x^{(k)}$, $k \geqslant 0$, contains the zero y;

(9) if every matrix $V \in \mathcal{V}^{(0)}$ is nonsingular, then we have $\lim_{k\to\infty} x^{(k)} = y$;

(13) if the sequence $\{x^{(k)}\}_{k=0}^{\infty}$ converges to y, then for some norm $\|\cdot\|$ it follows that

$$\|d(x^{(k+1)})\| \leqslant \gamma \|d(x^{(k)})\|^2, \qquad \gamma \geqslant 0,$$

that is, the R order of the method (12) satisfies

$$O_R((12), y) \geqslant 2$$

(see Appendix A, Theorem 2).

Proof: The verification of (8) and (9) is completely analogous to the corresponding proof in Theorem 1. One only has to consider

$$\{F^{-1} \mid F \in \mathcal{F}^{(k)}\} \subseteq \mathcal{V}^{(k)} \subseteq \mathcal{V}^{(0)}, \qquad k \geqslant 0.$$

Of (13): From the fact that $x^{(k+1)} \subseteq m(x^{(k)}) - \mathcal{V}^{(k)} f(m(x^{(k)}))$ it follows that

$$\begin{aligned}
d(x^{(k+1)}) &\leqslant d(m(x^{(k)}) - \mathcal{V}^{(k)} f(m(x^{(k)}))) \\
&= d(\mathcal{V}^{(k)})\,|f(m(x^{(k)}))| \\
&= d(\mathcal{V}^{(k)})\,|f(m(x^{(k)})) - f(y)| \\
&= d(\mathcal{V}^{(k)})\,|J(m(x^{(k)}))(m(x^{(k)}) - y)| \\
&\leqslant d(\mathcal{V}^{(k)})\,|J(m(x^{(k)}))|\tfrac{1}{2}d(x^{(k)}) \\
&\leqslant d(\mathcal{V}^{(k)})\,|\mathcal{F}^{(0)}|\tfrac{1}{2}d(x^{(k)}).
\end{aligned}$$

We now use a monotonic vector norm with a compatible monotonic matrix norm and obtain

$$\begin{aligned}
\|d(x^{(k+1)})\| &\leqslant \|d(\mathcal{V}^{(k)})\|\tfrac{1}{2}\|\,|\mathcal{F}^{(0)}|\,\|\,\|d(x^{(k)})\| \\
&= c\|d(\mathcal{V}^{(k)})\|\,\|d(x^{(k)})\|.
\end{aligned}$$

We finally have to estimate $\|d(\mathcal{V}^{(k)})\|$. $\mathcal{V}^{(k)}$ satisfies the equation

$$\mathcal{V}^{(k)} = \{(\mathcal{I} - (m(\mathcal{F}^{(k)}))^{-1}\mathcal{F}^{(k)})\mathcal{V}^{(k)} + (m(\mathcal{F}^{(k)}))^{-1}\} \cap \mathcal{V}^{(k)},$$

and therefore using (10.21) and (10.19) it follows that

$$\begin{aligned}
d(\mathcal{V}^{(k)}) &\leqslant d((\mathcal{I} - (m(\mathcal{F}^{(k)}))^{-1}\mathcal{F}^{(k)})\mathcal{V}^{(k)} + (m(\mathcal{F}^{(k)}))^{-1}) \\
&= d((\mathcal{I} - (m(\mathcal{F}^{(k)}))^{-1}\mathcal{F}^{(k)})\mathcal{V}^{(k)}) \\
&= d(((m(\mathcal{F}^{(k)}))^{-1}(m(\mathcal{F}^{(k)}) - \mathcal{F}^{(k)}))\mathcal{V}^{(k)}) \\
&= |m(\mathcal{F}^{(k)})^{-1}|\,d(\mathcal{F}^{(k)})|\mathcal{V}^{(k)}|.
\end{aligned}$$

Using the above monotonic matrix norm and the relations

$$|\mathscr{V}^{(k)}| \leqslant |\mathscr{V}^{(0)}| \qquad \text{and} \qquad |(m(\mathscr{F}^{(k)}))^{-1}| \leqslant |\mathscr{V}^{(0)}|,$$

we get that

$$\|d(\mathscr{V}^{(k)})\| \leqslant \tilde{\gamma}\|d(\mathscr{F}^{(k)})\|$$

where $\tilde{\gamma}$ is a constant independent of k.

The individual elements of the matrix $d(\mathscr{F}^{(k)})$ are

$$d\left(\frac{\partial}{\partial x_j} f_i(X_1^{(k)}, \ldots, X_n^{(k)})\right),$$

and they may be estimated using the relation (3.5′) by

$$d\left(\frac{\partial}{\partial x_j} f_i(X_1^{(k)}, \ldots, X_n^{(k)})\right) \leqslant \sum_{r=1}^{n} \gamma_{ijr}^{(k)} d(X_r^{(k)}) \leqslant \sum_{r=1}^{n} \gamma_{ijr}^{(0)} d(X_r^{(k)}).$$

Consequently we have

$$d(\mathscr{F}^{(k)}) \leqslant \mathscr{C}\, d(x^{(k)}),$$

where $\mathscr{C} = (\gamma_{ijl}^{(0)})$ is a bilinear operator from $V_n(\mathbb{R}) \times V_n(\mathbb{R})$ to $V_n(\mathbb{R})$. If one defines a norm for $\mathscr{C}$ as usual by

$$\|\mathscr{C}\| = \sup_{\|x\|=1} \sup_{\|y\|=1} \|\mathscr{C}xy\|,$$

it follows that

$$\|d(\mathscr{F}^{(k)})\| \leqslant \|\mathscr{C}\|\, \|d(x^{(k)})\|;$$

that is,

$$\|d(\mathscr{V}^{(k)})\| \leqslant \bar{c}\|d(x^{(k)})\| \qquad \text{with} \quad \bar{c} = \|\mathscr{C}\|\tilde{\gamma}.$$

Putting this into the above estimate for $d(x^{(k+1)})$ we get

$$\|d(x^{(k+1)})\| \leqslant \tilde{c}\|d(x^{(k)})\|^2 \qquad \text{with} \quad \tilde{c} = c\bar{c}$$

for a monotonic vector norm. Using the concept of norm equivalence in the same manner as in the proof of (18.8) in Theorem 18.1 we can use an arbitrary norm and we have relation (13) with a γ independent of k. ■

We now wish to consider another possibility for the calculation of a solution y of the equation $f(x) = o$ using repeated inclusions. This method is suitable for certain classes of nonlinear systems of equations.

We proceed from the system (4) with $x := m(x^{(0)})$ according to (18.1). The following considerations may also be realized using (6). If, for example, we assume that $\mathscr{F}^{(0)} = f'(x^{(0)})$ satisfies the assumptions of Theorem 15.3, then the Gaussian algorithm can be carried out on the interval matrix $\mathscr{F}^{(0)}$ and the vector $f(m(x^{(0)}))$. For the interval vector $w^{(0)}$ calculated from this elimination it follows that $m(x^{(0)}) - w^{(0)}$, that is, also

$$\tilde{x}^{(0)} = \{m(x^{(0)}) - w^{(0)}\} \cap x^{(0)},$$

contains the solution y of the nonlinear system of equations.

Following (2) the equation

$$f(x) = \mathcal{J}(x)(x - y)$$

is written. In this equation the matrix $\mathcal{J}(x)$ is decomposed into

$$\mathcal{J}(x) = \mathcal{D}(x) - \mathcal{B}(x),$$

where $\mathcal{D}(x)$ contains the diagonal of $\mathcal{J}(x)$. From the inclusion monotonicity setting $x = m(x^{(0)})$ one has that

$$\begin{aligned} y &= m(x^{(0)}) - \mathcal{D}(m(x^{(0)}))^{-1}\{\mathcal{B}(m(x^{(0)}))(m(x^{(0)}) - y) + f(m(x^{(0)}))\} \\ &\in \{m(x^{(0)}) - \mathcal{D}(x^{(0)})^{-1}\{\mathcal{B}(x^{(0)})(m(x^{(0)}) - \tilde{x}^{(0)}) + f(m(x^{(0)}))\}\} \cap \tilde{x}^{(0)} \\ &=: x^{(1)}. \end{aligned}$$

Here we have decomposed $\mathcal{F}^{(0)} = f'(x^{(0)})$ into

$$\mathcal{F}^{(0)} = \mathcal{D}(x^{(0)}) - \mathcal{B}(x^{(0)})$$

in the same manner as $\mathcal{J}(x)$. Using $\mathcal{D}(x^{(0)}) = \operatorname{diag}(D_i(x^{(0)}))$ we define $\mathcal{D}(x^{(0)})^{-1} = \operatorname{diag}(1/D_i(x^{(0)}))$. If now $\mathcal{F}^{(1)} = f'(x^{(1)})$ again satisfies the conditions of Theorem 15.3 then one may calculate, starting with $x^{(1)}$, in an analogous manner a new inclusion $x^{(2)}$ for y and so on. The repeated application of this step leads to the following iteration method

(14) (a) $\mathcal{F}^{(k)} = f'(x^{(k)}) = \mathcal{D}(x^{(k)}) - \mathcal{B}(x^{(k)})$, $m(x^{(k)})$ according to (18.1).

(b) The application of the Gaussian algorithm on $(\mathcal{F}^{(k)}, f(m(x^{(k)})))$ gives $w^{(k)}$.

(c) $\tilde{x}^{(k)} = \{m(x^{(k)}) - w^{(k)}\} \cap x^{(k)}$.

(d) $$\begin{aligned} x^{(k+1)} = \{m(x^{(k)}) - \mathcal{D}(x^{(k)})^{-1}\{\mathcal{B}(x^{(k)})(m(x^{(k)}) - \tilde{x}^{(k)}) \\ + f(m(x^{(k)}))\}\} \cap \tilde{x}^{(k)}, \; k \geqslant 0. \end{aligned}$$

Step (d) of the algorithm requires $\sim n^2$ operations (interval multiplications or divisions) for a full matrix. Since step (b) requires $\sim n^3/3$ operations it follows that the work required in step (d) is insignificant for larger n.

In order to investigate when $\lim_{k\to\infty} x^{(k)} = y$ is valid in (14) and in order to estimate the R order of convergence, we now introduce some preliminary lemmas.

Lemma 3: Let $\mathscr{A} = (A_{ij})$ be a real interval matrix satisfying the assumptions of Theorem 15.3. If $\mathscr{A}$ is decomposed into $\mathscr{A} = \mathscr{D} - \mathscr{B}$ such that $\mathscr{D}$ contains the diagonal of $\mathscr{A}$ then the matrix $\tilde{\mathscr{D}} = \text{diag}(1/A_{ii})$ satisfies the inequality

$$\rho(|\tilde{\mathscr{D}}|\,|\mathscr{B}|) < 1.$$

Proof: We again represent the elements of $\mathscr{A}$ by the midpoint and the half-width as $A_{ij} = \langle a_{ij}, r_{ij} \rangle$, $1 \leqslant i, j \leqslant n$. Now consider the definition of the matrix $\mathscr{B}$ in Theorem 15.3 as well as the relation

$$|1/A_{ii}| = 1/(a_{ii} - r_{ii}).$$

It is immediately clear that the real matrix $|\tilde{\mathscr{D}}|\,|\mathscr{B}|$ is the total step matrix for the total step method belonging to the matrix $\mathscr{B}$. According to Theorem 3.10 in Varga [26] we therefore have $\rho(|\tilde{\mathscr{D}}|\,|\mathscr{B}|) < 1$. ■

Let now $\mathscr{A} = (A_{ij})$ be a real interval matrix and x a real interval vector for which

$$(15) \qquad d(A_{ij}) \leqslant \alpha_{ij}\|d(x)\|, \qquad \alpha_{ij} \geqslant 0, \quad 1 \leqslant i, j \leqslant n,$$

is valid. If $0 \notin A_{ij}$, then (15) implies the relation

$$(16) \qquad d(1/A_{ij}) \leqslant \hat{\alpha}_{ij}\|d(x)\|, \qquad \hat{\alpha}_{ij} \geqslant 0, \quad 1 \leqslant i, j \leqslant n.$$

Let $h = (H_i)$ be a real interval vector satisfying

$$(17) \qquad |H_i| \leqslant \beta_i\|d(x)\|, \qquad \beta_i \geqslant 0, \quad 1 \leqslant i \leqslant n,$$

$$(18) \qquad d(H_i) \leqslant \gamma_i\|d(x)\|^2, \qquad \gamma_i \geqslant 0, \quad 1 \leqslant i \leqslant n.$$

Lemma 4: Let the inequalities (15)–(18) be satisfied for $\mathscr{A} = (A_{ij})$ and $h = (H_i)$. If the Gaussian algorithm can be carried out on the interval matrix $\mathscr{A}$ and the interval vector h, then the resulting upper triangular matrix $\tilde{\mathscr{A}} = (\tilde{A}_{ij})$ and the corresponding vector $\tilde{h} = (\tilde{H}_i)$ satisfy the relations (15)–(18) with constants $\tilde{\alpha}_{ij} \geqslant 0$, $\tilde{\beta}_i \geqslant 0$, $\tilde{\gamma}_i \geqslant 0$.

Proof: The proof follows using complete induction on the $n - 1$ steps of the Gaussian algorithm. In each step it is shown that from the assumption that the matrix–vector pair in the previous step satisfies the above conditions it follows that the matrix–vector pair in the current step also satisfies the conditions. We carry through the proof for the first step. In this step, starting with the matrix $\mathscr{A}$ and the vector h, a new matrix $\mathscr{A}'$ and a vector h' are computed according to the formulas (see Chapter 15)

$$\begin{aligned} A'_{1j} &= A_{1j}, && 1 \leqslant j \leqslant n, \quad H'_1 = H_1, \\ A'_{i1} &= 0, && 2 \leqslant i \leqslant n, \\ A'_{ij} &= A_{ij} - (A_{i1}/A_{11})A_{1j}, && 2 \leqslant i, j \leqslant n, \\ H'_i &= H_i - (A_{i1}/A_{11})H_1, && 2 \leqslant i \leqslant n. \end{aligned}$$

For $2 \leqslant i, j \leqslant n$, using (15)–(18) and (2.12), we then get

$$\begin{aligned} d(A'_{ij}) &= d(A_{ij}) + d((A_{i1}/A_{11})A_{1j}) \\ &\leqslant \alpha_{ij}\|d(\mathscr{x})\| + \left|\frac{A_{i1}}{A_{11}}\right| d(A_{1j}) + \left(d(A_{i1})\left|\frac{1}{A_{11}}\right| + d\left(\frac{1}{A_{11}}\right)|A_{i1}|\right)|A_{1j}| \\ &\leqslant \tilde{\alpha}_{ij}\|d(\mathscr{x})\|, \qquad \tilde{\alpha}_{ij} \geqslant 0. \end{aligned}$$

Correspondingly one obtains for $2 \leqslant i \leqslant n$ that

$$|H'_i| \leqslant |H_i| + |A_{i1}/A_{11}|\,|H_1| \leqslant \tilde{\beta}_i\|d(\mathscr{x})\|, \qquad \tilde{\beta}_i \geqslant 0,$$

as well as

$$\begin{aligned} d(H'_i) &= d(H_i) + |H_1|\left(d(A_{i1})\left|\frac{1}{A_{11}}\right| + d\left(\frac{1}{A_{11}}\right)|A_{i1}|\right) + d(H_1)\left|\frac{A_{i1}}{A_{11}}\right| \\ &\leqslant \tilde{\gamma}_i\|d(\mathscr{x})\|^2, \qquad \tilde{\gamma}_i \geqslant 0. \end{aligned}$$

The inequalities (15)–(18) are trivially satisfied for the other elements of $\mathscr{A}'$ and $\mathscr{h}'$. ■

Lemma 5: Let $\tilde{\mathscr{A}} = (\tilde{A}_{ij})$ be an upper triangular matrix for which $0 \notin \tilde{A}_{ii}$, $1 \leqslant i \leqslant n$, and $\tilde{\mathscr{h}} = (\tilde{H}_i)$ be a vector satisfying the conditions (15)–(18). The interval vector $\mathscr{y} = (Y_i)$ calculated according to the formulas

$$Y_n = \frac{\tilde{H}_n}{\tilde{A}_{nn}}, \qquad Y_i = \frac{1}{\tilde{A}_{ii}}\left(\tilde{H}_i - \sum_{j=i+1}^{n} \tilde{A}_{ij}Y_j\right), \qquad 1 \leqslant i \leqslant n-1,$$

then satisfies

$$\|d(\mathscr{y})\| \leqslant c\|d(\mathscr{x})\|^2, \qquad c \geqslant 0.$$

Proof: It first follows from (2.12) that

$$d(Y_n) \leqslant d(\tilde{H}_n)\left|\frac{1}{\tilde{A}_{nn}}\right| + d\left(\frac{1}{\tilde{A}_{nn}}\right)|\tilde{H}_n| \leqslant \delta_n\|d(\mathscr{x})\|^2, \qquad \delta_n \geqslant 0,$$

and that

$$|Y_n| = |1/\tilde{A}_{nn}|\,|\tilde{H}_n| \leqslant \kappa_n\|d(\mathscr{x})\|.$$

If it now holds that

$$d(Y_i) \leqslant \delta_i\|d(\mathscr{x})\|^2, \qquad |Y_i| \leqslant \kappa_i\|d(\mathscr{x})\|, \qquad 1 < i_0 + 1 \leqslant i \leqslant n,$$

then it follows that

$$\begin{aligned} d(Y_{i_0}) &\leqslant d\left(\frac{1}{\tilde{A}_{i_0i_0}}\right)\left|\tilde{H}_{i_0} - \sum_{j=i_0+1}^{n} \tilde{A}_{i_0j}Y_j\right| + \left|\frac{1}{\tilde{A}_{i_0i_0}}\right| d\left(\tilde{H}_{i_0} - \sum_{j=i_0+1}^{n} \tilde{A}_{i_0j}Y_j\right) \\ &\leqslant \delta_{i_0}\|d(\mathscr{x})\|^2, \\ |Y_{i_0}| &\leqslant \left|\frac{1}{\tilde{A}_{i_0i_0}}\right|\left(|\tilde{H}_{i_0}| + \sum_{j=i_0+1}^{n} |\tilde{A}_{i_0j}|\,|Y_j|\right) \\ &\leqslant \kappa_{i_0}\|d(\mathscr{x})\|. \end{aligned}$$

From this it follows that

$$d(Y_i) \leqslant \delta_i \|d(x)\|^2, \qquad 1 \leqslant i \leqslant n,$$

$$\max_{1 \leqslant i \leqslant n} \{d(Y_i)\} \leqslant \max_{1 \leqslant i \leqslant n} \{\delta_i\} \|d(x)\|^2.$$

From the theorem of norm equivalence the assertion now follows. ■

Combining Lemmas 4 and 5 we now have the following:

Lemma 6: Let the inequalities (15)–(18) be satisfied for $\mathcal{A} = (A_{ij})$ and $h = (H_i)$. If the Gaussian algorithm can be carried out on $\mathcal{A}$ and h, then the interval vector y computed by this algorithm satisfies the inequality

$$\|d(y)\| \leqslant c \|d(x)\|^2, \qquad c \geqslant 0. \quad ■$$

After these preliminary lemmas we now prove the following statement about the method (14).

Theorem 7: Let $x^{(0)}$ be an interval vector, and let $y \in x^{(0)}$ be a zero of the function $f(x)$. Let the interval evaluation $f'(x)$ of the Frechét derivative $f'(x)$ satisfy the assumptions of Theorem 15.3 for all $x \subseteq x^{(0)}$. Let the inequality (3.5′) be furthermore satisfied componentwise for $f'(x^{(0)})$. The sequence $\{x^{(k)}\}_{k=0}^{\infty}$ computed according to (14) then satisfies:

(8) each iterate $x^{(k)}$, $k \geqslant 0$, contains the zero y;

(19) the iteration method (14) may be carried out for each $k \geqslant 0$, and $\lim_{k\to\infty} x^{(k)} = y$ is satisfied;

(20) the inequality

$$\|d(x^{(k+1)})\| \leqslant c \|d(x^{(k)})\|^2, \qquad c \geqslant 0,$$

holds, that is, the R order of the method (14) satisfies

$$O_R((14), y) \geqslant 2$$

(see Appendix A, Theorem 2).

Proof: The verification of (8) was carried out for $k = 1$ in the development of the method (14). The general proof may be done using complete induction.

Of (19): Theorem 15.3 guarantees that the Gaussian algorithm can be carried out. The assumptions of the theorem guarantee that the diagonal of $f'(x^{(k)})$ does not contain any zeros, which implies that (d) in (14) can always be carried out. The interval vectors computed according to (14) satisfy

$$x^{(0)} \supseteq x^{(1)} \supseteq \cdots \supseteq x^{(k)} \supseteq x^{(k+1)} \supseteq \cdots.$$

This sequence converges to a limiting value x according to Corollary 10.8. It then follows that $\lim_{k\to\infty} m(x^{(k)}) = m(x)$, $\lim_{k\to\infty} f(m(x^{(k)})) = f(m(x))$,

$\lim_{k\to\infty} f'(x^{(k)}) = f'(x)$. Since $w^{(k)}$ depends continuously on $f(m(x^{(k)}))$ and $f'(x^{(k)})$ it also follows that $\lim_{k\to\infty} w^{(k)} = w$ as well as

$$\lim_{k\to\infty} \tilde{x}^{(k)} = \tilde{x} = \{m(x) - w\} \cap x,$$

$$\lim_{k\to\infty} x^{(k)} = x = \{m(x) - \mathcal{D}(x)^{-1}\{\mathcal{B}(x)(m(x) - \tilde{x}) + f(m(x))\}\} \cap \tilde{x}.$$

From these equations it follows that

$$\tilde{x} \subseteq x,$$

$$x \subseteq m(x) - \mathcal{D}(x)^{-1}\{\mathcal{B}(x)(m(x) - \tilde{x}) + f(m(x))\};$$

that is,

$$x \subseteq m(x) - \mathcal{D}(x)^{-1}\{\mathcal{B}(x)(m(x) - x) + f(m(x))\}. \tag{21}$$

In particular it then holds that

$$m(x) \in m(x) - \mathcal{D}(x)^{-1}\{\mathcal{B}(x)(m(x) - x) + f(m(x))\},$$

from which

$$o \in \mathcal{D}(x)^{-1}\{\mathcal{B}(x)(m(x) - x) + f(m(x))\}.$$

It therefore follows that

$$o \in \mathcal{B}(x)(m(x) - x) + f(m(x)).$$

From (21), applying (2.19) componentwise, it now follows that

$$\begin{aligned} d(x) &\leqslant |\mathcal{D}(x)^{-1}|\, d(\mathcal{B}(x)(m(x) - x) + f(m(x))) \\ &= |\mathcal{D}(x)^{-1}|\,|\mathcal{B}(x)|\, d(x) \\ &\leqslant |\mathcal{D}(x^{(0)})^{-1}|\,|\mathcal{B}(x^{(0)})|\, d(x). \end{aligned}$$

According to Lemma 3 it holds that $\rho(|\mathcal{D}(x^{(0)})^{-1}|\,|\mathcal{B}(x^{(0)})|) < 1$; that is, $d(x) = o$. Since $y \in x^{(k)}$, $k \geqslant 0$, it follows using (10.27) that $\lim_{k\to\infty} x^{(k)} = y$, which was to be proved.

Of (20): We show first that the inequalities (15)–(18) are satisfied with $x := x^{(0)}$ for

$$\mathcal{A} := f'(x^{(0)}) \qquad \text{and} \qquad h := f(m(x^{(0)})) = (f_i(m(x^{(0)}))).$$

From the assumptions, each element of $f'(x^{(0)})$ satisfies an inequality (3.5′) from which it follows, using the norm equivalence, that

$$d(A_{ij}) \leqslant \alpha_{ij}\|d(x^{(0)})\|, \qquad 1 \leqslant i, j \leqslant n;$$

that is, (15) follows. From the mean-value theorem it follows with $0 < \theta_i < 1$

that

$$|H_i| = |f_i(m(x^{(0)}))| \leqslant |f'_i(y + \theta_i(y - m(x^{(0)})))|\,|m(x^{(0)}) - y|$$
$$\leqslant |f'_i(x^{(0)})|\,|m(x^{(0)}) - y|.$$

From this it follows since $m(x^{(0)})$, $y \in x^{(0)}$, and using the norm equivalence theorem that

$$|H_i| \leqslant \beta_i \|d(x^{(0)})\|, \qquad 1 \leqslant i \leqslant n;$$

that is, (17) follows. Since $d(f(m(x^{(0)}))) = o$ it follows trivially that (18) is valid. According to Lemma 6 it then follows that

$$\|d(w^{(0)})\| \leqslant c\|d(x^{(0)})\|^2,$$

and from parts (c) and (d) of (14) one then obtains

$$\|d(x^{(1)})\| \leqslant c\|d(x^{(0)})\|^2.$$

The above steps may be repeated for all $k \geqslant 0$ by replacing the relevant constants by those of the first step since $x^{(k)} \supseteq x^{(k+1)}$, $k \geqslant 0$. This therefore proves the inequality

$$\|d(x^{(k+1)})\| \leqslant c\|d(x^{(k)})\|^2, \qquad k \geqslant 0.$$

The theorem follows using Theorem 2, Appendix A. ■

We now discuss a class of problems where the assumptions of Theorem 7 always can be satisfied. For this we consider the boundary value problem

$$y'' = f(t, y), \qquad y(0) = \bar{a}, \qquad y(1) = \bar{b}, \qquad f_y(t, y) \geqslant 0, \quad t \in (0, 1).$$

The application of the usual difference method on this problem leads to a nonlinear system of equations that satisfies all the conditions of Theorem 7. (The explicit form of the system is given in Chapter 21.) We choose $n = 25$ points in the interval $(0, 1)$ and in Table 1 we give the results for $f(t, y) = y + \sin(y)$, $y(0) = 0$, $y(1) = 1$, of the iterates for the approximation for $y(\frac{1}{2})$. The interval vector $x^{(0)}$ was calculated as in Chapter 21. All rounding errors were included in the computations.

Table 1

k	Lower bound	Upper bound
0	−0.5	0.5
1	0.3966601342385	0.4046693140542
2	0.3986876258099	0.3986882685536
3	0.3986880255411	0.3986880255471
4	0.3986880255420	0.3986880255466

We now return to Lemma 10 of Chapter 11, and we use this lemma to prove some existence statements for the solution of nonlinear systems of equations.

Theorem 8: Let $\underset{\cdot}{f}: \mathfrak{B} \subseteq V_n(\mathbb{R}) \rightarrow V_n(\mathbb{R})$ be continuously differentiable in $\mathfrak{B}$ and assume that an interval evaluation of the derivative exists for some $x \subseteq \mathfrak{B}$. Suppose also that the Gaussian algorithm is feasible for $\underset{\cdot}{f}'(x)$ and that applying the Gaussian algorithm to $\underset{\cdot}{f}'(x)$ with the right-hand side $\underset{\cdot}{f}(\underset{\cdot}{y})$ for a fixed $\underset{\cdot}{y} \in x$ results in an interval vector z. If $\underset{\cdot}{y} - z \subseteq x$, then it follows that $\underset{\cdot}{f}$ has a zero in x. If on the other hand $x \cap \{\underset{\cdot}{y} - z\} = \emptyset$, then there is no zero of $\underset{\cdot}{f}$ in x.

Proof: We start with Eq. (2) written in matrix form:

$$\underset{\cdot}{f}(\underset{\cdot}{x}) - \underset{\cdot}{f}(\underset{\cdot}{y}) = \underset{\cdot}{J}(\underset{\cdot}{x})(\underset{\cdot}{x} - \underset{\cdot}{y}).$$

Since the Gaussian algorithm can be carried out on the matrix $\underset{\cdot}{f}'(x)$ it follows that all the point matrices from $\underset{\cdot}{f}'(x)$, in particular, the matrix $\underset{\cdot}{J}(\underset{\cdot}{x})$, are nonsingular. We consider the mapping

$$\underset{\cdot}{p}: x \subset V_n(\mathbb{R}) \rightarrow V_n(\mathbb{R})$$

with

$$\underset{\cdot}{p}(\underset{\cdot}{x}) = \underset{\cdot}{x} - \underset{\cdot}{J}(\underset{\cdot}{x})^{-1}\underset{\cdot}{f}(\underset{\cdot}{x}).$$

It follows that

$$\begin{aligned}\underset{\cdot}{p}(\underset{\cdot}{x}) &= \underset{\cdot}{x} - \underset{\cdot}{J}(\underset{\cdot}{x})^{-1}\underset{\cdot}{f}(\underset{\cdot}{y}) + \underset{\cdot}{J}(\underset{\cdot}{x})^{-1}(\underset{\cdot}{f}(\underset{\cdot}{y}) - \underset{\cdot}{f}(\underset{\cdot}{x}))\\ &= \underset{\cdot}{y} - \underset{\cdot}{J}(\underset{\cdot}{x})^{-1}\underset{\cdot}{f}(\underset{\cdot}{y}) \in \underset{\cdot}{y} - z.\end{aligned}$$

If we therefore have $\underset{\cdot}{y} - z \subseteq x$, then it follows that $\underset{\cdot}{p}(\underset{\cdot}{x}) \in x$ for $\underset{\cdot}{x} \in x$, and from Lemma 11.10 it follows that $\underset{\cdot}{f}$ has a zero in x.

In order to prove the second part we suppose that $\underset{\cdot}{f}$ has a zero $\underset{\cdot}{x}^*$ in x. It then follows for the map $\underset{\cdot}{p}$ defined above for $\underset{\cdot}{x} = \underset{\cdot}{x}^*$ that

$$\underset{\cdot}{p}(\underset{\cdot}{x}^*) = \underset{\cdot}{x}^* \in \underset{\cdot}{y} - z.$$

Since $\underset{\cdot}{x}^* \in x$ we have $x \cap \{\underset{\cdot}{y} - z\} \neq \emptyset$, which is a contradiction. ■

The proof of the previous theorem shows that one may avoid using the Gaussian algorithm if one knows an interval matrix $\mathcal{V}$ with the property that

$$\underset{\cdot}{J}(\underset{\cdot}{x})^{-1} \in \mathcal{V} \qquad \text{for} \quad \underset{\cdot}{x} \in x.$$

Using the map $\underset{\cdot}{p}$ defined in the previous theorem one then obtains the relation

$$\underset{\cdot}{p}(\underset{\cdot}{x}) = \underset{\cdot}{x} - \underset{\cdot}{J}(\underset{\cdot}{x})^{-1}\underset{\cdot}{f}(\underset{\cdot}{x}) \in \underset{\cdot}{y} - \mathcal{V}\underset{\cdot}{f}(\underset{\cdot}{y}).$$

We then have

Theorem 9: Let $f: \mathfrak{B} \subseteq V_n(\mathbb{R}) \to V_n(\mathbb{R})$ be continuously differentiable in $\mathfrak{B}$ and assume that the derivative has an interval arithmetic evaluation for some $x \subseteq \mathfrak{B}$. Let the interval matrix $\mathcal{V}$ have the property $\mathcal{J}(x)^{-1} \in \mathcal{V}$ for $x \in x$. If now

$$y - \mathcal{V}f(y) \subseteq x$$

for a fixed $y \in x$, then there exists a zero of f in x. If, on the other hand,

$$\{y - \mathcal{V}f(y)\} \cap x = \emptyset,$$

then there is no zero of f in x. ■

In general one will compute the matrix $\mathcal{V}$ with the property required by this theorem by applying the Gaussian algorithm n times on the interval matrix $f'(x)$ with the columns of the unit matrix as the right-hand sides. A corresponding statement to the previous theorem, where the Gaussian algorithm for an interval matrix need not be applied, was proved by Moore in [12]. See also Moore [13].

Theorem 10: Let $f: \mathfrak{B} \subseteq V_n(\mathbb{R}) \to V_n(\mathbb{R})$ be continuously differentiable in $\mathfrak{B}$ and assume that the derivative has an interval arithmetic evaluation for some $x \subseteq \mathfrak{B}$. If the inclusion

$$k(x) = y - \mathcal{Y} \cdot f(y) + (\mathcal{J} - \mathcal{Y} \cdot f'(x))(x - y) \subseteq x$$

holds for a $y \in x$, a nonsingular point matrix $\mathcal{Y} \in M_{nn}(\mathbb{R})$ and the unit matrix $\mathcal{J}$, then it follows that f has a zero in x. On the other hand, if

$$\{y - \mathcal{Y} \cdot f(y) + (\mathcal{J} - \mathcal{Y} \cdot f'(x))(x - y)\} \cap x = \emptyset,$$

then there is no zero of f in x. ■

The obvious choice for $\mathcal{Y}$ is $m(f'(x))^{-1}$. This requires the inversion of a point matrix as well as some matrix–matrix multiplications and some matrix–vector multiplications in order to check the inclusion relation. A substantial part of these multiplications can be saved if one applies the Gaussian algorithm on a linear system of equations with a point matrix as the matrix of coefficients and an interval vector as the right-hand side. This is made more precise in the following theorem.

Theorem 11: Let $f: \mathfrak{B} \subseteq V_n(\mathbb{R}) \to V_n(\mathbb{R})$ be continuously differentiable in $\mathfrak{B}$ and assume that the derivative has an interval arithmetic evaluation for some $x \subseteq \mathfrak{B}$. The real point matrix $\tilde{\mathcal{Y}}$ is furthermore assumed to be nonsingular. The Gaussian algorithm is applied to the point matrix $\tilde{\mathcal{Y}}$ with the right-hand side $f(y) - (\tilde{\mathcal{Y}} - f'(x))(x - y)$ for a $y \in x$ such that an interval vector z is computed. If now $y - z \subseteq x$ is satisfied, then f has a zero in x. If on the other hand $x \cap \{y - z\} = \emptyset$, then there is no zero of f in x.

Proof: Let us consider the map

$$p: x \subseteq \mathfrak{B} \subseteq V_n(\mathbb{R}) \to V_n(\mathbb{R})$$

defined by

$$p(x) = x - \tilde{Y}^{-1} f(x).$$

It then follows that the equation

$$\begin{aligned} p(x) &= y - \tilde{Y}^{-1} f(y) + x - y - \tilde{Y}^{-1}(f(x) - f(y)) \\ &= y - \tilde{Y}^{-1} f(y) + \tilde{Y}^{-1}(\tilde{Y} - J(x))(x - y) \end{aligned}$$

is valid or that the equation

$$\tilde{Y}(p(x) - y) = -f(y) + (\tilde{Y} - J(x))(x - y)$$

is satisfied.

Since $x \in x$ and since $J(x) \in f'(x)$ we get that

$$-f(y) + (\tilde{Y} - J(x))(x - y) \in -f(y) + (\tilde{Y} - f'(x))(x - y).$$

If one now applies the Gaussian algorithm on the linear system of equations given by the matrix $\tilde{Y}$ and the right-hand side $f(y) - (\tilde{Y} - f'(x))(x - y)$, then one obtains an interval vector z, for which $p(x) - y \in -z$ holds for $x \in x$. If now $y - z \subseteq x$, then it follows that $p(x) \in x$ for $x \in x$. The proof may now be completed as in the previous theorem. ■

The main interest in this theorem is not the savings in computational work as compared to the procedure in Theorem 10. In comparison to the analogous result from Theorem 8, the main gain is that it is always possible to carry out the Gaussian algorithm on a nonsingular point matrix. On the other hand the assumption $k(x) \subseteq x$ is at least for the choice of $y = m(x)$ (midpoint of x) and $Y^{-1} = m(f'(x)) = \tilde{Y}$ not as strong as the requirement $y - z \subseteq x$ from Theorem 11. One may show this in the following manner: Using the notation introduced in Chapter 15 for the vector z computed by Gaussian algorithm we have for an arbitrary nonsingular Y

$$z = g(Y, f(y) - (Y - f'(x))(x - y)).$$

As a special case of the general statement proven in Chapter 15 (property (3) of Gaussian algorithm) we obtain

$$Y^{-1}(f(y) - (Y - f'(x))(x - y)) \subseteq g(Y, f(y) - (Y - f'(x))(x - y));$$

that is, we have the inclusion

$$y - Y^{-1}(f(y) - (Y - f'(x))(x - y)) \subseteq y - z.$$

We now show that for $y = m(x)$, $Y = m(f'(x))$ the left-hand side of this relation is equal to $\underset{\cdot}{k}(x)$ from Theorem 10. This follows from the fact that both $m(\underset{\cdot}{f}'(x)) - \underset{\cdot}{f}'(x)$ and $x - m(x)$ are symmetric as well as using (10.8) and the last formula in (10.9) obtaining

$$
\begin{aligned}
&m(x) - m(\underset{\cdot}{f}'(x))^{-1}(\underset{\cdot}{f}(m(x)) - (m(\underset{\cdot}{f}'(x)) - \underset{\cdot}{f}'(x))(x - m(x))) \\
&\quad = m(x) - m(\underset{\cdot}{f}'(x))^{-1}\underset{\cdot}{f}(m(x)) - (\mathcal{I} - m(\underset{\cdot}{f}'(x))^{-1}\underset{\cdot}{f}'(x))(x - m(x)) \\
&\quad = \underset{\cdot}{k}(x).
\end{aligned}
$$

We now consider some existence statements where the interval evaluation of the second derivative appears. For this we assume that the map $\underset{\cdot}{f}: \mathfrak{B} \subseteq V_n(\mathbb{R}) \to V_n(\mathbb{R})$ is twice continuously differentiable and that the second derivative $\underset{\cdot}{f}''(\underset{\cdot}{x}) \in M_{n^3}(\mathbb{R})$ has an interval arithmetic evaluation. For some $x \subseteq \mathfrak{B}$ and for an initially arbitrary real vector $m(x) \in x$ we consider the following interval vectors:

(22) $\underset{\cdot}{k}_1(x) := m(x) - \underset{\cdot}{\mathcal{Y}} \cdot \underset{\cdot}{f}(m(x)) - \underset{\cdot}{\mathcal{Y}}(\underset{\cdot}{f}''(x)(x - m(x))(x - m(x)))$

with $\underset{\cdot}{\mathcal{Y}} = \underset{\cdot}{f}'(m(x))^{-1}$,

(23) $\underset{\cdot}{k}_2(x) := m(x) - \underset{\cdot}{\mathcal{Y}} \cdot \underset{\cdot}{f}(m(x)) - \frac{1}{2}(\underset{\cdot}{\mathcal{Y}} \cdot \underset{\cdot}{f}''(x))(x - m(x))(x - m(x))$

with $\underset{\cdot}{\mathcal{Y}} = \underset{\cdot}{f}'(m(x))^{-1}$,

(24) $\underset{\cdot}{k}_3(x) := m(x) - \underset{\cdot}{\mathcal{Y}} \cdot \underset{\cdot}{f}(m(x))$

$$+ \{\mathcal{I} - \underset{\cdot}{\mathcal{Y}} \cdot \underset{\cdot}{f}'(m(x)) - \tfrac{1}{2}(\underset{\cdot}{\mathcal{Y}} \cdot \underset{\cdot}{f}''(x))(x - m(x))\}(x - m(x))$$

with an arbitrary nonsingular matrix $\underset{\cdot}{\mathcal{Y}}$.

Analogous to the previous theorem we first show

Theorem 12: Let $\underset{\cdot}{f}: \mathfrak{B} \subseteq V_n(\mathbb{R}) \to V_n(\mathbb{R})$ be twice continuously differentiable in $\mathfrak{B}$. If for some i, $1 \leqslant i \leqslant 3$, with an $x \subseteq \mathfrak{B}$, it holds that

$$\underset{\cdot}{k}_i(x) \subseteq x,$$

then it follows that $\underset{\cdot}{f}$ has a zero in x. If, on the other hand, $\underset{\cdot}{k}_i(x) \cap x = \varnothing$ for some i, $1 \leqslant i \leqslant 3$, then there is no zero of $\underset{\cdot}{f}$ in x.

Proof: Of (22): Consider again the map

$$\underset{\cdot}{p}(\underset{\cdot}{x}) = \underset{\cdot}{x} - \underset{\cdot}{\mathcal{Y}} \cdot \underset{\cdot}{f}(\underset{\cdot}{x})$$

and use the representation

$$\underset{\cdot}{p}(\underset{\cdot}{x}) = m(x) - \underset{\cdot}{\mathcal{Y}} \cdot \underset{\cdot}{f}(m(x)) + (\underset{\cdot}{x} - m(x)) - \underset{\cdot}{\mathcal{Y}}(\underset{\cdot}{f}(\underset{\cdot}{x}) - \underset{\cdot}{f}(m(x)))$$

for $\underset{\cdot}{x} \in x$. Using the mean-value theorem one obtains

$$\underset{\cdot}{f}(\underset{\cdot}{x}) - \underset{\cdot}{f}(m(x)) = \underset{\cdot}{\mathcal{J}}(\underset{\cdot}{x}, m(x))(\underset{\cdot}{x} - m(x)).$$

Here the ith row of $\mathcal{J}(x, m(x))$ is equal to $f_i(z^{(i)})$ with $z^{(i)} = m(x) + \theta_i(x - m(x))$, and $0 < \theta_i < 1$, $1 \leqslant i \leqslant n$. Furthermore we have that

$$f_i'(z^{(i)}) = f_i'(m(x)) + \int_0^1 f_i''(m(x) + t(z^{(i)} - m(x)))(z^{(i)} - m(x))\,dt;$$

that is,

$$\mathcal{J}(x, m(x)) = f'(m(x)) + \mathcal{B},$$

with a matrix $\mathcal{B}$ having the ith row equal to

$$\int_0^1 f_i''(m(x) + t(z^{(i)} - m(x)))(z^{(i)} - m(x))\,dt.$$

It therefore follows that

$$p(x) = m(x) - \mathcal{Y} \cdot f(m(x)) + \{\mathcal{I} - \mathcal{Y}(f'(m(x)) + \mathcal{B})\}(x - m(x)).$$

Setting

$$\mathcal{Y} = f'(m(x))^{-1}$$

we finally obtain

$$p(x) = m(x) - \mathcal{Y} \cdot f(m(x)) - \mathcal{Y}\mathcal{B}(x - m(x)).$$

Since $t \in [0, 1]$ and since $z^{(i)} \in x$ we have that $\mathcal{B} \in f''(x)(x - m(x))$, and therefore we get

$$p(x) \in m(x) - \mathcal{Y} \cdot f(m(x)) - \mathcal{Y}(f''(x)(x - m(x))(x - m(x))) = k_1(x).$$

Of (23) and (24): For the case of $k_2(x)$ and $k_3(x)$ one may proceed in a similar manner: For $1 \leqslant i \leqslant n$ we have

$$\begin{aligned} f_i(x) - f_i(m(x)) &= f_i'(m(x))(x - m(x)) \\ &\quad + \tfrac{1}{2} f_i''(m(x) + \theta_i(x - m(x)))(x - m(x))(x - m(x)), \\ &\qquad 0 < \theta_i < 1; \end{aligned}$$

that is,

$$f(x) - f(m(x)) = f'(m(x))(x - m(x)) + \tfrac{1}{2}\mathcal{B}(x - m(x))(x - m(x))$$

with a bilinear operator $\mathcal{B}$. It furthermore follows that

$$\mathcal{Y}(f(x) - f(m(x))) = \mathcal{Y} \cdot f'(m(x))(x - m(x)) + \tfrac{1}{2}\mathcal{Y}\mathcal{B}(x - m(x))(x - m(x))$$

and therefore since $\mathcal{B} \in f''(x)$ we get

$$\begin{aligned} p(x) &= m(x) - \mathcal{Y} \cdot f(m(x)) + \{\mathcal{I} - \mathcal{Y} \cdot f'(m(x)) \\ &\quad - \tfrac{1}{2}(\mathcal{Y}\mathcal{B})(x - m(x))\}(x - m(x)) \\ &\in m(x) - \mathcal{Y} \cdot f(m(x)) + \{\mathcal{I} - \mathcal{Y} \cdot f'(m(x)) \\ &\quad - \tfrac{1}{2}(\mathcal{Y} \cdot f''(x))(x - m(x))\}(x - m(x)) \\ &= k_3(x). \end{aligned}$$

With $\mathcal{Y} = f'(m(x))^{-1}$ one obtains $p(x) \in k_2(x)$.

Using Brouwer's fixed point theorem the first part of the theorem is now proved. The second part is shown as in Theorem 8. ■

We shall now compare some of these existence statements with some existence statements treated in the literature. For this we define the closed ball with midpoint y and radius r as

$$\zeta(y, r) := \{x \in V_n(\mathbb{R}) \mid \|x - y\| \leqslant r\}.$$

The norm is *always* the ∞ norm. Therefore one may also consider $\zeta(y, r)$ to be an interval vector whose components all have the width $2r$. The norms for linear and bilinear operators are defined as natural norms by using the ∞-vector norm.

The first theorem represents the existence statement of the Newton–Kantorovich theorem (see, for example Kantorovich and Akilov [8, p. 708, Theorem 6]).

Theorem 13: Let $x = \zeta(m(x), r) \in V_n(I(\mathbb{R}))$, let $f\colon V_n(\mathbb{R}) \to V_n(\mathbb{R})$ be a twice continuously differentiable mapping and assume that $\mathcal{Y} := f'(m(x))^{-1}$ exists.

If the conditions

(25) $$\|\mathcal{Y} \cdot f(m(x))\| \leqslant \eta,$$

(26) $$\|\mathcal{Y} \cdot f''(x)\| \leqslant k \qquad \text{for} \quad x \in \zeta(m(x), r),$$

(27) $$h := k\eta \leqslant \tfrac{1}{2},$$

(28) $$r \geqslant ((1 - \sqrt{1 - 2h})/h)\eta$$

are satisfied, then there is a zero of f in $\zeta(m(x), r)$. ■

The following theorems discuss some relationships between the existence statements of Theorems 12 and 13. In Theorem 12 $m(x)$ is then *always* chosen to be the midpoint.

Theorem 14: (a) Let $x = \zeta(m(x), r)$ with $r := \tfrac{1}{2}\|d(x)\|$ and assume that for $r_i := \tfrac{1}{2}\|d(k_i(x))\|$ it follows that $\zeta(m(k_i(x)), r_i) \subseteq \zeta(m(x), r) = x$ for $i = 1$ or 2.

Then the assumptions (25)–(28) of Theorem 13 are satisfied. (b) The converse is also valid for $i = 2$: If the conditions (25)–(28) are satisfied with the additional restriction

$$r \leqslant ((1 + \sqrt{1 - 2h_2})/h_2)\eta,$$

then it follows that $\zeta(m(k_2(x)), r_2) \subseteq \zeta(m(x), r) = x$. In particular it follows that $k_2(x) \subseteq x$.

Proof: (a) The inclusion $\zeta(m(k_i(x)), r_i) \subseteq \zeta(m(x), r)$ is valid if and only if $\|m(k_i(x)) - m(x)\| + r_i \leqslant r$.

From the fact that

$$m(k_i(x)) = m(x) - \mathscr{Y} \cdot f(m(x))$$

it follows that

$$\eta := \|\mathscr{Y} \cdot f(m(x))\| = \|m(k_i(x)) - m(x)\|.$$

According to Theorem 10.12, (b) and (c), the widths of $k_1(x)$ and $k_2(x)$ satisfy

$$d(k_1(x)) = \tfrac{1}{2}|\mathscr{Y}|\,|f''(x)|\,d(x)\,d(x),$$
$$d(k_2(x)) = \tfrac{1}{4}|\mathscr{Y} \cdot f''(x)|\,d(x)\,d(x).$$

We define

$$\mathscr{C} = |\mathscr{Y}|\,|f''(x)| = (c_{ijk}), \qquad k_1 = \max_{1 \leqslant i \leqslant n} \sum_{j,k=1}^{n} c_{ijk},$$

and

$$\mathscr{D} = |\mathscr{Y} \cdot f''(x)| = (d_{ijk}), \qquad k_2 = \max_{1 \leqslant i \leqslant n} \sum_{j,k=1}^{n} d_{ijk},$$

and obtain

$$\|d(k_1(x))\| = 2k_1r^2 \qquad \text{and} \qquad \|d(k_2(x))\| = k_2r^2,$$

since $d(x) = 2r(1, 1, \ldots, 1)^{\mathrm{T}}$. It therefore follows that

$$r_1 = \tfrac{1}{2}\|d(k(x))\| = k_1r^2, \qquad r_2 = \tfrac{1}{2}\|d(k_2(x))\| = \tfrac{1}{2}k_2r^2.$$

The assumption $\zeta(m(k_i(x)), r_i) \subseteq \zeta(m(x), r)$ finally gives

$$\eta + k_1r^2 \leqslant r \qquad \text{for} \quad i = 1$$

and

$$\eta + \tfrac{1}{2}k_2r^2 \leqslant r \qquad \text{for} \quad i = 2.$$

Since

$$\eta + \tfrac{1}{2}k_1r^2 \leqslant \eta + k_1r^2$$

we obtain

$$\eta + \tfrac{1}{2}k_i r^2 \leqslant r, \qquad i = 1 \text{ or } 2.$$

This implies necessarily that

$$h_i = k_i \eta \leqslant \tfrac{1}{2}, \qquad i = 1 \text{ or } 2,$$

and

$$\frac{1 - \sqrt{1 - 2h_i}}{h_i}\eta \leqslant r \leqslant \frac{1 + \sqrt{1 - 2h_i}}{h_i}\eta, \qquad i = 1 \text{ or } 2.$$

This shows that assumptions (25), (27), and (28) of Theorem 13 are satisfied. It remains to verify (26). Since $\underset{\cdot}{x} \in \zeta(m(x), r) = x$ it follows that

$$\underset{\cdot}{\mathscr{Y}} \cdot \underset{\cdot}{f}''(\underset{\cdot}{x}) \in \underset{\cdot}{\mathscr{Y}} \cdot \underset{\cdot}{f}''(x)$$

and therefore that

$$|\underset{\cdot}{\mathscr{Y}} \cdot \underset{\cdot}{f}''(\underset{\cdot}{x})| \leqslant |\underset{\cdot}{\mathscr{Y}} \cdot \underset{\cdot}{f}''(x)| \leqslant |\underset{\cdot}{\mathscr{Y}}|\,|\underset{\cdot}{f}''(x)|.$$

From this it follows that

$$\|\underset{\cdot}{\mathscr{Y}} \cdot \underset{\cdot}{f}''(\underset{\cdot}{x})\| = \|\,|\underset{\cdot}{\mathscr{Y}} \cdot \underset{\cdot}{f}''(\underset{\cdot}{x})|\,\| \leqslant \|\,|\underset{\cdot}{\mathscr{Y}} \cdot \underset{\cdot}{f}''(x)|\,\| = \|\mathscr{D}\| \leqslant \max_{1 \leqslant i \leqslant n} \sum_{j,k=1}^{n} d_{ijk}$$

and

$$\|\underset{\cdot}{\mathscr{Y}} \cdot \underset{\cdot}{f}''(\underset{\cdot}{x})\| \leqslant \|\,|\mathscr{Y}|\,|\underset{\cdot}{f}''(x)|\,\| = \|\mathscr{C}\| \leqslant \max_{1 \leqslant i \leqslant n} \sum_{j,k=1}^{n} c_{ijk}.$$

This completes the proof of (a).

(b) We note that under the assumptions of the theorem it follows that

$$\eta + \tfrac{1}{2}k_2 r^2 \leqslant r,$$

which is equivalent to

$$\zeta(m(\underset{\cdot}{k}_2(x)), r_2) \subseteq \zeta(m(x), r) = x.$$

Since $\underset{\cdot}{k}_2(x) \subseteq \zeta(m(\underset{\cdot}{k}_2(x)), r_2)$ it follows that $\underset{\cdot}{k}_2(x) \subseteq x$. ■

The statement of Theorem 10 with $\underset{\cdot}{k}(x)$ and the statement of Newton–Kantorovich was compared by Rall [21]. The assumption $\underset{\cdot}{k}(x) \subseteq x$ for the existence of a zero leads in the case of

$$x = \zeta(m(x), r)$$

to

$$((1 - \sqrt{1 - 4h})/2h)\eta \leqslant r \leqslant ((1 + \sqrt{1 - 4h})/2h)\eta.$$

Here it is assumed that $h := b\lambda\eta \leqslant \frac{1}{4}$. λ is a certain Lipschitz constant for the first derivative, and

$$\|\mathcal{Y}\| \leqslant b, \qquad \|\mathcal{Y} \cdot f(m(x))\| \leqslant \eta \qquad (\mathcal{Y} = f'(m(x))^{-1}).$$

The question of under which conditions the inclusion $k_1(x) \subseteq x$ is valid for $x = \zeta(m(x), r)$ leads essentially to the same conditions. $k_1(x) \subseteq x$ is certainly true if $\zeta(m(k_1(x)), r_1) \subseteq \zeta(m(x), r)$ is satisfied. As in the proof of part (a) of the previous theorem this is equivalent to

$$\eta + k_1 r^2 \leqslant r,$$

which leads to

$$h_1 = k_1 \eta \leqslant \tfrac{1}{4}$$

and

$$((1 - \sqrt{1 - 4h_1})/2h_1)\eta \leqslant r \leqslant ((1 + \sqrt{1 - 4h_1})/2h_1)\eta.$$

If one now notes that λb is essentially an upper bound for k_1, then one obtains for $k_1(x) \subseteq x$ the same sufficient conditions as for $k(x) \subseteq x$.

We finally remark that in the case of one equation in one unknown the condition $k_2(x) \subseteq x$ is satisfied if and only if

$$\zeta(m(k_2(x)), r_2) \subseteq \zeta(m(x), r)$$

is valid. Together with the preceding theorem we then have

Corollary 15: In the case of one equation in one unknown the conditions of the Newton–Kantorovich theorem are satisfied if and only if $k_2(x) \subseteq x = \zeta(m(x), r)$. ■

We now cite a result that we shall compare later to the existence statement from Theorem 12 based on $k_3(x)$.

Theorem 16: Let $x = \zeta(m(x), r) \in V_n(I(\mathbb{R}))$ and let $f: V_n(\mathbb{R}) \to V_n(\mathbb{R})$ be a twice differentiable map. For $\mathcal{Y} \in M_{nn}(\mathbb{R})$ it is assumed that the following conditions hold:

(29) $$\|\mathcal{Y} \cdot f(m(x))\| \leqslant \bar{\eta},$$

(30) $$\|\mathcal{I} - \mathcal{Y} \cdot f'(m(x))\| \leqslant \delta < 1,$$

(31) $$\|\mathcal{Y} \cdot f''(x)\| \leqslant \bar{k} \qquad \text{for all} \quad x \in \zeta(m(x), r),$$

(32) $$h := \bar{k}\bar{\eta}/(1 - \delta)^2 \leqslant \tfrac{1}{2},$$

(33) $$r \geqslant \bar{r}_0 = ((1 - \sqrt{1 - 2h})/h)(\bar{\eta}/(1 - \delta)).$$

The function f then has a zero in $\zeta(m(x), r)$. ■

The proof given in Kantorovich and Akilov [8] essentially consists of verifying the assumptions of Theorem 13.

In the formulation of the next theorem we use the fact that $k_3(x)$ may be written

$$k_3(x) = m(x) - \mathscr{Y} \cdot f(m(x)) + (\mathscr{I} - \mathscr{Y} \cdot f'(m(x)))(x - m(x)) + \tfrac{1}{2}(\mathscr{Y} \cdot f''(x))(x - m(x))(x - m(x))$$

provided $m(x)$ is the midpoint of x. This is shown by noting that $x - m(x)$ is symmetric, which means that the original representation of $k_3(x)$ may be rewritten

$$k_3(x) = m(x) - \mathscr{Y} \cdot f(m(x)) + |\mathscr{I} - \mathscr{Y} \cdot f'(m(x)) - \tfrac{1}{2}(\mathscr{Y} \cdot f''(x))(x - m(x))|(x - m(x)).$$

Furthermore we have that

$$\begin{aligned} &|\mathscr{I} - \mathscr{Y} \cdot f'(m(x)) - \tfrac{1}{2}(\mathscr{Y} \cdot f''(x))(x - m(x))| \\ &= |\mathscr{I} - \mathscr{Y} \cdot f'(m(x))| + \tfrac{1}{4}d((\mathscr{Y} \cdot f''(x))(x - m(x))) \\ &= |\mathscr{I} - \mathscr{Y} \cdot f'(m(x))| + \tfrac{1}{4}|\mathscr{Y} \cdot f''(x)|\, d(x). \end{aligned}$$

Since both matrices that occur in the sum are nonnegative we may write $k_3(x)$ as

$$k_3(x) = m(x) - \mathscr{Y} \cdot f(m(x)) + |\mathscr{I} - \mathscr{Y} \cdot f'(m(x))|(x - m(x)) + \tfrac{1}{4}|\mathscr{Y} \cdot f''(x)|\, d(x)(x - m(x))$$

finally obtaining

$$k_3(x) = m(x) - \mathscr{Y} \cdot f(m(x)) + (\mathscr{I} - \mathscr{Y} \cdot f'(m(x)))(x - m(x)) + \tfrac{1}{2}(\mathscr{Y} \cdot f''(x))(x - m(x))(x - m(x)).$$

With this we may now formulate

Theorem 17: Define

$$\begin{aligned} r' &= \tfrac{1}{2}\|d(m(x) - \mathscr{Y} \cdot f(m(x)) + \tfrac{1}{2}(\mathscr{Y} \cdot f''(x))(x - m(x))(x - m(x)))\|, \\ r'' &= \tfrac{1}{2}\|d((\mathscr{I} - \mathscr{Y} \cdot f'(m(x)))(x - m(x)))\|, \\ r_3 &= r' + r'', \\ r &= \tfrac{1}{2}\|d(x)\|, \\ \delta &= \|\mathscr{I} - \mathscr{Y} \cdot f'(m(x))\| < 1, \end{aligned}$$

and assume that

$$\zeta(m(k_3(x)), r_3) \subseteq \zeta(m(x), r) = x$$

holds.

Then the assumptions (29)–(33) of Theorem 16 (and therefore also those of Theorem 13) are satisfied. (It is to be noted that because of $\frac{1}{2}\|d(k_3(x))\| \leqslant r' + r''$ the hypothesis is stronger than the requirements for $k_1(x)$ or $k_2(x)$ in Theorem 14.)

Proof: We have

$$d(m(x) - \mathscr{Y} \cdot f(m(x)) + \tfrac{1}{2}(\mathscr{Y} \cdot f''(x))(x - m(x))(x - m(x)))$$
$$= \tfrac{1}{4}|\mathscr{Y} \cdot f''(x)|\, d(x)\, d(x).$$

Since $d(x) = 2r(1, 1, \ldots, 1)^{\mathrm{T}}$ it follows that

$$r' = \tfrac{1}{2}\bar{k}r^2$$

using $\mathscr{C} := |\mathscr{Y} \cdot f''(x)|$ and $\bar{k} = \max_{1 \leqslant i \leqslant n} \sum_{j,k} c_{ijk}$.

Furthermore we have

$$d((\mathscr{I} - \mathscr{Y} \cdot f'(m(x)))(x - m(x))) = |\mathscr{I} - \mathscr{Y} \cdot f'(m(x))|\, d(x),$$

and therefore

$$r'' = \delta r.$$

The assumption $\zeta(m(k_3(x)), r_3) \subseteq \zeta(m(x), r)$ is satisfied if and only if

$$\|m(k_3(x)) - m(x)\| + r_3 \leqslant r$$

holds. Setting $\bar{\eta} = \|\mathscr{Y} \cdot f(m(x))\|$ it follows that this is equivalent to

$$\bar{\eta} + r' + r'' \leqslant r$$

since

$$m(k_3(x)) = m(x) - \mathscr{Y} \cdot f(m(x));$$

in other words it is equivalent to

$$\bar{\eta} + \tfrac{1}{2}\bar{k}r^2 \leqslant r(1 - \delta).$$

From this it follows necessarily that

$$h := \bar{k}\bar{\eta}/(1 - \delta)^2 \leqslant \tfrac{1}{2}$$

and

$$\frac{1 - \sqrt{1 - 2h}}{h}\,\frac{\bar{\eta}}{1 - \delta} \leqslant r \leqslant \frac{1 + \sqrt{1 - 2h}}{h}\,\frac{\bar{\eta}}{1 - \delta}.$$

This means that conditions (29), (30) and (32), (33) of Theorem 16 have been verified.

It remains to prove that

$$\|\mathscr{Y} \cdot f''(x)\| \leqslant \bar{k} \qquad \text{for} \quad x \in x.$$

Since $\underset{\cdot}{x} \in x$ it follows that

$$|\underset{\cdot}{\mathcal{Y}} \cdot \underset{\cdot}{f}''(\underset{\cdot}{x})| \leqslant |\underset{\cdot}{\mathcal{Y}} \cdot \underset{\cdot}{f}''(x)|$$

and therefore

$$\|\underset{\cdot}{\mathcal{Y}} \cdot \underset{\cdot}{f}''(\underset{\cdot}{x})\| \leqslant \|\,|\underset{\cdot}{\mathcal{Y}} \cdot \underset{\cdot}{f}''(x)|\,\| = \|\mathcal{C}\| \leqslant \max_{1 \leqslant i \leqslant n} \sum_{j,k} c_{ijk}. \quad ■$$

As already remarked, it does not necessarily follow from $\underset{\cdot}{k}_i(x) \subseteq x$ that the assumptions of Theorem 13 are satisfied. We therefore consider the following simple *example*:

Let $\underset{\cdot}{f}(\underset{\cdot}{x}) = \binom{x^2 - 1}{y^2 - 0.01}$, $\underset{\cdot}{x} = \binom{x}{y}$, and $x = \binom{[0.98,\ 1.18]}{[0,\ 0.2]}$, then $\underset{\cdot}{f}$ has the zero $\underset{\cdot}{x}^* = \binom{1}{0.1}$ in x. With $m(x) = \binom{1.08}{0.1}$ it follows that $\underset{\cdot}{f}(m(x)) = \binom{0.1664}{0}$. Furthermore we obtain $\underset{\cdot}{f}'(\underset{\cdot}{x}) = \begin{pmatrix} 2x & 0 \\ 0 & 2y \end{pmatrix}$, $\underset{\cdot}{f}'(m(x)) = \begin{pmatrix} 2.16 & 0 \\ 0 & 0.2 \end{pmatrix}$, $\underset{\cdot}{f}'(m(x))^{-1} = \begin{pmatrix} 1/2.16 & 0 \\ 0 & 5 \end{pmatrix}$, $\underset{\cdot}{f}'(m(x))^{-1}\underset{\cdot}{f}(m(x)) = \binom{0.1664/2.16}{0}$; that is, $\eta = 0.1664/2.16$. The second derivative is

$$\underset{\cdot}{f}''(\underset{\cdot}{x}) = \left(\begin{array}{cc|cc} 2 & 0 & 0 & 0 \\ 0 & 0 & 0 & 2 \end{array}\right)$$

from which we get

$$\underset{\cdot}{f}'(m(x))^{-1}\underset{\cdot}{f}''(\underset{\cdot}{x}) = \left(\begin{array}{cc|cc} 2/2.16 & 0 & 0 & 0 \\ 0 & 0 & 0 & 10 \end{array}\right)$$

for $\underset{\cdot}{x} \in x$. A simple calculation gives $\|\underset{\cdot}{f}'(m(x))^{-1}\underset{\cdot}{f}''(\underset{\cdot}{x})\|_\infty = 10 =: k$. It therefore follows that $h = k\eta = 1.664/2.16 > \frac{1}{2}$ which shows that Theorem 13 cannot be applied. On the other hand, for $\underset{\cdot}{k}_1(x)$ from Theorem 12 setting $\underset{\cdot}{\mathcal{Y}} = m(\underset{\cdot}{f}'(x))^{-1}$, we get that

$$\begin{aligned} \underset{\cdot}{k}_1(x) &= m(x) - \underset{\cdot}{\mathcal{Y}} \cdot \underset{\cdot}{f}(m(x)) - \underset{\cdot}{\mathcal{Y}} \cdot (\underset{\cdot}{f}''(x)(x - m(x))(x - m(x))) \\ &\subseteq \begin{pmatrix} [0.9937, & 1.0123] \\ [0, & 0.2] \end{pmatrix} \subseteq x. \end{aligned}$$

This means that Theorem 12 with $i = 1$ (Eq. (22)) is applicable. A similar calculation shows that for this example both $\underset{\cdot}{k}_2(x) \subseteq x$ as well as $\underset{\cdot}{k}(x) \subseteq x$ (see Theorem 10) holds for $\underset{\cdot}{\mathcal{Y}} = m(\underset{\cdot}{f}'(x))^{-1}$.

When one considers the advantages and disadvantages of the existence statements of Theorems 12 and 13 then the statements of Theorem 12 seem to be superior since they are valid for arbitrary interval vectors rather than only for the balls that are defined by the ∞ norm. The result of Theorem 13 is, on the other hand, valid for arbitrary norms. These remarks show that the existence statements found in this chapter may be considered to be important supplements to the theorem of Newton–Kantorovich.

This is also true for the existence statements of Theorems 8–11. In comparison to the classical statement of Theorem 13 they have the important advantage of neither needing the second derivative nor a Lipschitz constant.

Remarks: Rokne and Lancaster also calculate an inclusion for the solution of a system $f(x) = o$ of equations analogous to the case of one equation in one unknown. One step of Newton's method is carried out for this and with the aid of estimates for f, f', and f'' one determines an inclusion. This is in fact an application of the theorem of Kantorovich. This procedure in Rokne and Lancaster [23] is then applied to the inclusion of eigenvalues of real matrices in Rokne [22]. The eigenvalue problem is there in a known manner carried over to the solution of a nonlinear system of equations on which the method from Rokne and Lancaster [23] is applied. A modification of the interval Newton method (7) is given by Krawczyk [10]. In this modification there is no need to compute inclusions for the inverses of all the matrices contained in a matrix. The iteration procedure

$$(34)\quad \begin{cases} x^{(k+1)} = \{m(x^{(k)}) - \mathscr{Y}\cdot f(m(x^{(k)})) + (\mathscr{I} - \mathscr{Y}\cdot f'(x^{(k)})) \\ \qquad\qquad \times (x^{(k)} - m(x^{(k)}))\} \cap x^{(k)} \\ m(x^{(k)}) \in x^{(k)}, \end{cases}$$

does only require a matrix $\mathscr{Y}$. If

$$\| |\mathscr{I} - \mathscr{Y}\cdot f'(x^{(0)})| \| < \tfrac{1}{2}$$

is satisfied, then the convergence of the inclusions toward the zero is proved. Superlinear convergence is proved if $\mathscr{Y}$ is replaced by $\mathscr{Y}^{(k)} = f'(m(x^{(k)}))^{-1}$. This procedure is used to include eigenvalues of real matrices in Krawczyk [B161]. The expression on the left of the intersection in the above iteration method (34) is the expression $k(x)$ from Theorem 10.

A recent paper by Moore [14] is concerned with the convergence of the above method (34) (see Theorem 1 in Moore [14]). We prove here a generalization of the results by Moore. In fact we wish to show that if the relation

$$k(x^{(0)}) = m(x^{(0)}) - \mathscr{Y}\cdot f(m(x^{(0)})) + (\mathscr{I} - \mathscr{Y}\cdot f'(x^{(0)}))(x^{(0)} - m(x^{(0)})) \\ \subset x^{(0)}$$

holds for some $\mathscr{Y} \in M_{nn}(\mathbb{R})$ and an interval vector $x^{(0)}$ with the midpoint $m(x^{(0)})$, then the iteration method (34) with the midpoint of $x^{(k)}$ being $m(x^{(k)})$ for $k \geqslant 0$ converges toward the unique zero y of f in the interval $x^{(0)}$.

The relation $k(x^{(0)}) \subset x^{(0)}$ is defined in the same manner as in Theorem 14.6. The above assumption $k(x^{(0)}) \subset x^{(0)}$ therefore has in particular the implication that $d(x^{(0)}) > o$ as a consequence. We have $k(x^{(0)}) \subset x^{(0)}$ if and only if the distance from $k(x^{(0)})$ to $m(x^{(0)})$ satisfies

$$q(k(x^{(0)}), m(x^{(0)})) < \tfrac{1}{2} d(x^{(0)})$$

or equivalently

$$|k(x^{(0)}) - m(x^{(0)})| < \tfrac{1}{2} d(x^{(0)}).$$

Because of

$$|k(x^{(0)}) - m(x^{(0)})| = |\mathcal{Y} \cdot f(m(x^{(0)}))| + \tfrac{1}{2} |\mathcal{I} - \mathcal{Y} \cdot f'(x^{(0)})| \, d(x^{(0)}),$$

it certainly follows that

$$|\mathcal{I} - \mathcal{Y} \cdot f'(x^{(0)})| \, d(x^{(0)}) < d(x^{(0)}).$$

From this it follows that the spectral radius of the nonnegative matrix $|\mathcal{I} - \mathcal{Y} \cdot f'(x^{(0)})|$ is smaller than 1. (The row sums weighted with the positive vector $d(x^{(0)})$ are all less than 1.) The limiting value x^* computed from (34) satisfies

$$x^* = \{m(x^*) - \mathcal{Y} \cdot f(m(x^*)) + (\mathcal{I} - \mathcal{Y} \cdot f'(x^*))(x^* - m(x^*))\} \cap x^*;$$

that is,

$$d(x^*) \leqslant |\mathcal{I} - \mathcal{Y} \cdot f'(x^*)| \, d(x^*).$$

From inclusion monotonicity it follows that $f'(x^*) \subseteq f'(x^{(0)})$, and we therefore can write

$$|\mathcal{I} - \mathcal{Y} \cdot f'(x^*)| \leqslant |\mathcal{I} - \mathcal{Y} \cdot f'(x^{(0)})|.$$

From the theorem of Perron and Frobenius on the monotonicity of the spectral radius of nonnegative matrices it follows that $\rho(|\mathcal{I} - \mathcal{Y} \cdot f'(x^*)|) < 1$ also.

It now follows that $d(x^*) = o$ and that $\lim_{k \to \infty} x^{(k)} \doteq y$. The result of Moore, quoted above, contains the special case of intervals for which $d(X_1^{(0)}) = \cdots = d(X_n^{(0)})$ ("hypercubes").

In the thesis of Rump [24] there is an algorithm using $k(x)$ (including FORTRAN programs) for the determination of an including vector for the solution of a set of linear equations. Numerous numerical examples and further applications to concrete problems of numerical mathematics are also contained in his work. See also Kaucher and Rump [9].

The existence statements (Theorem 12) and the comparison theorems (Theorems 14 and 17 as well as Corollary 15) were proven by Platzöder [20].

Analogous to the method (34), motivated by the existence statement of Theorem 10, one may also, starting with Theorem 11, define the iteration method

$$(35)\quad \begin{cases} \tilde{\mathscr{x}}^{(k+1)} = m(\mathscr{x}^{(k)}) - \mathscr{g}(\underset{\cdot}{\mathscr{Y}}, \underset{\cdot}{\mathscr{f}}(m(\mathscr{x}^{(k)})) - (\underset{\cdot}{\mathscr{Y}} - \mathscr{f}'(\mathscr{x}^{(k)}))(\mathscr{x}^{(k)} - m(\mathscr{x}^{(k)})) \\ \qquad\qquad\qquad\qquad \text{with} \quad m(\mathscr{x}^{(k)}) \in \mathscr{x}^{(k)}, \text{ midpoint of } \mathscr{x}^{(k)}, \\ \mathscr{x}^{(k+1)} = \tilde{\mathscr{x}}^{(k+1)} \cap \mathscr{x}^{(k)}, \qquad k \geqslant 0. \end{cases}$$

The computational effort required in this method is significantly lower per step than in method (34). (The interval vector $\mathscr{g}(\mathscr{A}, \mathscr{b})$ denotes again the interval vector calculated from the interval matrix $\mathscr{A}$ and the right-hand side $\mathscr{b}$ by applying the Gaussian algorithm. See Chapter 15.)

We use the abbreviation

$$\tilde{\underset{\cdot}{\mathscr{k}}}(\mathscr{x}) = m(\mathscr{x}) - \mathscr{g}(\underset{\cdot}{\mathscr{Y}}, \underset{\cdot}{\mathscr{f}}(m(\mathscr{x})) - (\underset{\cdot}{\mathscr{Y}} - \mathscr{f}'(\mathscr{x}))(\mathscr{x} - m(\mathscr{x}))).$$

Analogously to the statement for (34) we now show: If the relation $\tilde{\underset{\cdot}{\mathscr{k}}}(\mathscr{x}^{(0)}) \subset \mathscr{x}^{(0)}$ holds for some $\underset{\cdot}{\mathscr{Y}} \in M_{nn}(\mathbb{R})$ and an interval vector $\mathscr{x}^{(0)}$ with the midpoint $m(\mathscr{x}^{(0)})$, then the iteration method (35) with the midpoint of $\mathscr{x}^{(k)}$ being $m(\mathscr{x}^{(k)})$ for $k \geqslant 0$ converges to the unique zero $\lim_{k\to\infty} \mathscr{x}^{(k)} = \underset{\cdot}{x}^*$.

In order to *prove* this we note from Theorem 11 that since $\tilde{\underset{\cdot}{\mathscr{k}}}(\mathscr{x}^{(0)}) \subset \mathscr{x}^{(0)}$ it follows that there is a zero $\underset{\cdot}{x}^*$ of $\underset{\cdot}{\mathscr{f}}$ in $\mathscr{x}^{(0)}$. Using the relations given for the Gaussian algorithm at the start of Chapter 15 we may then write

$$\begin{aligned} \underset{\cdot}{x}^* &= \underset{\cdot}{x}^* - \underset{\cdot}{\mathscr{Y}}^{-1}\underset{\cdot}{\mathscr{f}}(\underset{\cdot}{x}^*) \\ &= m(\mathscr{x}^{(0)}) - \underset{\cdot}{\mathscr{Y}}^{-1}\underset{\cdot}{\mathscr{f}}(m(\mathscr{x}^{(0)})) + \underset{\cdot}{x}^* - m(\mathscr{x}^{(0)}) + \underset{\cdot}{\mathscr{Y}}^{-1}(\underset{\cdot}{\mathscr{f}}(m(\mathscr{x}^{(0)})) - \underset{\cdot}{\mathscr{f}}(\underset{\cdot}{x}^*)) \\ &\in m(\mathscr{x}^{(0)}) - \underset{\cdot}{\mathscr{Y}}^{-1}(\underset{\cdot}{\mathscr{f}}(m(\mathscr{x}^{(0)})) - (\underset{\cdot}{\mathscr{Y}} - \mathscr{f}'(\mathscr{x}^{(0)}))(\mathscr{x}^{(0)} - m(\mathscr{x}^{(0)}))) \\ &\subseteq m(\mathscr{x}^{(0)}) - \mathscr{g}(\underset{\cdot}{\mathscr{Y}}, \underset{\cdot}{\mathscr{f}}(m(\mathscr{x}^{(0)})) - (\underset{\cdot}{\mathscr{Y}} - \mathscr{f}'(\mathscr{x}^{(0)}))(\mathscr{x}^{(0)} - m(\mathscr{x}^{(0)}))) = \tilde{\underset{\cdot}{\mathscr{k}}}(\mathscr{x}^{(0)}). \end{aligned}$$

Therefore $\mathscr{x}^{(1)}$ exists and it holds that $\underset{\cdot}{x}^* \in \mathscr{x}^{(1)}$. Using complete induction it follows that $\underset{\cdot}{x}^* \in \mathscr{x}^{(k)}$, $k \geqslant 0$.

The assumption that $\tilde{\underset{\cdot}{\mathscr{k}}}(\mathscr{x}^{(0)}) \subset \mathscr{x}^{(0)}$ is equivalent to componentwise proper inclusion; that is, there exists a real number $0 \leqslant \alpha < 1$ for which

$$d(\tilde{\underset{\cdot}{\mathscr{k}}}(\mathscr{x}^{(0)})) \leqslant \alpha\, d(\mathscr{x}^{(0)}).$$

From

$$\tilde{\mathscr{x}}^{(1)} = \tilde{\underset{\cdot}{\mathscr{k}}}(\mathscr{x}^{(0)}) \subset \mathscr{x}^{(0)}$$

it follows that

$$\mathscr{x}^{(1)} = \tilde{\underset{\cdot}{\mathscr{k}}}(\mathscr{x}^{(0)}) \cap \mathscr{x}^{(0)} = \tilde{\underset{\cdot}{\mathscr{k}}}(\mathscr{x}^{(0)}) = \tilde{\mathscr{x}}^{(1)},$$

and therefore

$$d(\tilde{\mathscr{x}}^{(1)}) = d(\mathscr{x}^{(1)}) \leqslant \alpha\, d(\mathscr{x}^{(0)}).$$

From $\mathscr{x}^{(1)} \subset \mathscr{x}^{(0)}$ it follows from the inclusion monotonicity that

$$|\underset{\cdot}{\mathscr{Y}} - \mathscr{f}'(\mathscr{x}^{(1)})| \leqslant |\underset{\cdot}{\mathscr{Y}} - \mathscr{f}'(\mathscr{x}^{(0)})|$$

and therefore

$$\begin{aligned} d(\underset{\cdot}{f}(m(\boldsymbol{x}^{(1)})) &- (\underset{\cdot}{\mathscr{Y}} - \boldsymbol{f}'(\boldsymbol{x}^{(1)}))(\boldsymbol{x}^{(1)} - m(\boldsymbol{x}^{(1)}))) \\ &= |\underset{\cdot}{\mathscr{Y}} - \boldsymbol{f}'(\boldsymbol{x}^{(1)})|\, d(\boldsymbol{x}^{(1)}) \\ &\leqslant \alpha\, |\underset{\cdot}{\mathscr{Y}} - \boldsymbol{f}'(\boldsymbol{x}^{(0)})|\, d(\boldsymbol{x}^{(0)}) \\ &= \alpha\, d(\underset{\cdot}{f}(m(\boldsymbol{x}^{(0)})) - (\underset{\cdot}{\mathscr{Y}} - \boldsymbol{f}'(\boldsymbol{x}^{(0)}))(\boldsymbol{x}^{(0)} - m(\boldsymbol{x}^{(0)}))). \end{aligned}$$

As proved generally in Chapter 15 we therefore have that

$$d(\tilde{\boldsymbol{x}}^{(2)}) \leqslant \alpha\, d(\tilde{\boldsymbol{x}}^{(1)});$$

that is,

$$d(\tilde{\boldsymbol{x}}^{(2)}) \leqslant \alpha^2\, d(\boldsymbol{x}^{(0)}).$$

From

$$d(\boldsymbol{x}^{(2)}) \leqslant \inf\{d(\tilde{\boldsymbol{x}}^{(2)}), d(\boldsymbol{x}^{(1)})\}$$

it follows that

$$d(\boldsymbol{x}^{(2)}) \leqslant \alpha^2\, d(\boldsymbol{x}^{(0)}).$$

By using complete induction one proves that

$$d(\boldsymbol{x}^{(k+1)}) \leqslant \alpha^{k+1}\, d(\boldsymbol{x}^{(0)}), \qquad k \geqslant 0.$$

With this it follows that $\lim_{k\to\infty} d(\boldsymbol{x}^{(k)}) = \underset{\cdot}{o}$ and because of the fact that $\underset{\cdot}{x}^* \in \boldsymbol{x}^{(k)}$ it follows that $\lim_{k\to\infty} \boldsymbol{x}^{(k)} = \underset{\cdot}{x}^*$. This completes the proof.

We now wish to add some further remarks regarding the method (35). (The same remarks are also valid for (34).) If one carries out the method (35) starting with an arbitrary interval vector $\boldsymbol{x}^{(0)}$, that is, one not necessarily satisfying the assumption $\tilde{\boldsymbol{k}}(\boldsymbol{x}^{(0)}) \subset \boldsymbol{x}^{(0)}$, and with a nonsingular matrix $\underset{\cdot}{\mathscr{Y}}$, then one of the two following cases will occur:

(i) For some $k_0 \geqslant 0$ it holds that $\tilde{\boldsymbol{x}}^{(k_0+1)} \cap \boldsymbol{x}^{(k_0)} = \varnothing$. In this case there is no zero of $\boldsymbol{f}$ in $\boldsymbol{x}^{(0)}$. If there were a zero of $\boldsymbol{f}$ in $\boldsymbol{x}^{(0)}$, then one could show as in the case of $\tilde{\boldsymbol{k}}(\boldsymbol{x}^{(0)}) \subset \boldsymbol{x}^{(0)}$ that the intersection is nonempty for all k.

(ii) The method (35) is defined for all $k \geqslant 0$. That is, $\lim_{k\to\infty} \boldsymbol{x}^{(k)} = \boldsymbol{x}^*$ holds. If now $d(\boldsymbol{x}^*) \neq \underset{\cdot}{o}$, then it is not possible to make a statement about the existence or nonexistence of a zero. If, on the other hand, $d(\boldsymbol{x}^*) = \underset{\cdot}{o}$, that is, $\lim_{k\to\infty} \boldsymbol{x}^{(k)} = \underset{\cdot}{x}^*$, then it follows that $\underset{\cdot}{x}^*$ is a zero of $\underset{\cdot}{f}$. This follows from (35) by passing to the limit $k \to \infty$.

It is now remarkable that the assumption $d(\tilde{\boldsymbol{k}}(\boldsymbol{x}^{(0)})) \leqslant \alpha\, d(\boldsymbol{x}^{(0)})$, $0 \leqslant \alpha < 1$, is already sufficient for either (i) or (ii) with $d(\boldsymbol{x}^*) = \underset{\cdot}{o}$ to occur: Under the assumption $d(\tilde{k}(x^{(0)})) \leqslant \alpha\, d(x^{(0)})$, $0 \leqslant \alpha < 1$, one shows as above in the case of the stronger assumption $\tilde{k}(x^{(0)}) \subset x^{(0)}$ that the inequality $d(x^{(k+1)}) \leqslant \alpha^{k+1} d(x^{(0)})$ is valid for all $k \geqslant 0$ for which $\tilde{x}^{(k+1)} \cap x^{(k)} \neq \emptyset$ holds. If f does not have a zero in $x^{(0)}$, then the intersection will be empty for some $k_0 \geqslant 0$. Otherwise one obtains the contradiction $\lim_{k\to\infty} x^{(k)} = x^*$ with $f(x^*) = o$. If f does have a zero x^* in $x^{(0)}$, then one shows as in the above case of $\tilde{k}(x^{(0)}) \subset x^{(0)}$ that the intersection cannot be empty for *any* $k \geqslant 0$ and therefore that $\lim_{k\to\infty} x^{(k)} = x^*$ with $f(x^*) = o$ holds.

An inclusion for the solutions of a set of linear equations of, for example, a system of the form (6), was used in the development of methods (7) (resp., (12)). This inclusion could also basically be determined with the help of the Gaussian algorithm of Chapter 15. Such a method would certainly converge for a sufficiently small width of $x^{(0)}$. No verifiable conditions guaranteeing the convergence has been found as of yet. It would be extremely useful to know such criteria, since one would then be able to avoid the inversion of $m(\mathcal{F}^{(k)})$ as well as the use of (10). One would then be able to carry out the repeated inclusion of the solution y with an acceptable amount of calculation. Even though in (14) the Gaussian algorithm and step (d) have to be executed, this procedure is substantially cheaper computationally than other methods considered in this chapter. It is also important, as shown in Theorem 7, that there are no limitations on the size of the width of $x^{(0)}$. The convergence statement of Theorem 7 is therefore practically a global convergence statement. The numerical example for method (14) demonstrates the importance of this statement. It is well known that solving the discretized boundary value problem using Newton's method presents no problems if the function $f(t, y)$ is convex with respect to y. In the case of nonconvexity, however, it is only possible to prove convergence for starting values sufficiently close to the solution. No convexity assumptions are needed for the method (14).

In Chapter 21 a method is considered that also solves the above discretized boundary value problem where one avoids the interval Gaussian algorithm. [See method (21.8′).] Together with the interval evaluation of the derivative one requires, however, two function evaluations in each step in order to achieve quadratic convergence.

Using the Gaussian algorithm for including the set of solutions described by (4) (resp., (6)) leads to the algorithm

$$x^{(k+1)} = \{m(x^{(k)}) - g(f'(x^{(k)}), f(m(x^{(k)})))\} \cap x^{(k)}. \tag{36}$$

Here $g(\cdot, \cdot)$ denotes the mapping defined by applying the Gaussian algorithm. For a single equation in one unknown this method is identical to method (7.11). Unfortunately the contents of Theorem 7.4 cannot be carried over to systems of equations. This was demonstrated by Schwandt [25], who discussed an example for which $x^* \in x^{(0)}$, $f(x^*) = o$ and $x^{(1)} = x^{(0)}$ with $d(x^{(1)}) \neq o$. In Alefeld [27] the matrix $\mathcal{Y}$ appearing in (35) is allowed to vary with k. Using $\mathcal{Y}^{(k)} = m(f'(x^{(k)}))$ a modification of (35) was given that is quadratically convergent. In Alefeld [28] one can find some sufficient conditions for the convergence of (36) to a solution of $f(x) = o$.

REFERENCES

[1] G. Alefeld, Intervallanalytische Methoden bei nichtlinearen Gleichungen. Jahrbuch Überblicke Mathematik 1979, 63–78, Bibliograph. Inst., Mannheim (1979).

[2] G. Alefeld and J. Herzberger, Über das Newton-Verfahren bei nichtlinearen Gleichungssystemen. *Z. Angew. Math. Mech.* **50**, 773–774 (1970).

[3] G. Alefeld and J. Herzberger, Ein Verfahren zur monotonen Einschließung von Lösungen nichtlinearer Gleichungssysteme. *Z. Angew. Math. Mech.* **53**, T 176–T 177 (1973).

[4] L. Collatz, "Funktionalanalysis und numerische Mathematik." Springer-Verlag, Berlin and New York, 1964.

[5] J. A. Grant and G. D. Hitchins, The solution of polynomial equations in interval arithmetic. *Comput. J.* **16**, 69–72 (1973).

[6] E. R. Hansen, On solving systems of equations using interval arithmetic. *Math. Comput.* **22**, 374–384 (1968).

[7] J. Herzberger, Zur Konvergenz intervallmäßiger Iterationsverfahren. *Z. Angew. Math. Mech.* **51**, T 56 (1971).

[8] L. V. Kantorovich and G. P. Akilov, "Functional Analysis in Normed Spaces" (English transl.). Pergamon, Oxford, 1964.

[9] E. Kaucher and S. M. Rump, Small bounds for the solution of systems of linear equations. *In* "Computing Supplementum 2" (G. Alefeld and R. D. Grigorieff, eds.). Springer-Verlag, Berlin and New York, 1980.

[10] R. Krawczyk, Newton-Algorithmen zur Bestimmung von Nullstellen mit Fehlerschranken. *Computing* **4**, 187–201 (1969).

[11] R. E. Moore, "Interval Analysis." Prentice-Hall, Englewood Cliffs, New Jersey, 1966.

[12] R. E. Moore, A test for existence of solutions to non-linear systems. *SIAM J. Numer. Anal.* **14**, 611–615 (1977).

[13] R. E. Moore, Methods and Applications of Interval Analysis. SIAM Studies in Applied Mathematics. SIAM, Philadelphia, Pennsylvania (1979).

[14] R. E. Moore, A computational test for convergence of iterative methods for non-linear systems. *SIAM J. Numer. Anal.* **15**, 1194–1196 (1978).

[15] R. E. Moore, Interval methods for nonlinear systems. *In* "Computing Supplementum 2" (G. Alefeld and R. D. Grigorieff, eds.). Springer-Verlag, Berlin and New York, 1980.

[16] K. Nickel, Zeros of polynomials and other topics. *In* "Topics in Interval Analysis" (E. Hansen, ed.). Oxford Univ. Press (Clarendon), London and New York, 1969.

[17] K. Nickel, On the Newton-method in Interval Analysis. MRC Technical Summary Rep. 1136, Univ. of Madison, Madison, Wisconsin (1971).

[18] K. Nickel, A globally convergent ball Newton method. *SIAM J. Numer. Anal.* **18**, 988–1003 (1981).

[19] J. M. Ortega, "Numerical Analysis: A Second Course." Academic Press, New York, 1972.

[20] L. Platzöder, Einige Beiträge über die Existenz von Lösungen nichtlinearer Gleichungssysteme und Verfahren zu ihrer Berechnung. Ph.D. Thesis, Techn. Univ. Berlin (1981).

[21] L. Rall, A Comparison of the Existence Theorems of Kantorovich and Moore. MRC Technical Summary Report #1944. Mathematics Research Center, Madison, Wisconsin, (1979).

[22] J. Rokne, Fehlererfassung bei Eigenwertproblemen von Matrizen. *Computing* **7**, 145–152 (1971).

[23] J. Rokne and P. Lancaster, Automatic error bounds for the approximate solutions of equations. *Computing* **4**, 294–303 (1969).

[24] S. M. Rump, Kleine Fehlerschranken bei Matrixproblemen. Ph.D. Thesis, Univ. Karlsruhe Fakultät für Mathematik, Karlsruhe, Germany, 1980.

[25] H. Schwandt, Schnelle fast global konvergente Verfahren für die Fünf-Punkte-Diskretisierung der Poissongleichung mit Dirichletschen Randbedingungen auf Rechtecksgebieten. Ph. D. Thesis, Technische Univ. Berlin, Fachbereich Mathematik (1981).

[26] R. S. Varga, "Matrix Iterative Analysis." Prentice-Hall, Englewood Cliffs, New Jersey, 1962.

Added in proof:

[27] G. Alefeld and L. Platzöder, A quadratically convergent Krawczyk-like algorithm. *SIAM J. Numer. Anal.* (to be published).

[28] G. Alefeld, On the convergence of some interval arithmetic modifications of Newton's method. *SIAM J. Numer. Anal.* (to be published).

Chapter 20 / NEWTON-LIKE METHODS WITHOUT MATRIX INVERSIONS

In the last chapter it was shown that the iteration (19.12) converges at least quadratically in the width of the iterates. At each step both a real matrix $m(\mathscr{F}^{(k)})$ has to be inverted as well as (19.10) calculated until a fixed point has been reached. We shall now modify (19.12) in such a manner that only one step of the iterative determination of an inclusion $\mathscr{V}^{(k)}$ for the set $\{\mathfrak{F}^{-1} \mid \mathfrak{F} \in \mathscr{F}^{(k)}\}$ is performed. We recall that in (19.12) we had to perform the iteration (19.10). For this purpose we choose another method for the iterative inclusion of the inverse of an interval matrix. This algorithm is a generalization of algorithm (18.9) with $r = 2$ for interval matrices. We therefore introduce the following.

Theorem 1: Let a real nonsingular matrix $\mathfrak{A}$ and a sequence of interval matrices $\{\mathscr{A}^{(k)}\}_{k=0}^{\infty}$ for which $\mathfrak{A} \in \mathscr{A}^{(k)}$, $k \geqslant 0$, be given. Let $\lim_{k\to\infty} \mathscr{A}^{(k)} = \mathfrak{A}$, and let $\mathscr{W}^{(0)}$ be an interval matrix for which $\mathfrak{A}^{-1} \in \mathscr{W}^{(0)}$. For the iteration

$$(1) \qquad \mathscr{W}^{(k+1)} = \{m(\mathscr{W}^{(k)}) + \mathscr{W}^{(k)}(\mathfrak{I} - \mathscr{A}^{(k)} m(\mathscr{W}^{(k)}))\} \cap \mathscr{W}^{(k)},$$

$m(\mathscr{W}^{(k)})$ according to (18.1), $k \geqslant 0$, it holds that

(2) (a) $\mathfrak{A}^{-1} \in \mathscr{W}^{(k)}$, $k \geqslant 0$;

(b) if every matrix $\mathfrak{W} \in \mathscr{W}^{(0)}$ is nonsingular, then it follows that

$$\lim_{k\to\infty} \mathscr{W}^{(k)} = \mathfrak{A}^{-1}.$$

Proof: Of (a): From the assumptions we have that $\mathscr{A}^{-1} \in \mathscr{W}^{(0)}$ and $\mathscr{A} \in \mathscr{A}^{(0)}$. With the aid of (10.10′) it therefore follows that

$$\begin{aligned}\mathscr{A}^{-1} &= m(\mathscr{W}^{(0)}) + \mathscr{A}^{-1}(\mathscr{I} - \mathscr{A} \cdot m(\mathscr{W}^{(0)}))\\ &\in \{m(\mathscr{W}^{(0)}) + \mathscr{W}^{(0)}(\mathscr{I} - \mathscr{A}^{(0)} m(\mathscr{W}^{(0)}))\} \cap \mathscr{W}^{(0)}.\end{aligned}$$

A complete induction completes the proof.

Of (b): The monotonically descending sequence of matrices

$$\mathscr{W}^{(0)} \supseteq \mathscr{W}^{(1)} \supseteq \mathscr{W}^{(2)} \supseteq \cdots$$

converges to an interval matrix $\mathscr{W}$ (compare with Corollary 10.8). From the continuous dependence of the midpoint $m(\mathscr{W}^{(k)})$ on $\mathscr{W}^{(k)}$ it follows that

$$\lim_{k\to\infty} m(\mathscr{W}^{(k)}) = m(\mathscr{W}).$$

From this and the continuity of the iteration (1) it follows that

$$\begin{aligned}\mathscr{W} &= \lim_{k\to\infty} (\{m(\mathscr{W}^{(k)}) + \mathscr{W}^{(k)}(\mathscr{I} - \mathscr{A}^{(k)} m(\mathscr{W}^{(k)}))\} \cap \mathscr{W}^{(k)})\\ &= \{m(\mathscr{W}) + \mathscr{W}(\mathscr{I} - \mathscr{A} \cdot m(\mathscr{W}))\} \cap \mathscr{W},\end{aligned}$$

and therefore

$$m(\mathscr{W}) \in m(\mathscr{W}) + \mathscr{W}(\mathscr{I} - \mathscr{A} \cdot m(\mathscr{W})).$$

Let the jth column of the matrix $m(\mathscr{W})$ be denoted by $(m(\mathscr{W}))_j$. Because of (10.1) we get the representation

$$(m(\mathscr{W}))_j = (m(\mathscr{W}))_j + \mathscr{W}^{(j)}((\mathscr{I} - \mathscr{A} \cdot m(\mathscr{W})))_j$$

with a matrix $\mathscr{W}^{(j)} \in \mathscr{W} \subseteq \mathscr{W}^{(0)}$. From this it follows that

$$\mathscr{W}^{(j)}((\mathscr{I} - \mathscr{A} \cdot m(\mathscr{W})))_j = o.$$

The matrices $\mathscr{A}$ and $\mathscr{W}^{(j)}$ are nonsingular by assumption and together with

$$(\mathscr{A} m(\mathscr{W}))_j = \mathscr{A}(m(\mathscr{W}))_j,$$

we conclude that

$$(m(\mathscr{W}))_j = (\mathscr{A}^{-1})_j,$$

which shows that $m(\mathscr{W}) = \mathscr{A}^{-1}$. But this implies $\mathscr{W} = \mathscr{A}^{-1}$, which completes the proof of the theorem. ■

The iteration method (18.9) with $r = 2$ is obtained as a special case when $\mathscr{A}^{(k)} = \mathscr{A}$. If one assumes in (1) that the sequence $\{\mathscr{A}^{(k)}\}_{k=0}^{\infty}$ satisfies

$$\mathscr{A}^{(0)} \supseteq \mathscr{A}^{(1)} \supseteq \mathscr{A}^{(2)} \supseteq \cdots \qquad \text{with} \quad \lim_{k\to\infty} \mathscr{A}^{(k)} = \mathscr{A},$$

then it follows, analogously to (2a) that the inclusion

$$\{\underset{\cdot}{\mathscr{A}}^{-1} \mid \underset{\cdot}{\mathscr{A}} \in \mathscr{A}\} \subseteq \mathscr{W}^{(0)}$$

always implies

$$\{\underset{\cdot}{\mathscr{A}}^{-1} \mid \underset{\cdot}{\mathscr{A}} \in \mathscr{A}\} \subseteq \mathscr{W}^{(k)}, \qquad k \geqslant 0.$$

If one additionally choses $\mathscr{A}^{(k)} = \mathscr{A}$, then one obtains a method from (1) for the iterative inclusion of an interval matrix. A similar iteration method was already considered in (19.10).

The iteration (1) is now used to construct a method similar to (19.12). Therefore let $\mathfrak{x}^{(0)}$ be an interval vector containing a zero of the function $\underset{\cdot}{\mathfrak{f}}(\underset{\cdot}{\mathfrak{x}})$. Assume furthermore that $\mathscr{V}^{(0)}$ is an interval matrix with the property that $\underset{\cdot}{\mathscr{J}}(\underset{\cdot}{\mathfrak{x}})^{-1} \in \mathscr{V}^{(0)}$ for $\underset{\cdot}{\mathfrak{x}} \in \mathfrak{x}^{(0)}$, where $\underset{\cdot}{\mathscr{J}}(\underset{\cdot}{\mathfrak{x}})$ is defined according to (19.3). Let again $\mathscr{F}^{(k)} = \underset{\cdot}{\mathfrak{f}}'(\mathfrak{x}^{(k)})$.

Now consider the iteration

$$(3)\quad \begin{cases} \mathfrak{x}^{(k+1)} = \{m(\mathfrak{x}^{(k)}) - \mathscr{V}^{(k)}\underset{\cdot}{\mathfrak{f}}(m(\mathfrak{x}^{(k)}))\} \cap \mathfrak{x}^{(k)}, \\ \mathscr{V}^{(k+1)} = \{m(\mathscr{V}^{(k)}) + \mathscr{V}^{(k)}(\underset{\cdot}{\mathscr{J}} - \mathscr{F}^{(k+1)}m(\mathscr{V}^{(k)}))\} \cap \mathscr{V}^{(k)}, \qquad k \geqslant 0. \end{cases}$$

The method (3) consists of a simultaneous execution of (19.7) and (1). It differs from the iteration (19.12) mainly in the calculation of a new interval matrix $\mathscr{V}^{(k+1)}$. Here only one step is executed in an appropriate method.

In (3) sequences of interval matrices and interval vectors

$$\mathfrak{x}^{(0)} \supseteq \mathfrak{x}^{(1)} \supseteq \mathfrak{x}^{(2)} \supseteq \cdots \qquad \text{and} \qquad \mathscr{V}^{(0)} \supseteq \mathscr{V}^{(1)} \supseteq \mathscr{V}^{(2)} \supseteq \cdots$$

are calculated. We shall now prove some properties for these.

Theorem 2: Let $\mathfrak{x}^{(0)}$ be an interval vector and let $\underset{\cdot}{\mathfrak{y}} \in \mathfrak{x}^{(0)}$ be a zero of the function $\underset{\cdot}{\mathfrak{f}}(\underset{\cdot}{\mathfrak{x}})$. Let the derivative of $\underset{\cdot}{\mathfrak{f}}(\underset{\cdot}{\mathfrak{x}})$ satisfy a Lipschitz condition in $\mathfrak{x}^{(0)}$. Furthermore let $\mathscr{V}^{(0)}$ be an interval matrix containing the matrices $\underset{\cdot}{\mathscr{J}}(\underset{\cdot}{\mathfrak{x}})^{-1}$ for each $\underset{\cdot}{\mathfrak{x}} \in \mathfrak{x}^{(0)}$. For the interval evaluation $\underset{\cdot}{\mathfrak{f}}'(\mathfrak{x})$ of the Frechét derivative $\underset{\cdot}{\mathfrak{f}}'(\underset{\cdot}{\mathfrak{x}})$ the condition

$$\|d(\underset{\cdot}{\mathfrak{f}}'(\mathfrak{x}))\|' \leqslant c\|d(\mathfrak{x})\|'$$

holds for $\mathfrak{x} \subseteq \mathfrak{x}^{(0)}$. (This is certainly the case if every element of the matrix $\underset{\cdot}{\mathfrak{f}}'(\mathfrak{x})$ satisfies the condition (3.5′).)

The interval vectors $\{\mathfrak{x}^{(k)}\}_{k=0}^{\infty}$ and the interval matrices $\{\mathscr{V}^{(k)}\}_{k=0}^{\infty}$ calculated according to (3) then satisfy the following:

(4) each interval vector $\mathfrak{x}^{(k)}$, $k \geqslant 0$, contains the zero $\underset{\cdot}{\mathfrak{y}}$;

(5) if all the matrices $\underset{\cdot}{\mathscr{V}} \in \mathscr{V}^{(0)}$ are nonsingular, then it follows that $\lim_{k\to\infty} \mathfrak{x}^{(k)} = \underset{\cdot}{\mathfrak{y}}$ and $\lim_{k\to\infty} \mathscr{V}^{(k)} = \underset{\cdot}{\mathfrak{f}}'(\underset{\cdot}{\mathfrak{y}})^{-1}$;

(6) let $\mathscr{H}^{(k)} \in M_{n,n+1}(I(\mathbb{R}))$, $k \geq 0$, be the interval matrix whose first column is equal to the interval vector $x^{(k)}$ and whose other columns are the columns of the matrix $\mathscr{V}^{(k)}$; that is, $\mathscr{H}^{(k)} = (x^{(k)}, \mathscr{V}^{(k)})$. It then follows that

$$O_R((3), (y, f'(y)^{-1})) \geq 2.$$

(See Appendix A, Theorem 2.)

Proof: For the proof we again set $\mathscr{F}^{(k)} = f'(x^{(k)})$.

Of (4): As in the proof of the corresponding statement (19.8) of Theorem 19.1 one shows that $y \in x^{(1)}$. In addition the fact that $x^{(1)} \subseteq x^{(0)}$ implies that for $x \in x^{(1)}$ it follows that

$$\mathscr{J}(x)^{-1} \in \mathscr{V}^{(0)} \qquad \text{and} \qquad \mathscr{J}(x) \in \mathscr{F}^{(1)}.$$

Therefore using (10.10′)

$$\begin{aligned} \mathscr{J}(x)^{-1} &= m(\mathscr{V}^{(0)}) + \mathscr{J}(x)^{-1}(\mathscr{I} - \mathscr{J}(x) m(\mathscr{V}^{(0)})) \\ &\in \{m(\mathscr{V}^{(0)}) + \mathscr{V}^{(0)}(\mathscr{I} - \mathscr{F}^{(1)} m(\mathscr{V}^{(0)}))\} \cap \mathscr{V}^{(0)} = \mathscr{V}^{(1)}. \end{aligned}$$

Therefore we have

$$\{\mathscr{J}(x)^{-1} \mid x \in x^{(1)}\} \subseteq \mathscr{V}^{(1)}.$$

Using this result we prove that $y \in x^{(2)}$ and the proof of the statement follows by complete induction.

Of (5): Since $\mathscr{V}^{(k)} \subseteq \mathscr{V}^{(0)}$ we can prove that $\lim x^{(k)} = y$ in the same way as it was done in (19.9).

The second convergence statement follows from Theorem 1 since $f(y) \in \mathscr{F}^{(k)}$, $k \geq 0$.

Of (6): We use the abbreviation $\hat{\mathscr{J}} := f'(y)$. It then follows from (3) using (10.6), (10.7), and (10.9) that

$$\begin{aligned} x^{(k+1)} - y &\subseteq -\mathscr{V}^{(k)}\{f(m(x^{(k)})) - f(y) - f'(y)(m(x^{(k)}) - y)\} \\ &\quad - \mathscr{V}^{(k)}(\hat{\mathscr{J}} \cdot (m(x^{(k)}) - y)) + m(x^{(k)}) - y \\ &= -\mathscr{V}^{(k)}\{f(m(x^{(k)}) - f(y) - f'(y)(m(x^{(k)}) - y)\} \\ &\quad + (\hat{\mathscr{J}}^{-1} - \mathscr{V}^{(k)}) \cdot (\hat{\mathscr{J}} \cdot (m(x^{(k)}) - y)). \end{aligned}$$

We now take absolute values (Definition 10.6) and consider (10.15) and (10.27) and Theorem 3.2.12 in Ortega and Rheinboldt [6] as well as the relations

$$\begin{aligned} q(x^{(k)}, y) &= q(x^{(k)} - y, o) = |x^{(k)} - y|, \\ q(\mathscr{V}^{(k)}, \hat{\mathscr{J}}^{-1}) &= q(\mathscr{V}^{(k)} - \hat{\mathscr{J}}^{-1}, \mathscr{O}) = |\mathscr{V}^{(k)} - \hat{\mathscr{J}}^{-1}|. \end{aligned}$$

It then follows from the fact that $m(x^{(k)}) \in x^{(k)}$ using a monotonic vector norm and a subordinate monotonic matrix norm, as well as the theorem of norm equivalence, that

$$\|q(\underset{\cdot}{x}^{(k+1)}, \underset{\cdot}{y})\| \leqslant \| |\mathscr{V}^{(k)}| \| c_1 \|q(\underset{\cdot}{x}^{(k)}, \underset{\cdot}{y})\|^2 + \|q(\mathscr{V}^{(k)}, \hat{\underset{\cdot}{\mathscr{J}}}^{-1})\| \cdot \| |\hat{\mathscr{J}}| \| \cdot \|q(\underset{\cdot}{x}^{(k)}, \underset{\cdot}{y})\|.$$

Similarly one obtains from the second part of (3) that

$$\mathscr{V}^{(k+1)} - \hat{\underset{\cdot}{\mathscr{J}}}^{-1} \subseteq -((\hat{\underset{\cdot}{\mathscr{J}}}^{-1} - \mathscr{V}^{(k)}) \cdot \hat{\underset{\cdot}{\mathscr{J}}}) \cdot (\hat{\underset{\cdot}{\mathscr{J}}}^{-1} - m(\mathscr{V}^{(k)})) - \mathscr{V}^{(k)}((\mathscr{F}^{(k+1)} - \hat{\underset{\cdot}{\mathscr{J}}}) m(\mathscr{V}^{(k)})).$$

Again taking absolute values, considering the fact that $m(\mathscr{V}^{(k)}) \in \mathscr{V}^{(k)}$, as well as (10.15),

$$|\mathscr{F}^{(k+1)} - \hat{\underset{\cdot}{\mathscr{J}}}| = q(\mathscr{F}^{(k+1)}, \hat{\underset{\cdot}{\mathscr{J}}}) \leqslant d(\mathscr{F}^{(k+1)}),$$

$$\|d(\mathscr{F}^{(k+1)})\|' \leqslant c\|d(\underset{\cdot}{x}^{(k+1)})\|',$$

and (10.27) using a monotonic matrix norm and the theorem of norm equivalence, it follows that

$$\|q(\mathscr{V}^{(k+1)}, \hat{\underset{\cdot}{\mathscr{J}}}^{-1})\| \leqslant c_3 \| |\hat{\underset{\cdot}{\mathscr{J}}}| \| \cdot \|q(\mathscr{V}^{(k)}, \hat{\underset{\cdot}{\mathscr{J}}}^{-1})\|^2 + \| |\mathscr{V}^{(k)}| \|^2 c_4 \|q(\underset{\cdot}{x}^{(k+1)}, \underset{\cdot}{y})\|.$$

For $\mathscr{A} = (\ell, \mathscr{B}) \in M_{n,n+1}(\mathbb{R})$ a norm $\|\mathscr{A}\| = \max\{\|\ell\|, \|\mathscr{B}\|\}$ is introduced. Setting $r^{(k)} = \|(q(\underset{\cdot}{x}^{(k)}, \underset{\cdot}{y}), q(\mathscr{V}^{(k)}, \hat{\underset{\cdot}{\mathscr{J}}}^{-1}))\|$ and noting that $\lim_{k\to\infty} \mathscr{V}^{(k)} = \hat{\underset{\cdot}{\mathscr{J}}}^{-1}$, $\| |\mathscr{V}^{(k)}| \| \leqslant s$, as well as setting $\| |\hat{\underset{\cdot}{\mathscr{J}}}| \| = p$ it follows from the above inequalities that

$$\|q(\underset{\cdot}{x}^{(k+1)}, \underset{\cdot}{y})\| \leqslant (sc_1 + pc_2)(r^{(k)})^2,$$

$$\|q(\mathscr{V}^{(k+1)}, \hat{\underset{\cdot}{\mathscr{J}}}^{-1})\| \leqslant (c_3 p + s^2 c_4(sc_1 + pc_2))(r^{(k)})^2.$$

If we therefore define

$$\gamma = \max\{sc_1 + pc_2, c_3 p + s^2 c_4(sc_1 + pc_2)\},$$

it then follows that

$$r^{(k+1)} \leqslant \gamma (r^{(k)})^2, \qquad k \geqslant 0.$$

From Theorem 2, Appendix A, it now follows that we have at least quadratic convergence of the sequence

$$\{(\underset{\cdot}{x}^{(k)}, \mathscr{V}^{(k)})\}_{k=0}^{\infty} \quad \text{toward} \quad (\underset{\cdot}{y}, \underset{\cdot}{f}'(\underset{\cdot}{y})^{-1}). \quad \blacksquare$$

Remarks: Theorem 2 is found in Alefeld and Herzberger [2]. It is an improvement of the result in Alefeld and Herzberger [1], where only the superlinear convergence was shown. A modification of (3) was given in Madsen [5], where the sequence of widths was majorized by a quadratically convergent sequence. In this case an additional point equation system had to be solved at each step. Theorem 2 shows that this extra effort does not have any influence on the order of convergence.

The method (1) without taking intersections was considered in Adams and Ames [B344] for a special class of interval matrices.

REFERENCES

[1] G. Alefeld and J. Herzberger, Nullstelleneinschließung mit dem Newton-Verfahren ohne Invertierung von Intervallmatrizen. *Numer. Math.* **19**, 56–64 (1972).

[2] G. Alefeld and J. Herzberger, Über die Konvergenzordnung intervallarithmetischer Iterationsverfahren. *Kolloq. Konstrukt. Theorie Funktionen.* Tagungsbericht, Cluj (1973).

[3] W. Burmeister, Inversionsfreie Verfahren zur Lösung nichtlinearer Operatorgleichungen. *Z. Angew. Math. Mech.* **52**, 101–110 (1972).

[4] F. Krückeberg, Inversion von Matrizen mit Fehlererfassung. *Z. Angew. Math. Mech.* **46**, T 69 (1966).

[5] K. Madsen, On the solution of nonlinear equations in interval arithmetics. *BIT* **13**, 428–433 (1973).

[6] J. M. Ortega and W. C. Rheinboldt, "Iterative Solution of Nonlinear Equations in Several Variables". Academic Press, New York, 1970.

[7] J. W. Schmidt and D. Leder, Ableitungsfreie Verfahren ohne Auflösung linearer Gleichungen. *Computing* **5**, 71–81 (1970).

[8] S. Ulm, On iteration methods with successive approximations of the inverse operator (Russian). *Izv. Akad. Nauk. Est. SSR* **16**, 403–411 (1967).

Chapter 21 / NEWTON-LIKE METHODS FOR PARTICULAR SYSTEMS OF NONLINEAR EQUATIONS

We shall now consider the same problem formulation as in the two previous sections while assuming particular properties for the system

$$f(x) = o$$

under consideration. The statements that are proved in this section closely correspond to results in Ortega and Rheinboldt [6, 7]. It is, however, important to note that there are no assumptions made on the convexity of f. The Frechét derivative of f at the point x is denoted by $f'(x) = (\partial f_i/\partial x_j)$. The natural partial order is used both for $V_n(\mathbb{R})$ as well as the set $M_{nn}(\mathbb{R})$ consisting of the real $n \times n$ point matrices. We first prove a general theorem and then we show how it can be applied to particular systems.

Theorem 1: Let $f\colon \mathfrak{D} \subseteq V_n(\mathbb{R}) \to V_n(\mathbb{R})$ be a Frechét-differentiable map and assume that the set $\{x \mid x^{(0)} \leqslant x \leqslant y^{(0)}\}$ is contained in $\mathfrak{D}$.

Let the following hold true:

(0) $$f(x^{(0)}) \leqslant o \leqslant f(y^{(0)}).$$

$$\mathcal{A}\colon \{x \mid x^{(0)} \leqslant x \leqslant y^{(0)}\} \times \{x \mid x^{(0)} \leqslant x \leqslant y^{(0)}\} \to M_{nn}(\mathbb{R}),$$

$$\mathcal{A} = \mathcal{A}(x, y),$$

is a map for which

(1) $$f(x) - f(y) = \mathscr{A}(x, y)(x - y) \qquad \text{for} \quad x^{(0)} \leqslant x \leqslant y \leqslant y^{(0)}.$$

$$\mathscr{B}\colon \{x \mid x^{(0)} \leqslant x \leqslant y^{(0)}\} \times \{x \mid x^{(0)} \leqslant x \leqslant y^{(0)}\} \to M_{nn}(\mathbb{R}),$$
$$\mathscr{B} = \mathscr{B}(x, y),$$

is a continuous map for which

(2) $$\mathscr{B}(\bar{x}, \bar{y}) \leqslant \mathscr{B}(x, y) \qquad \text{for} \quad x^{(0)} \leqslant x \leqslant \bar{x} \leqslant \bar{y} \leqslant y \leqslant y^{(0)},$$

(3) $$\mathscr{A}(x, y) \leqslant \mathscr{B}(x, y) \qquad \text{for} \quad x^{(0)} \leqslant x \leqslant y \leqslant y^{(0)},$$

(4) $\mathscr{B}(x, y)^{-1}$ exists and $\mathscr{B}(x, y)^{-1} \geqslant \mathscr{O}$ for $x^{(0)} \leqslant x \leqslant y \leqslant y^{(0)}$.

$\mathscr{P}^{(0)} \in M_{nn}(\mathbb{R})$ is nonsingular and it holds that

(5) $$\mathscr{B}(x^{(0)}, y^{(0)})\mathscr{P}^{(0)} \leqslant \mathscr{I} \qquad (\mathscr{I} \text{ unit matrix}),$$

(6) $$\mathscr{P}^{(0)}\mathscr{B}(x^{(0)}, y^{(0)}) \leqslant \mathscr{I},$$

(7) $$\mathscr{P}^{(0)} \geqslant \mathscr{O}.$$

Then the iteration

(8) $$\begin{cases} y^{(k+1)} = y^{(k)} - \mathscr{P}^{(k)} f(y^{(k)}), \\ x^{(k+1)} = x^{(k)} - \mathscr{P}^{(k)} f(x^{(k)}), \\ \mathscr{P}^{(k+1)} = \mathscr{P}^{(k)} - \mathscr{P}^{(k)}(\mathscr{B}(x^{(k+1)}, y^{(k+1)})\mathscr{P}^{(k)} - \mathscr{I}), \qquad k \geqslant 0, \end{cases}$$

can be performed without restrictions and it holds that

(9) $$x^{(0)} \leqslant x^{(1)} \leqslant \cdots \leqslant x^{(k)} \leqslant x^{(k+1)} \leqslant y^{(k+1)} \leqslant y^{(k)} \leqslant \cdots \leqslant y^{(1)} \leqslant y^{(0)},$$

(10) $$f(x^{(k)}) \leqslant o \leqslant f(y^{(k)}),$$

(11) $$\mathscr{P}^{(0)} \leqslant \mathscr{P}^{(1)} \leqslant \cdots \leqslant \mathscr{P}^{(k)} \leqslant \mathscr{P}^{(k+1)} \leqslant \cdots,$$

(12) $$\mathscr{P}^{(k)} \text{ nonsingular},$$

(13) $$\mathscr{P}^{(k)}\mathscr{B}(x^{(k)}, y^{(k)}) \leqslant \mathscr{I},$$

(14) $$\mathscr{B}(x^{(k)}, y^{(k)})\mathscr{P}^{(k)} \leqslant \mathscr{I},$$

(15) $$\lim_{k \to \infty} x^{(k)} = x^* \leqslant y^* = \lim_{k \to \infty} y^{(k)},$$

(16) $$\mathscr{P}^* := \lim_{k \to \infty} \mathscr{P}^{(k)} = \mathscr{B}(x^*, y^*)^{-1}.$$

If f is continuous, then it follows that $f(x^*) = f(y^*) = o$. All solutions z^* of $f(x) = o$ from the set $\{x \mid x^{(0)} \leqslant x \leqslant y^{(0)}\}$ satisfy

$$x^* \leqslant z^* \leqslant y^*.$$

If $\mathscr{A}(x, y)$ is nonsingular for $x^{(0)} \leqslant x \leqslant y \leqslant y^{(0)}$, then it follows that $x^* = y^*$.

If, furthermore, the Frechét derivative $f'(x)$ satisfies a Lipschitz condition of the form

$$\|f'(x) - f'(y)\| \leqslant c_1\|x - y\| \qquad \text{for} \quad x^{(0)} \leqslant x, y \leqslant y^{(0)} \tag{17}$$

and if the map $\mathscr{B}(x, y)$ satisfies the condition

$$\|f'(z) - \mathscr{B}(x, y)\| \leqslant c_2\|x - y\| \qquad \text{for} \quad x^{(0)} \leqslant x \leqslant z \leqslant y \leqslant y^{(0)}, \tag{18}$$

then it follows that

$$\lim_{k \to \infty} \mathscr{P}^{(k)} = f'(x^*)^{-1} \tag{19}$$

and the sequence

$$\{(([x_i^{(k)}, y_i^{(k)}]), ([p_{ij}^{(k)}, p_{ij}^{(k)}]))\}_{k=0}^{\infty} \in M_{n,n+1}(I(\mathbb{R}))$$

then converges with an R order of at least 2 toward $(x^*, \mathscr{P}^*)$.

Proof: Statements (9)–(14) are proven by complete induction. The ideas for the proof of (9) and (10) are already found in Ortega and Rheinboldt [6] in the proof of Theorem 4.1, p. 177.

Let $k \geqslant 0$ and $x^{(k)} \leqslant x \leqslant y^{(k)}$. It then follows from (2) and (3) that

$$\mathscr{A}(x, y^{(k)}) \leqslant \mathscr{B}(x, y^{(k)}) \leqslant \mathscr{B}(x^{(k)}, y^{(k)}).$$

Since $\mathscr{P}^{(k)} \geqslant \mathscr{O}$ from (11) it follows that

$$\mathscr{P}^{(k)}\mathscr{B}(x^{(k)}, y^{(k)}) \geqslant \mathscr{P}^{(k)}\mathscr{A}(x, y^{(k)}),$$

and therefore using (13) we get

$$\mathscr{I} - \mathscr{P}^{(k)}\mathscr{A}(x, y^{(k)}) \geqslant \mathscr{I} - \mathscr{P}^{(k)}\mathscr{B}(x^{(k)}, y^{(k)}) \geqslant \mathscr{O}.$$

From this, using the fact that $x - y^{(k)} \leqslant o$ and using (1) one obtains the inequalities

$$\begin{aligned} x - \mathscr{P}^{(k)}f(x) &= y^{(k+1)} + (x - y^{(k)}) - \mathscr{P}^{(k)}(f(x) - f(y^{(k)})) \\ &= y^{(k+1)} + (x - y^{(k)}) - \mathscr{P}^{(k)}\mathscr{A}(x, y^{(k)})(x - y^{(k)}) \\ &= y^{(k+1)} + (\mathscr{I} - \mathscr{P}^{(k)}\mathscr{A}(x, y^{(k)}))(x - y^{(k)}) \\ &\leqslant y^{(k+1)}. \end{aligned} \tag{20}$$

Setting $x = x^{(k)}$, using (10) and the fact that $\mathscr{P}^{(k)} \geqslant \mathscr{O}$, one obtains the result

$$x^{(k)} \leqslant x^{(k)} - \mathscr{P}^{(k)}f(x^{(k)}) = x^{(k+1)} \leqslant y^{(k+1)} = y^{(k)} - \mathscr{P}^{(k)}f(y^{(k)}),$$

which proves (9).

From the fact that $x^{(k)} \leqslant y^{(k+1)} \leqslant y^{(k)}$ as well as from (2) and (3) it follows that

$$\mathscr{A}(y^{(k+1)}, y^{(k)}) \leqslant \mathscr{B}(y^{(k+1)}, y^{(k)}) \leqslant \mathscr{B}(x^{(k)}, y^{(k)}).$$

Since $\mathcal{P}^{(k)} \geqslant \mathcal{O}$ using (14) it then follows that

$$\mathcal{I} - \mathcal{A}(y^{(k+1)}, y^{(k)})\mathcal{P}^{(k)} \geqslant \mathcal{I} - \mathcal{B}(x^{(k)}, y^{(k)})\mathcal{P}^{(k)} \geqslant \mathcal{O}.$$

By (10) we have $f(y^{(k)}) \geqslant o$ and using (1) it follows that

$$\begin{aligned} f(y^{(k+1)}) &= f(y^{(k)}) + \mathcal{A}(y^{(k+1)}, y^{(k)})(y^{(k+1)} - y^{(k)}) \\ &= (\mathcal{I} - \mathcal{A}(y^{(k+1)}, y^{(k)})\mathcal{P}^{(k)})f(y^{(k)}) \geqslant o. \end{aligned}$$

Correspondingly we show that $f(x^{(k+1)}) \leqslant o$. This proves (10). From $x^{(k)} \leqslant x^{(k+1)} \leqslant y^{(k+1)} \leqslant y^{(k)}$ using (2) it follows that

$$\mathcal{B}(x^{(k+1)}, y^{(k+1)}) \leqslant \mathcal{B}(x^{(k)}, y^{(k)}).$$

Since $\mathcal{P}^{(k)} \geqslant \mathcal{O}$ and using (13) and (14), it therefore follows that

$$\begin{cases} \mathcal{B}(x^{(k+1)}, y^{(k+1)})\mathcal{P}^{(k)} \leqslant \mathcal{B}(x^{(k)}, y^{(k)})\mathcal{P}^{(k)} \leqslant \mathcal{I}, \\ \mathcal{P}^{(k)}\mathcal{B}(x^{(k+1)}, y^{(k+1)}) \leqslant \mathcal{P}^{(k)}\mathcal{B}(x^{(k)}, y^{(k)}) \leqslant \mathcal{I}. \end{cases} \tag{21}$$

Using (8) it follows that the equations

$$\mathcal{I} - \mathcal{B}(x^{(k+1)}, y^{(k+1)})\mathcal{P}^{(k+1)} = (\mathcal{I} - \mathcal{B}(x^{(k+1)}, y^{(k+1)})\mathcal{P}^{(k)})^2 \geqslant \mathcal{O}$$

and

$$\mathcal{I} - \mathcal{P}^{(k+1)}\mathcal{B}(x^{(k+1)}, y^{(k+1)}) = (\mathcal{I} - \mathcal{P}^{(k)}\mathcal{B}(x^{(k+1)}, y^{(k+1)}))^2 \geqslant \mathcal{O} \tag{22}$$

as well as

$$\mathcal{P}^{(k+1)} = \mathcal{P}^{(k)} - \mathcal{P}^{(k)}(\mathcal{B}(x^{(k+1)}, y^{(k+1)})\mathcal{P}^{(k)} - \mathcal{I}) \geqslant \mathcal{P}^{(k)} \geqslant \mathcal{O}$$

are valid. This proves (11), (13), and (14). In order to prove (12) we have to show that $\mathcal{P}^{(k+1)}$ is nonsingular. Since $\mathcal{P}^{(k)}$ is nonsingular we have that

$$\mathcal{B}(x^{(k+1)}, y^{(k+1)}) = \mathcal{M} - \mathcal{N}$$

with

$$\mathcal{M} = (\mathcal{P}^{(k)})^{-1}, \qquad \mathcal{N} = (\mathcal{P}^{(k)})^{-1} - \mathcal{B}(x^{(k+1)}, y^{(k+1)})$$

is valid. The assumption shows that $\mathcal{M}^{-1} = \mathcal{P}^{(k)} \geqslant \mathcal{O}$ is valid and using (21) it follows that

$$\begin{aligned} \mathcal{M}^{-1}\mathcal{N} &= \mathcal{I} - \mathcal{P}^{(k)}\mathcal{B}(x^{(k+1)}, y^{(k+1)}) \geqslant \mathcal{O}, \\ \mathcal{N}\mathcal{M}^{-1} &= \mathcal{I} - \mathcal{B}(x^{(k+1)}, y^{(k+1)})\mathcal{P}^{(k)} \geqslant \mathcal{O}; \end{aligned}$$

that is, $\mathcal{M} - \mathcal{N}$ is a weakly regular splitting of $\mathcal{B}(x^{(k+1)}, y^{(k+1)})$ (see Ortega and Rheinboldt [7, Definition 2.4.15]). According to (4) we have that $\mathcal{B}(x^{(k+1)}, y^{(k+1)})^{-1} \geqslant \mathcal{O}$. From Theorem 2.4.17 in Ortega and Rheinboldt [7, p. 56] it then follows that

$$\rho(\mathcal{I} - \mathcal{P}^{(k)}\mathcal{B}(x^{(k+1)}, y^{(k+1)})) < 1.$$

(ρ denotes the spectral radius.) This means that

$$\mathcal{I} - (\mathcal{I} - \mathcal{P}^{(k)}\mathcal{B}(x^{(k+1)}, y^{(k+1)}))^2$$

can be inverted. This again implies from (22) that $\mathcal{P}^{(k+1)}$ can be inverted. The relation (15) follows from (9) using standard arguments.

From (13) or (14) using (4) it follows that

$$\mathcal{P}^{(k)} \leqslant \mathcal{B}(x^{(k)}, y^{(k)})^{-1}.$$

The continuity of $\mathcal{B}(x, y)$ implies that $\lim_{k\to\infty} \mathcal{B}(x^{(k)}, y^{(k)}) = \mathcal{B}(x^*, y^*)$ and since $x^{(k)} \leqslant x^* \leqslant y^* \leqslant y^{(k)}$ it follows from (2) that

$$\mathcal{B}(x^*, y^*) \leqslant \mathcal{B}(x^{(k)}, y^{(k)}),$$

from which it follows using (4) that

$$\mathcal{B}(x^{(k)}, y^{(k)})^{-1} \leqslant \mathcal{B}(x^*, y^*)^{-1};$$

that is,

$$\mathcal{P}^{(k)} \leqslant \mathcal{B}(x^*, y^*)^{-1}$$

is valid. The sequence (11) is therefore bounded above and hence convergent. From the iteration (8) using (21) and (11) it follows that

$$\begin{aligned}\mathcal{P}^{(k)} - \mathcal{P}^{(k+1)} &= \mathcal{P}^{(k)}(\mathcal{B}(x^{(k+1)}, y^{(k+1)})\mathcal{P}^{(k)} - \mathcal{I}) \\ &\leqslant \mathcal{P}^{(0)}(\mathcal{B}(x^{(k+1)}, y^{(k+1)})\mathcal{P}^{(k)} - \mathcal{I}) \leqslant \mathcal{O},\end{aligned}$$

which implies that

$$\begin{aligned}\mathcal{O} = \lim_{k\to\infty} (\mathcal{P}^{(k)} - \mathcal{P}^{(k+1)}) &\leqslant \lim_{k\to\infty} \mathcal{P}^{(0)}[\mathcal{B}(x^{(k+1)}, y^{(k+1)})\mathcal{P}^{(k)} - \mathcal{I}] \\ &= \mathcal{P}^{(0)}(\mathcal{B}(x^*, y^*)\mathcal{P}^* - \mathcal{I}) \leqslant \mathcal{O}.\end{aligned}$$

Since $\mathcal{P}^{(0)}$ is nonsingular it follows that we have verified (16).

Let $x^{(k)} \leqslant z \leqslant y^{(k)}$ for some $k \geqslant 0$ with $f(z) = o$. Then from (20) we obtain

$$z = z - \mathcal{P}^{(k)}f(z) \leqslant y^{(k+1)}.$$

From the relation

$$x - \mathcal{P}^{(k)}f(x) \geqslant x^{(k+1)},$$

which holds for $x^{(k)} \leqslant x \leqslant y^{(k)}$ and which can be proved in the same manner as (20), it follows setting $x := z$ that

$$z = z - \mathcal{P}^{(k)}f(z) \geqslant x^{(k+1)};$$

that is,

$$x^{(k)} \leqslant x^{(k+1)} \leqslant z \leqslant y^{(k+1)} \leqslant y^{(k)}$$

for all $k \geqslant 0$. When we take limits in these inequalities it follows that all the solutions of $f(x) = o$ that are enclosed by $x^{(0)}$ and $y^{(0)}$ are also enclosed by x^* and y^*.

The iteration (8) also implies that if f is continuous then it follows that

$$o = \lim_{k\to\infty} (y^{(k)} - y^{(k+1)}) = \lim_{k\to\infty} \mathscr{P}^{(k)} f(y^{(k)}) = \mathscr{P}^* f(y^*).$$

Since $\mathscr{P}^* = \mathscr{B}(x^*, y^*)^{-1}$ is nonsingular it follows that $f(y^*) = o$. Correspondingly one shows that $f(x^*) = o$.

If $\mathscr{A}(x, y)$ is nonsingular for $x^{(0)} \leqslant x \leqslant y \leqslant y^{(0)}$, then it follows from (1) that

$$o = f(x^*) - f(y^*) = \mathscr{A}(x^*, y^*)(x^* - y^*);$$

that is, $x^* = y^*$.

If we now assume (18) then it follows that

$$\lim_{k\to\infty} x^{(k)} = x^* = y^* = \lim_{k\to\infty} y^{(k)} \quad \text{implies} \quad \lim_{k\to\infty} \mathscr{B}(x^{(k)}, y^{(k)}) = f'(x^*).$$

Using (16) we therefore prove (19).

The proof of the quadratic convergence follows analogously to the proof in Schmidt and Leder [11, Theorem 2]. We abbreviate

$$\mathscr{B}^{(k)} := \mathscr{B}(x^{(k)}, y^{(k)}), \qquad p := \|f'(x^*)\|, \qquad \mathscr{P}^* = f'(x^*)^{-1}.$$

Because of $\lim_{k\to\infty} \mathscr{P}^{(k)} = \mathscr{P}^* = f'(x^*)^{-1}$ there exists a constant s, such that $\|\mathscr{P}^{(k)}\| \leqslant s$, $k \geqslant 0$ is valid.

From the iteration (8) it follows using (17) that

$$\begin{aligned}(23)\quad \|y^{(k+1)} - x^*\| &= \|y^{(k)} - x^* - \mathscr{P}^{(k)} f(y^{(k)})\| \\ &= \|-\mathscr{P}^{(k)}(f(y^{(k)}) - f(x^*) - f'(x^*)(y^{(k)} - x^*)) \\ &\quad + (f'(x^*)^{-1} - \mathscr{P}^{(k)}) f'(x^*)(y^{(k)} - x^*)\| \\ &\leqslant (c_1/2)\|\mathscr{P}^{(k)}\|\,\|y^{(k)} - x^*\|^2 + p\|\mathscr{P}^* - \mathscr{P}^{(k)}\|\,\|y^{(k)} - x^*\|.\end{aligned}$$

Correspondingly one shows that

$$\begin{aligned}(24)\qquad \|x^{(k+1)} - x^*\| &\leqslant (c_1/2)\|\mathscr{P}^{(k)}\|\,\|x^{(k)} - x^*\|^2 \\ &\quad + p\|\mathscr{P}^* - \mathscr{P}^{(k)}\|\,\|x^{(k)} - x^*\|.\end{aligned}$$

Using a monotonic matrix norm as well as the theorem of norm equivalence we may write (23) and (24) as

$$(23')\quad \|y^{(k+1)} - x^*\| \leqslant \gamma_1\|y^{(k)} - x^{(k)}\|^2 + \gamma_2\|\mathscr{P}^* - \mathscr{P}^{(k)}\|\,\|y^{(k)} - x^{(k)}\|,$$

$$(24')\quad \|x^{(k+1)} - x^*\| \leqslant \gamma_3\|y^{(k)} - x^{(k)}\|^2 + \gamma_4\|\mathscr{P}^* - \mathscr{P}^{(k)}\|\,\|y^{(k)} - x^{(k)}\|.$$

From the iteration (8) using $\mathscr{I} = f'(x^*)\mathscr{P}^*$ we obtain

$$\begin{aligned}\mathscr{P}^{(k+1)} - \mathscr{P}^* &= \mathscr{P}^{(k)} - \mathscr{P}^* - \mathscr{P}^{(k)}(\mathscr{B}^{(k+1)}\mathscr{P}^{(k)} - \mathscr{I}) \\ &= (\mathscr{P}^* - \mathscr{P}^{(k)}) f'(x^*)(\mathscr{P}^{(k)} - \mathscr{P}^*) + \mathscr{P}^{(k)}(f'(x^*) - \mathscr{B}^{(k+1)})\mathscr{P}^{(k)}.\end{aligned}$$

Using (18) we now obtain

$$\text{(25)} \qquad \begin{aligned} \|\mathcal{P}^{(k+1)} - \mathcal{P}^*\| &\leq p\|\mathcal{P}^{(k)} - \mathcal{P}^*\|^2 + c_2\|\mathcal{P}^{(k)}\|^2 \|x^{(k+1)} - y^{(k+1)}\| \\ &\leq \gamma_5\|\mathcal{P}^{(k)} - \mathcal{P}^*\|^2 + \gamma_6\|y^{(k+1)} - x^{(k+1)}\|. \end{aligned}$$

We introduce a norm for $\mathcal{A} = (\ell, \mathcal{B}) \in M_{n,n+1}(\mathbb{R})$ by setting

$$\|\mathcal{A}\| = \max\{\|\ell\|, \|\mathcal{B}\|\}$$

and we let

$$d^{(k)} = \|(y^{(k)} - x^{(k)}, \mathcal{P}^* - \mathcal{P}^{(k)})\|.$$

From (23′), (24′), and (25) we now get

$$\|y^{(k+1)} - x^{(k+1)}\| \leq \left(\sum_{i=1}^{4} \gamma_i\right)(d^{(k)})^2$$

$$\|\mathcal{P}^* - \mathcal{P}^{(k+1)}\| \leq \left(\gamma_5 + \gamma_6\left(\sum_{i=1}^{4} \gamma_i\right)\right)(d^{(k)})^2;$$

that is,

$$d^{(k+1)} \leq \gamma(d^{(k)})^2$$

with

$$\gamma = \max\left\{\sum_{i=1}^{4} \gamma_i, \gamma_5 + \gamma_6 \sum_{i=1}^{4} \gamma_i\right\}.$$

Using Theorem 2, Appendix A, we now get the theorem. ■

The contents of Theorem 1 may readily be carried over to methods with a higher rate of convergence (see Schmidt and Leder [11]). For this we keep the matrix $\mathcal{P}^{(k)}$ fixed over several steps. The approximate inversion of $\mathcal{B}(x^{(k+1)}, y^{(k+1)})$ is done using a higher order method. This results in the following iteration

$$\begin{cases} y^{(k,0)} = y^{(k)}, \\ y^{(k,r+1)} = y^{(k,r)} - \mathcal{P}^{(k)} f(y^{(k,r)}), \qquad 0 \leq r \leq m, \\ y^{(k+1)} = y^{(k,m+1)}, \\ x^{(k,0)} = x^{(k)}, \\ x^{(k,r+1)} = x^{(k,r)} - \mathcal{P}^{(k)} f(x^{(k,r)}), \qquad 0 \leq r \leq m, \\ x^{(k+1)} = x^{(k,m+1)}, \\ \mathcal{P}^{(k+1)} = \mathcal{P}^{(k)} + \mathcal{P}^{(k)} \sum_{\mu=1}^{m+1} (-1)^\mu [\mathcal{B}(x^{(k+1)}, y^{(k+1)})\mathcal{P}^{(k)} - \mathcal{I}]^\mu, \end{cases}$$

$k \geq 0$.

This method converges under the conditions of Theorem 1 with $m + 2$ being the order of convergence.

A special case of Theorem 1 is obtained when the matrices $\mathcal{B}(x^{(k)}, y^{(k)})$ are inverted exactly rather than approximately. This is contained in the following theorem.

Theorem 2: Let $\mathcal{P}^{(k)} = \mathcal{B}(x^{(k)}, y^{(k)})^{-1}$ in Theorem 1 and change the iteration (8) to

$$(8') \qquad \begin{cases} y^{(k+1)} = y^{(k)} - \mathcal{P}^{(k)} f(y^{(k)}), \\ x^{(k+1)} = x^{(k)} - \mathcal{P}^{(k)} f(x^{(k)}), \qquad k \geqslant 0. \end{cases}$$

Then all the statements of Theorem 1 are valid.

Proof: (5), (6), and (7) as well as (12), (13), and (14) are trivially satisfied because of (4). Inequality (11) also follows using (4). The other parts of the proof can be repeated without changes. ■

In the numerical execution of the method (8′) it is to be noted that the triangular decomposition of $\mathcal{B}(x^{(k)}, y^{(k)})$ in the Gaussian algorithm only has to be performed once.

We now consider a class of systems of nonlinear equations where we may apply the preceding theorems.

Let

$$(26) \qquad f: \mathfrak{D} \subseteq V_n(\mathbb{R}) \to V_n(\mathbb{R}), \qquad f(x) = \mathcal{H} x + \varepsilon h(x) + b$$

with $\varepsilon \geqslant 0$ and $b \in V_n(\mathbb{R})$. The matrix $\mathcal{H}$ is an M matrix; that is, $h_{ij} \leqslant 0$ for $i \neq j$ and $\mathcal{H}^{-1} \geqslant \mathcal{O}$. The map

$$h: \mathfrak{D} \subseteq V_n(\mathbb{R}) \to V_n(\mathbb{R}), \qquad h(x) = (h_i(x))$$

is continuously differentiable and

$$h_i(x) = \sum_{j=1}^{n} \alpha_{ij} g_j(x_j), \qquad 1 \leqslant i \leqslant n,$$

where $g_i: \mathfrak{D}_i \subseteq \mathbb{R} \to \mathbb{R}$, $g_i = g_i(s)$, $1 \leqslant i \leqslant n$,

$$\frac{d}{ds} g_i(s) \geqslant 0, \qquad 1 \leqslant i \leqslant n,$$

and

$$\alpha_{ij} \geqslant 0, \qquad 1 \leqslant i, j \leqslant n,$$

is satisfied.

Setting

$$g(x) = (g_i(x_i)) \qquad \text{and} \qquad \mathcal{U} = (\alpha_{ij})$$

we may write f as

$$f(x) = \mathcal{H}x + \varepsilon\mathcal{U}g(x) + b.$$

Let $\mathcal{A}(x, y)$ be the matrix whose ith row is equal to

$$f_i'(x + \theta_i(y - x)) = \Big(h_{i1} + \varepsilon\alpha_{i1}\frac{d}{dx_1}g_1(x_1 + \theta_i(y_1 - x_1)), \ldots,$$
$$h_{in} + \varepsilon\alpha_{in}\frac{d}{dx_n}g_n(x_n + \theta_i(y_n - x_n))\Big)$$

for $\theta_i \in (0, 1)$, $1 \leqslant i \leqslant n$. If $\mathfrak{D}$ is convex then it follows from the mean-value theorem for maps from $\mathfrak{D} \subseteq V_n(\mathbb{R}) \to \mathbb{R}$ that

$$f(x) - f(y) = \mathcal{A}(x, y)(x - y)$$

for $x, y \in \mathfrak{D}$.

Let $v_j(x_j, y_j)$ be the upper bound of the interval evaluation of dg_j/dx_j for the interval $[x_j, y_j]$ with $x_j \leqslant y_j$. Inclusion monotonicity now implies that

$$\frac{d}{dx_j}g_j(\bar{x}_j + \bar{\theta}_i(\bar{y}_j - \bar{x}_j)) \in \frac{d}{dx_j}g_j([\bar{x}_j, \bar{y}_j]) \subseteq \frac{d}{dx_j}g_j([x_j, y_j])$$

for $[\bar{x}_j, \bar{y}_j] \subseteq [x_j, y_j]$; that is,

$$v_j(x_j, y_j) \geqslant \frac{d}{dx_j}g_j(\bar{x}_j + \bar{\theta}_i(\bar{y}_j - \bar{x}_j)) \geqslant 0.$$

If we now put

$$\bar{\mathcal{U}} = (\bar{u}_{ij}) \qquad \text{with} \quad \bar{u}_{ij} = \alpha_{ij}v_j(x_j, y_j), \quad 1 \leqslant i, j \leqslant n,$$

then we have with

$$\mathcal{B}(x, y) = (h_{ij} + \varepsilon\alpha_{ij}v_j(x_j, y_j))$$

that

$$\mathcal{B}(\bar{x}, \bar{y}) \leqslant \mathcal{B}(x, y) \qquad \text{for} \quad x \leqslant \bar{x} \leqslant \bar{y} \leqslant y$$

as well as

$$\mathcal{A}(x, y) \leqslant \mathcal{B}(x, y).$$

Assumptions (1), (2), and (3) of Theorem 1 are now satisfied. Since $\mathcal{H}$ is an M matrix it follows from Theorem 2.4.10 in Ortega and Rheinboldt [7] for sufficiently small $\varepsilon \geqslant 0$ and also from the fact that $\mathcal{U} \geqslant \mathcal{O}$ that $\mathcal{B}(x, y)$ is an M matrix. This means that (4) is satisfied (provided ε satisfies the relation

$$h_{ij} + \varepsilon\alpha_{ij}v_j(x_j^{(0)}, y_j^{(0)}) \leqslant 0, \qquad i \neq j, \quad 1 \leqslant i, j \leqslant n).$$

Let $\mathscr{D}$ denote the diagonal part of $\mathscr{B}(\underset{\cdot}{x}^{(0)}, \underset{\cdot}{y}^{(0)})$. Since $\mathscr{D}^{-1} \geqslant \mathscr{O}$ and since $\mathscr{B}(\underset{\cdot}{x}^{(0)}, \underset{\cdot}{y}^{(0)}) = \mathscr{D} - (\mathscr{D} - \mathscr{B}(\underset{\cdot}{x}^{(0)}, \underset{\cdot}{y}^{(0)}))$ setting $\mathscr{P}^{(0)} = \mathscr{D}^{-1}$ it follows that

$$\mathscr{B}(\underset{\cdot}{x}^{(0)}, \underset{\cdot}{y}^{(0)})\mathscr{P}^{(0)} = \mathscr{I} - (\mathscr{D} - \mathscr{B}(\underset{\cdot}{x}^{(0)}, \underset{\cdot}{y}^{(0)}))\mathscr{D}^{-1} \leqslant \mathscr{I},$$

$$\mathscr{P}^{(0)}\mathscr{B}(\underset{\cdot}{x}^{(0)}, \underset{\cdot}{y}^{(0)}) = \mathscr{I} - \mathscr{D}^{-1}(\mathscr{D} - \mathscr{B}(\underset{\cdot}{x}^{(0)}, \underset{\cdot}{y}^{(0)})) \leqslant \mathscr{I},$$

since $\mathscr{D} - \mathscr{B}(\underset{\cdot}{x}^{(0)}, \underset{\cdot}{y}^{(0)}) \geqslant \mathscr{O}$.

This choice of $\mathscr{P}^{(0)}$ ensures that the assumptions (5), (6), and (7) of Theorem 1 are satisfied.

If one partitions $\mathscr{B}(\underset{\cdot}{x}^{(0)}, \underset{\cdot}{y}^{(0)})$ into $\mathscr{D} - \mathscr{L} - \mathscr{R}$ where $\mathscr{L}$ is a strictly lower triangular matrix and $\mathscr{R}$ a strictly upper triangular matrix, then it follows that $\mathscr{P}^{(0)} = (\mathscr{D} - \mathscr{L})^{-1} \geqslant \mathscr{O}$ and because of the fact that $\mathscr{R} \geqslant \mathscr{O}$ it follows that

$$\mathscr{B}(\underset{\cdot}{x}^{(0)}\underset{\cdot}{y}^{(0)})\mathscr{P}^{(0)} = \mathscr{I} - \mathscr{R}(\mathscr{D} - \mathscr{L})^{-1} \leqslant \mathscr{I},$$

$$\mathscr{P}^{(0)}\mathscr{B}(\underset{\cdot}{x}^{(0)}, \underset{\cdot}{y}^{(0)}) = \mathscr{I} - (\mathscr{D} - \mathscr{L})^{-1}\mathscr{R} \leqslant \mathscr{I}.$$

This shows another possible choice for $\mathscr{P}^{(0)}$.

The assumption (17) is satisfied if the derivatives of the maps g_i satisfy a Lipschitz condition. The assumption (18) is then satisfied without further conditions. For

$$\underset{\cdot}{f}'(\underset{\cdot}{z}) = \left(h_{ij} + \varepsilon\alpha_{ij}\frac{d}{dx_j}g_j(z_j)\right)$$

and

$$\mathscr{B}(\underset{\cdot}{x}, \underset{\cdot}{y}) = (h_{ij} + \varepsilon\alpha_{ij}v_j(x_j, y_j))$$

it clearly holds that

$$\underset{\cdot}{f}'(\underset{\cdot}{z}) - \mathscr{B}(\underset{\cdot}{x}, \underset{\cdot}{y}) = \left(\varepsilon\alpha_{ij}\left(\frac{d}{dx_j}g_j(z_j) - v_j(x_j, y_j)\right)\right).$$

If we now define

$$[u_j(x_j, y_j), v_j(x_j, y_j)] := \frac{d}{dx_j}g_j([x_j, y_j]),$$

then from inclusion monotonicity it follows that

$$\frac{d}{dx_j}g_j(z_j) \in [u_j(x_j, y_j), v_j(x_j, y_j)].$$

From Theorem 3.5 it then follows that

$$v_j(x_j, y_j) - \frac{d}{dx_j}g_j(z_j) \leqslant v_j(x_j, y_j) - u_j(x_j, y_j) \leqslant c_j(y_j - x_j).$$

Therefore we get

$$\mathcal{B}(x, y) - f'(z) \leqslant \max_{1 \leqslant i \leqslant n} \{y_i - x_i\}\mathcal{C}$$

with a matrix $\mathcal{C} = (c_{ij})$, $c_{ij} = c_j \varepsilon \alpha_{ij} \geqslant 0, 1 \leqslant i, j \leqslant n$, from which (18) follows.

Systems of nonlinear equations of the form (26) result when one solves a boundary value problem of the form

$$y'' = f(t, y), \qquad y(0) = \bar{a}, \qquad y(1) = \bar{b}$$

numerically. The application of a finite difference method on this system leads to the n equations

$$x_{i-1} - 2x_i + x_{i+1} - h^2\{\alpha f(t_{i-1}, x_{i-1}) + \beta f(t_i, x_i) + \gamma f(t_{i+1}, x_{i+1})\} = 0, \qquad 1 \leqslant i \leqslant n,$$

where $t_i = ih$, $0 \leqslant i \leqslant n + 1$, $(n + 1)h = 1$, and $x_0 = \bar{a}$, $x_{n+1} = \bar{b}$. x_i is an approximation for $y(t_i)$. For $\alpha = \gamma = 0$, $\beta = 1$ one obtains the standard difference method. For $\alpha = \gamma = \frac{1}{2}$, $\beta = \frac{10}{12}$ one obtains the Hermitian method (see Zurmühl [13, p. 461]). Writing

$$x = \begin{pmatrix} x_1 \\ x_2 \\ \vdots \\ x_n \end{pmatrix}, \quad g(x) = \begin{pmatrix} f(t_1, x_1) \\ f(t_2, x_2) \\ \vdots \\ f(t_n, x_n) \end{pmatrix}, \quad b = \begin{pmatrix} -\bar{a} + \alpha h^2 f(0, \bar{a}) \\ 0 \\ \vdots \\ -\bar{b} + \gamma h^2 f(1, \bar{b}) \end{pmatrix},$$

$$\mathcal{H} = \begin{bmatrix} 2 & -1 & & & 0 \\ -1 & 2 & -1 & & \\ & \ddots & \ddots & \ddots & \\ & & -1 & 2 & -1 \\ 0 & & & -1 & 2 \end{bmatrix}, \quad \mathcal{U} = \begin{bmatrix} \beta & \gamma & & & 0 \\ \alpha & \beta & \gamma & & \\ & \ddots & \ddots & \ddots & \\ & & \alpha & \beta & \gamma \\ 0 & & & \alpha & \beta \end{bmatrix},$$

and setting $\varepsilon = h^2$, one may then write these n equations as

$$f(x) = \mathcal{H}x + \varepsilon \mathcal{U} g(x) - b.$$

Under the assumption that $\partial f/\partial y \geqslant 0$ for $t \in (0, 1)$ it follows that f satisfies all the conditions for systems of the form (26). In the case of the Hermitian method one must also choose h small enough.

We consider the boundary problem

$$y'' = \sin y + y, \qquad y(0) = 1, \qquad y(1) = 1$$

for a concrete example. Since $\partial^2 f/\partial y^2 = -\sin y$ is indefinite it is clear that the discretized problem is not convex in the whole of $V_n(\mathbb{R})$. The iterates as approximations to $y(\frac{1}{2})$ for $n = 5, 25, 51, 101$, are given in Table 1 for the usual difference method and in Table 2 for the Hermitian method. The starting

Table 1

Usual Difference Method, $y(\frac{1}{2})$

k	$n = 5$	
0	− 0.5	0.5
1	0.3940299983760	0.4000335866235
2	0.3989339526997	0.3989347005095
3	0.3989344659822	0.3989344659822
k	$n = 25$	
0	− 0.5	0.5
1	0.3935413781128	0.3997788906381
2	0.3986874733555	0.3986882610402
3	0.3986880255447	0.3986880255452
4	0.3986880255452	0.3986880255452
k	$n = 51$	
0	− 0.5	0.5
1	0.3935206369679	0.3997680696930
2	0.3986771185469	0.3986779080124
3	0.3986776725106	0.3986776725169
4	0.3986776725137	0.3986776725153
k	$n = 101$	
0	− 0.5	0.5
1	0.3935155168238	0.3997653993461
2	0.3986745645657	0.3986753545055
3	0.3986751189564	0.3986751189564

Table 2

Hermitian Method, $y(\frac{1}{2})$

k	$n = 5$	
0	− 0.5	0.5
1	0.3938048950831	0.3997635541509
2	0.3986758325059	0.3986765352929
3	0.3986763144021	0.3986763144021
k	$n = 25$	
0	− 0.5	0.5
1	0.3935292233327	0.3997644587939
2	0.3986736779243	0.3986744630652
3	0.3986742283178	0.3986742283191
k	$n = 51$	
0	− 0.5	0.5
1	0.3935175960669	0.3997644612118
2	0.3986736691791	0.3986744580066
3	0.3986742226748	0.3986742226762
4	0.3986742226762	0.3986742226762
k	$n = 101$	
0	− 0.5	0.5
1	0.3935147300836	0.3997644611468
2	0.3986736680718	0.3986744577855
3	0.3986742223087	0.3986742223447
4	0.3986742223155	0.3986742223174

vectors $x^{(0)}$ and $y^{(0)}$ were determined according to 13.4.6c in Ortega and Rheinboldt in [7]. For these values it also holds for the Hermitian method that

$$f(x^{(0)}) \leqslant \rho \leqslant f(y^{(0)}).$$

Remarks: As already mentioned in the text, the results in this section are closely connected to the corresponding results in Ortega and Rheinboldt [6, 7]. The corresponding statements are all proved under certain convexity assumptions on the map f. Simple examples show that the methods considered in Ortega and Rheinboldt [6, 7] cannot relax these convexity assumptions. The interval computations in this chapter are only used to calculate the matrices $\mathscr{B}(x^{(k)}, y^{(k)})$ from systems of the form (26) for the methods (8) and (8′).

The n-dimensional analogs of the regula falsi and the secant method are considered in Schmidt [8, 9] and Schmidt and Leonhardt [10] and applied to the inclusion of solutions of convex maps. It has not been completely clarified to what extent these methods can be modified by the aid of interval arithmetic in order that they also include solutions of nonconvex systems.

In the assumptions to Theorem 1 we notice that (17) may be proven from (18) using the triangle inequality.

REFERENCES

[1] G. Alefeld and J. Herzberger, Über das Newton-Verfahren bei nichtlinearen Gleichungssystemen. *Z. Angew. Math. Mech.* **50**, 773–774 (1970).

[2] G. Alefeld and J. Herzberger, Nullstelleneinschließung mit dem Newton-Verfahren ohne Invertierung von Intervallmatrizen. *Numer. Math.* **19**, 56–64 (1972).

[3] A. N. Baluev, Zur Methode von Chaplygin (Russian). *Dokl. Akad. Nauk SSSR* **83**, 781–789 (1952).

[4] L. Collatz, "Funktionalanalysis und numerische Mathematik." Springer-Verlag, Berlin and New York, 1964.

[5] R. E. Moore, "Interval Analysis." Prentice-Hall, Englewood Cliffs, New Jersey, 1966.

[6] J. M. Ortega and W. C. Rheinboldt, Monotone iterations for nonlinear equations with application to Gauss-Seidel methods. *SIAM J. Numer. Anal.* **4**, 171–190 (1967).

[7] J. M. Ortega and W. C. Rheinboldt, "Iterative Solution of Nonlinear Equations in Several Variables." Academic Press, New York 1970.

[8] J. W. Schmidt, Einschließung von Nullstellen bei Operatoren mit monoton zerlegbarer Steigung durch überlinear konvergente Iterationsverfahren. *Ann. Acad. Sci. Fenn. Ser. A. I. Math.* (1972).

[9] J. W. Schmidt, Eingrenzung von Lösungen nichtlinearer Gleichungen durch Verfahren mit höherer Konvergenzgeschwindigkeit. *Computing* **8**, 208–215 (1971).

[10] J. W. Schmidt and H. Leonhardt, Eingrenzung von Lösungen mit Hilfe der Regula falsi. *Computing* **6**, 318–329 (1970).

[11] J. W. Schmidt and D. Leder, Ableitungsfreie Verfahren ohne Auflösung linearer Gleichungen. *Computing* **5**, 71–81 (1970).

[12] J. S. Vandergraft, Newton's method for convex operators in partially ordered spaces. *SIAM J. Numer. Anal.* **4**, 406–432 (1967).

[13] R. Zurmühl, "Praktische Mathematik," 4th ed. Springer-Verlag, Berlin and New York, 1962.

Chapter 22 / NEWTON-LIKE TOTAL STEP AND SINGLE STEP METHODS

In this chapter we shall consider methods for the iterative inclusion of solutions of real nonlinear systems of equations where it is neither required to invert interval matrices as in Chapter 19 nor is it required to approximately perform this inversion as required in Chapter 20. We again start with the system of nonlinear equations

$$f(x) = o$$

and we assume that f is continuously differentiable over an interval vector $x^{(0)}$. Using the mean-value theorem, setting $f(x) = (f_i(x))$ it then follows for $f_i(x)$ that

$$f_i(x) = f_i(y) + \sum_{j=1}^{n} \frac{\partial}{\partial x_j} f_i(y + \theta_i(x - y))(x_j - y_j),$$

$$0 < \theta_i < 1, \qquad 1 \leqslant i \leqslant n, \qquad y, x \in x^{(0)},$$

or setting

$$J(x) = \left(\frac{\partial}{\partial x_j} f_i(y + \theta_i(x - y))\right)$$

we have the equation

$$f(x) - f(y) = J(x)(x - y).$$

If y is a solution of the system of equations in $\mathfrak{x}^{(0)}$ and if we decompose the matrix $\mathcal{J}(x)$ into

$$\mathcal{J}(x) = \mathcal{D}(x) - \mathcal{L}(x) - \mathcal{U}(x),$$

where $\mathcal{D}(x)$ is a diagonal matrix, $\mathcal{L}(x)$ a strictly lower triangular matrix, and $\mathcal{U}(x)$ an upper triangular matrix, then the last equation implies for nonsingular $\mathcal{D}(x)$ that the relation

$$y = x - \mathcal{D}(x)^{-1}\{\mathcal{B}(x)(x - y) + f(x)\}$$

is valid where $\mathcal{B}(x) = \mathcal{L}(x) + \mathcal{U}(x)$.

Since $y + \theta_i(x - y) \in \mathfrak{x}^{(0)}$, $1 \leqslant i \leqslant n$, it follows from the inclusion monotonicity that

$$\frac{\partial}{\partial x_j} f_i(y + \theta_i(x - y)) \in \frac{\partial}{\partial x_j} f_i(\mathfrak{x}^{(0)}), \qquad 1 \leqslant i, j \leqslant n.$$

We now decompose the interval matrix

$$f'(\mathfrak{x}^{(0)}) = \left(\frac{\partial}{\partial x_j} f_i(\mathfrak{x}^{(0)})\right)$$

in the same manner as the matrix $\mathcal{J}(x)$ into

$$f'(\mathfrak{x}^{(0)}) = \mathcal{D}(\mathfrak{x}^{(0)}) - \mathcal{L}(\mathfrak{x}^{(0)}) - \mathcal{U}(\mathfrak{x}^{(0)}).$$

We also define $\mathcal{B}(\mathfrak{x}^{(0)}) := \mathcal{L}(\mathfrak{x}^{(0)}) + \mathcal{U}(\mathfrak{x}^{(0)})$ and $x := m(\mathfrak{x}^{(0)})$ (see (18.1)). Provided no diagonal element of $f'(\mathfrak{x}^{(0)})$ contains zero and using inclusion monotonicity we obtain

$$y \in m(\mathfrak{x}^{(0)}) - \mathcal{D}(\mathfrak{x}^{(0)})^{-1}\{\mathcal{B}(\mathfrak{x}^{(0)})(m(\mathfrak{x}^{(0)}) - y) + f(m(\mathfrak{x}^{(0)}))\}.$$

It is understood that $\mathcal{D}(\mathfrak{x}^{(0)})^{-1}$ is the diagonal matrix formed from the diagonal matrix

$$\mathcal{D}(\mathfrak{x}^{(0)}) = \operatorname{diag}\left(\frac{\partial}{\partial x_i} f_i(\mathfrak{x}^{(0)})\right)$$

by inverting the diagonal elements.

If one now choses an interval vector $\mathfrak{z}^{(0)}$ containing y, then one shows using complete induction for the iterates of the iteration method

(1)
$$\mathfrak{z}^{(\nu+1)} = \{m(\mathfrak{x}^{(0)}) - \mathcal{D}(\mathfrak{x}^{(0)})^{-1}\{\mathcal{B}(\mathfrak{x}^{(0)})(m(\mathfrak{x}^{(0)}) - \mathfrak{z}^{(\nu)}) + f(m(\mathfrak{x}^{(0)}))\}\} \cap \mathfrak{z}^{(\nu)}$$

that the relation $y \in \mathfrak{z}^{(\nu)}$, $\nu \geqslant 0$ is valid. Since the sequence $\{\mathfrak{z}^{(\nu)}\}_{\nu=0}^{\infty}$ converges to a limit $\mathfrak{z}$ because of taking intersections it also follows that $y \in \mathfrak{z}$. Starting with $\mathfrak{x}^{(1)} := \mathfrak{z}$ one calculates the interval matrix $f'(\mathfrak{x}^{(1)})$ and with this a new inclusion $\mathfrak{x}^{(2)}$ for $\underset{\cdot}{y}$ and so on. We shall choose now, however, a different route. In the above iteration method we set $\mathfrak{z}^{(0)} := \mathfrak{x}^{(0)}$. From the considerations above it follows that

$$\underset{\cdot}{y} \in \mathfrak{z}^{(1)} = \{m(\mathfrak{x}^{(0)}) - \mathscr{D}(\mathfrak{x}^{(0)})^{-1}\{\mathscr{B}(\mathfrak{x}^{(0)})(m(\mathfrak{x}^{(0)}) - \mathfrak{x}^{(0)}) \\ + \mathfrak{f}(m(\mathfrak{x}^{(0)}))\}\} \cap \mathfrak{x}^{(0)}$$

and $\mathfrak{z}^{(1)} \subseteq \mathfrak{x}^{(0)}$. It then seems reasonable to put $\mathfrak{x}^{(1)} := \mathfrak{z}^{(1)}$ and using this interval vector, which encloses $\underset{\cdot}{y}$ at least as well as $\mathfrak{x}^{(0)}$, a new interval matrix $\mathfrak{f}'(\mathfrak{x}^{(1)})$ is calculated. In the above iteration (1) exactly one step is required in order to compute a new inclusion. Combining all this one finally obtains the following iteration procedure

$$(2)\quad \begin{cases} \mathfrak{u}^{(k+1)} = m(\mathfrak{x}^{(k)}) - \mathscr{D}(\mathfrak{x}^{(k)})^{-1}\{\mathscr{B}(\mathfrak{x}^{(k)})(m(\mathfrak{x}^{(k)}) - \mathfrak{x}^{(k)}) + \mathfrak{f}(m(\mathfrak{x}^{(k)}))\} \\ \mathfrak{x}^{(k+1)} = \mathfrak{x}^{(k)} \cap \mathfrak{u}^{(k+1)}, \qquad k \geqslant 0. \end{cases}$$

The given system of nonlinear equations is therefore transformed, as in Newton's method, into a linear system of equations with interval coefficients. From this a new inclusion for $\underset{\cdot}{y}$ is computed in a manner similar to the total step method. We therefore call this method the Newton total step method.

In a similar manner to the above method one arrives at the result that the following iteration method with $\mathfrak{u}^{(k)} = (U_i^{(k)})$, $\mathfrak{x}^{(k)} = (X_i^{(k)})$,

$$(3)\quad \begin{cases} U_i^{(k+1)} = m(X_i^{(k)}) - \dfrac{1}{(\partial/\partial x_i)f_i(\mathfrak{x}^{(k)})}\left\{\displaystyle\sum_{j=1}^{i-1} \frac{\partial}{\partial x_j} f_i(\mathfrak{x}^{(k)})(m(X_j^{(k)}) - X_j^{(k+1)})\right. \\ \qquad\qquad \left. + \displaystyle\sum_{j=i+1}^{n} \frac{\partial}{\partial x_j} f_i(\mathfrak{x}^{(k)})(m(X_j^{(k)}) - X_j^{(k)}) + f_i(m(\mathfrak{x}^{(k)}))\right\}, \\ X_i^{(k+1)} = U_i^{(k+1)} \cap X_i^{(k)}, \qquad 1 \leqslant i \leqslant n, \qquad k \geqslant 0, \end{cases}$$

generates interval vectors $\mathfrak{x}^{(k)}$ that contain the solution $\underset{\cdot}{y}$. We call this method the Newton single step method taking componentwise intersections.

We shall now investigate the conditions for convergence toward the solution $\underset{\cdot}{y} \in \mathfrak{x}^{(0)}$ for these two methods. Since we take intersections we have each of the above methods generating a sequence

$$\mathfrak{x}^{(0)} \supseteq \mathfrak{x}^{(1)} \supseteq \cdots \supseteq \mathfrak{x}^{(k)} \supseteq \mathfrak{x}^{(k+1)} \supseteq \cdots$$

that converges to an interval vector $\mathfrak{x}$ according to Corollary 10.8. Furthermore we have that

$$\underset{\cdot}{y} \in \mathfrak{x}^{(k)}, \qquad k \geqslant 0, \qquad \text{and} \qquad \underset{\cdot}{y} \in \mathfrak{x}.$$

Without further assumptions we do not have in general that $\lim_{k\to\infty} \mathfrak{x}^{(k)} = \underset{\cdot}{y}$. The following theorem gives sufficient conditions for $d(\mathfrak{x}) = \underset{\cdot}{o}$ and therefore, since $\underset{\cdot}{y} \in \mathfrak{x}$, for $\lim_{k\to\infty} \mathfrak{x}^{(k)} = \underset{\cdot}{y}$.

Theorem 1: Let the system of nonlinear equations $\mathbf{f}(\mathbf{x}) = \mathbf{o}$ have a solution $\mathbf{y}$ in the interval vector $\mathbf{x}^{(0)}$. The interval matrix

$$\mathbf{f}'(\mathbf{x}^{(0)}) = \left(\frac{\partial}{\partial x_j} f_i(\mathbf{x}^{(0)})\right)$$

is then decomposed into

$$\mathbf{f}'(\mathbf{x}^{(0)}) = \mathcal{D}(\mathbf{x}^{(0)}) - \mathcal{L}(\mathbf{x}^{(0)}) - \mathcal{U}(\mathbf{x}^{(0)}),$$

and assume it holds that

$$0 \notin \frac{\partial}{\partial x_i} f_i(\mathbf{x}^{(0)}), \qquad 1 \leqslant i \leqslant n.$$

Let $\mathcal{B}(\mathbf{x}^{(0)}) = \mathcal{L}(\mathbf{x}^{(0)}) + \mathcal{U}(\mathbf{x}^{(0)})$. If the condition

$$\rho(|\mathcal{D}(\mathbf{x}^{(0)})^{-1}|\,|\mathcal{B}(\mathbf{x}^{(0)})|) < 1$$

holds (ρ denotes the spectral radius) then the sequence $\{\mathbf{x}^{(k)}\}_{k=0}^{\infty}$ converges to the solution $\mathbf{y}$ in the methods (2) and (3).

Proof: We prove this fact for the Newton total step method (2). For $k \to \infty$ it follows from the iteration method (2) that

$$\mathbf{u} = m(\mathbf{x}) - \mathcal{D}(\mathbf{x})^{-1}\{\mathcal{L}(\mathbf{x})(m(\mathbf{x}) - \mathbf{x}) + \mathcal{U}(\mathbf{x})(m(\mathbf{x}) - \mathbf{x}) + \mathbf{f}(m(\mathbf{x}))\},$$

$$\mathbf{x} = \mathbf{x} \cap \mathbf{u}.$$

From $\mathbf{x} = \mathbf{x} \cap \mathbf{u}$ it follows that $\mathbf{x} \subseteq \mathbf{u}$, in particular,

$$m(\mathbf{x}) \in m(\mathbf{x}) - \mathcal{D}(\mathbf{x})^{-1}\{\mathcal{L}(\mathbf{x})(m(\mathbf{x}) - \mathbf{x}) + \mathcal{U}(\mathbf{x})(m(\mathbf{x}) - \mathbf{x}) + \mathbf{f}(m(\mathbf{x}))\};$$

that is,

$$\mathbf{o} \in \mathcal{D}(\mathbf{x})^{-1}\{\mathcal{L}(\mathbf{x})(m(\mathbf{x}) - \mathbf{x}) + \mathcal{U}(\mathbf{x})(m(\mathbf{x}) - \mathbf{x}) + \mathbf{f}(m(\mathbf{x}))\}.$$

Since the diagonal elements of $\mathcal{D}(\mathbf{x})^{-1}$ do not contain zero it follows that

$$\mathbf{o} \in \mathcal{L}(\mathbf{x})(m(\mathbf{x}) - \mathbf{x}) + \mathcal{U}(\mathbf{x})(m(\mathbf{x}) - \mathbf{x}) + \mathbf{f}(m(\mathbf{x})).$$

Using (2.19) it then follows that

$$\begin{aligned} d(\mathbf{x}) \leqslant d(\mathbf{u}) &= |\mathcal{D}(\mathbf{x})^{-1}|(|\mathcal{L}(\mathbf{x})|\, d(\mathbf{x}) + |\mathcal{U}(\mathbf{x})|\, d(\mathbf{x})) \\ &\leqslant |\mathcal{D}(\mathbf{x})^{-1}|(|\mathcal{L}(\mathbf{x})|\, d(\mathbf{u}) + |\mathcal{U}(\mathbf{x})|\, d(\mathbf{u})) \\ &= |\mathcal{D}(\mathbf{x})^{-1}|\,|\mathcal{B}(\mathbf{x})|\, d(\mathbf{u}). \end{aligned}$$

Since $\rho(|\mathcal{D}(\mathbf{x})^{-1}|\,|\mathcal{B}(\mathbf{x})|) \leqslant \rho(|\mathcal{D}(\mathbf{x}^{(0)})^{-1}|\,|\mathcal{B}(\mathbf{x}^{(0)})|) < 1$ it follows that $d(\mathbf{u}) = \mathbf{o}$; that is, $d(\mathbf{x}) = \mathbf{o}$. This proves that the sequence $\{\mathbf{x}^{(k)}\}_{k=0}^{\infty}$ converges to y. The proof for the other method may be carried out in a similar manner. ■

From the assumptions of Theorem 1 it also follows that the system of equations $f(x) = o$ can have no other solutions apart from y in $x^{(0)}$.

The above methods may be modified in various ways. These modifications are now described for the Newton total step method. In the case that the calculation of $f'(x^{(k)})$ is very expensive it is appropriate to execute more than one step of the calculation of an inclusion for y in method (1). The number of steps r executed after the calculation of $f'(x^{(k)})$ may also depend on k: $r = r_k$. This leads to the following iteration:

$$(4)\begin{cases} x^{(0,0)} = x^{(0)}, & \\ u^{(k+1,i)} = m(x^{(k)}) - \mathscr{D}(x^{(k)})^{-1}\{\mathscr{B}(x^{(k)})(m(x^{(k)}) - x^{(k,i-1)}) + f(m(x^{(k)}))\}, & \\ x^{(k,i)} = u^{(k+1,i)} \cap x^{(k,i-1)}, & 1 \leqslant i \leqslant r_k, \\ x^{(k+1)} = x^{(k,r_k)}, & \\ x^{(k+1,0)} = x^{(k+1)}, & k \geqslant 0. \end{cases}$$

The other iteration method above, the Newton single step method taking componentwise intersections (3), may also be modified in a similar manner. This method is convergent to y under the assumption of Theorem 1.

The modified methods are also of interest since they have superlinear convergence under appropriate conditions on the sequence $\{r_k\}_{k=0}^{\infty}$. We formulate and prove this only for the Newton total step method. Contents and proof of the following theorem is also valid for the other method.

Theorem 2: Let the assumptions of Theorem 1 be satisfied. If the modified Newton total step method (4) satisfies $r_k \to \infty$ for $k \to \infty$, then it follows that

$$\|d(x^{(k+1)})\| \leqslant c_k \|d(x^{(k)})\|, \qquad k \geqslant 0,$$

with $\lim_{k\to\infty} c_k = 0$ provided (3.5′) holds for the elements of $f'(x^{(k)})$.

Proof: Before we go into the details of the proof, we remark that under the conditions of Theorem 1 method (4) is convergent to y. This can be shown in the same manner as it was done in the proof of Theorem 1.

We now show that

$$d(x^{(k+1)}) = d(x^{(k,r_k)}) \leqslant (\mathscr{M}^{r_k} + \mathscr{R}_{r_k})d(x^{(k)}),$$

where

$$\mathscr{M} = |\mathscr{D}(x^{(0)})^{-1}|\,|\mathscr{B}(x^{(0)})|$$

and where $\mathscr{R}_{r_k}$ is a real matrix which depends on $x^{(k)}$ and for which we have $\mathscr{R}_{r_k} \to \mathscr{O}$ for $k \to \infty$. Because of $\rho(\mathscr{M}) < 1$ the claim then follows. In proving the above inequality we use the fact that because of (3.5′) it holds that

$$d(\mathcal{D}(x^{(k)})^{-1}\mathcal{B}(x^{(k)})) \leqslant \mathcal{C}\, d(x^{(k)})$$

as well as

$$\begin{aligned} d(\mathcal{D}(x^{(k)})^{-1})|f'(x^{(k)})| &\leqslant d(\mathcal{D}(x^{(k)})^{-1}|f'(x^{(0)})|) \\ &\leqslant \mathcal{E}\, d(x^{(k)}). \end{aligned}$$

Here $\mathcal{C}$ and $\mathcal{E}$ are symmetric bilinear operators that are independent of k.

Furthermore we use the fact that for interval matrices $\mathcal{A}$ and $\mathcal{B}$ and for a point vector c we have $\mathcal{A}(\mathcal{B}c) \subseteq (\mathcal{A}\mathcal{B})c$. (The proof uses Definition 10.3 and the distributive law $(A + B)c = Ac + Bc$ for $A, B \in I(\mathbb{R})$, $c \in \mathbb{R}$.)

Because of $x^{(k,0)} = x^{(k)}$ we get from (5)

$$\begin{aligned} d(x^{(k,1)}) &\leqslant d(u^{(k+1,1)}) \\ &\leqslant d(\mathcal{D}(x^{(k)})^{-1}\{\mathcal{B}(x^{(k)})(m(x^{(k)}) - x^{(k,0)})\}) + d(\mathcal{D}(x^{(k)})^{-1}f(m(x^{(k)}))) \\ &\leqslant (\mathcal{M} + \mathcal{R}_1)\, d(x^{(k)}), \end{aligned}$$

where

$$\mathcal{R}_1 = d(\mathcal{D}(x^{(k)})^{-1})|f'(x^{(k)})|$$

and $\mathcal{R}_1 \to \mathcal{O}$ for $k \to \infty$.

Assume now that for some $i \geqslant 1$ we have

$$d(x^{(k,i)}) \leqslant (\mathcal{M}^i + \mathcal{R}_i)\, d(x^{(k)})$$

and $\mathcal{R}_i \to \mathcal{O}$ for $k \to \infty$. Then we get from (5)

$$\begin{aligned} d(x^{(k,i+1)}) &\leqslant d(u^{(k+1,i+1)}) \\ &\leqslant d(\mathcal{D}(x^{(k)})^{-1}\{\mathcal{B}(x^{(k)})(m(x^{(k)}) - m(x^{(k,i)}) + m(x^{(k,i)}) - x^{(k,i)})\}) \\ &\quad + d(\mathcal{D}(x^{(k)})^{-1})|f'(x^{(k)})|\,|m(x^{(k)}) - m(x^{(k,i)}) + m(x^{(k,i)}) - y| \\ &\leqslant \mathcal{C}\, d(x^{(k)})|m(x^{(k)}) - m(x^{(k,i)})| + |\mathcal{D}(x^{(k)})^{-1}|\,|\mathcal{B}(x^{(k)})|\, d(x^{(k,i)}) \\ &\quad + \mathcal{E}\, d(x^{(k)})|m(x^{(k)}) - m(x^{(k,i)})| + \mathcal{R}_1\, d(x^{(k,i)}) \\ &\leqslant ((\mathcal{E} + \mathcal{C})|m(x^{(k)}) - m(x^{(k,i)})| + \mathcal{M}(\mathcal{M}^i + \mathcal{R}_i))\, d(x^{(k)}) \\ &\quad + \mathcal{R}_1(\mathcal{M}^i + \mathcal{R}_i)\, d(x^{(k)}) = (\mathcal{M}^{i+1} + \mathcal{R}_{i+1})\, d(x^{(k)}), \end{aligned}$$

where

$$\mathcal{R}_{i+1} = (\mathcal{E} + \mathcal{C})|m(x^{(k)}) - m(x^{(k,i)})| + \mathcal{M}\mathcal{R}_i + \mathcal{R}_1(\mathcal{M}^i + \mathcal{R}_i).$$

Since

$$|m(x^{(k)}) - m(x^{(k,i)})| \to o \qquad \text{for} \quad k \to \infty$$

and

$$\mathcal{R}_1 \to \mathcal{O} \qquad \text{for} \quad k \to \infty$$

as well as

$$\mathscr{R}_i \to \mathscr{O} \qquad \text{for} \quad k \to \infty,$$

it follows that

$$\mathscr{R}_{i+1} \to \mathscr{O} \qquad \text{for} \quad k \to \infty.$$

From this the assertion of the theorem can be deduced. ■

We shall now investigate a class of nonlinear systems of equations where the above methods can be applied.

A solution has to be found for the nonlinear boundary value problem

$$\Delta u \equiv u_{ss} + u_{tt} = f(s, t, u), \qquad (s, t) \in \Omega$$

with $u(s, t) = \phi(s, t)$ for $(s, t) \in \dot{\Omega}$. Here Ω is a simply connected, bounded, open domain in the (s, t) plane and $\dot{\Omega}$ denotes the boundary of Ω. We assume for the sake of simplicity that $\Omega = (0, 1) \times (0, 1)$. Under relatively weak conditions on ϕ and if f: $\Omega \times \mathbb{R} \to \mathbb{R}$ is continuously differentiable with $f_u(s, t, u) \geqslant 0$, $(s, t) \in \Omega$, $u \in \mathbb{R}$, then it follows that the boundary value problem has a unique solution. In order to determine numerical approximations for this solution one may for example transform the boundary value problem into a system of nonlinear equations using a difference method. If one chooses the same grid size both in the s and the t directions, then the replacement of the partial derivatives by the appropriate difference quotients leads to n^2 nonlinear equations

$$4x_{ij} - x_{i-1,j} - x_{i+1,j} - x_{i,j+1} - x_{i,j-1} + h^2 f(ih, jh, x_{ij}) = 0, \qquad 1 \leqslant i, j \leqslant n.$$

Here $h = 1/(n + 1)$, and x_{ij} is an approximate value for $u(ih, jh)$. For $i = 0, n + 1$ and $j = 0, n + 1$, the values of x_{ij} are determined by $\phi(s, t)$. In Ortega and Rheinboldt [2] it is shown that, under the above assumptions, the nonlinear system has a unique solution.

The n^2 equations may be written in the form

$$(5) \qquad \mathscr{f}(\mathscr{x}) = \mathscr{A}\mathscr{x} + \mathscr{g}(\mathscr{x}) - \mathscr{b} = \mathscr{o},$$

where $\mathscr{x} = (x_{ij}) \in V_{n^2}(\mathbb{R})$. As is well known, $\mathscr{A}$ is a block tridiagonal matrix and $\mathscr{g}_{ij}(\mathscr{x}) = g_{ij}(x_{ij})$. Under the assumption $f_u(s, t, u) \geqslant 0$ one may show, without much difficulty, that independently of the width $d(\mathscr{x}^{(0)})$ of an inclusion $\mathscr{x}^{(0)}$ for the unique solution of (5), the assumptions of Theorem 1 hold.

We consider the equation

$$\begin{aligned} u_{ss} + u_{tt} &= u^3/(s^2 + t^2 + 1), && (s, t) \in (0, 1) \times (0, 1). \\ u(s, 0) &= 1, && s \in [0, 1], \\ u(0, t) &= 1, && t \in [0, 1], \\ u(1, t) &= 2 - e^t, && t \in [0, 1], \\ u(s, 1) &= 2 - e^s, && s \in [0, 1], \end{aligned}$$

with $n = 5$ as a concrete example. One may easily convince oneself that the two vectors

$$\underset{\cdot}{y} = (y_{ij}), \qquad \underset{\cdot}{z} = (z_{ij}) \in V_{n^2}(\mathbb{R})$$

with $y_{ij} = -1$, $z_{ij} = 2$, $1 \leq i, j \leq n$, satisfy

$$\underset{\cdot}{f}(\underset{\cdot}{y}) \leq \underset{\cdot}{o} \leq \underset{\cdot}{f}(\underset{\cdot}{z})$$

for (5). From this it follows since $\underset{\cdot}{f}$ is an M function that the interval vector $\mathfrak{x}^{(0)} = (X_{ij}^{(0)}) \in V_{n^2}(I(\mathbb{R}))$ with $X_{ij}^{(0)} \doteq [-1, 2]$ contains the solution of (5).

We now present the results of the computations of approximations at the point $u(\frac{1}{2}, \frac{1}{2})$, computed using the corresponding modification of the Newton single step method, taking componentwise intersections, that was considered in Theorem 2 for the Newton total step method.

We choose $r_1 = 1$, $r_{k+1} = r_k + 1$, $k \geq 1$, and obtain the results in Table 1.

Table 1

k	Lower bound	Upper bound
0	[− 1	2]
1	[− 0.8683495081969,	1.932431741809]
2	[− 0.4334368762048,	1.662891859166]
3	[0.1747052174969,	1.110238786145]
4	[0.4964693372903,	0.7889353835863]
5	[0.6094825964828,	0.6756608548531]
6	[0.6370682656242,	0.6480110082876]
7	[0.6418668139331,	0.6432014513712]
8	[0.6424732262048,	0.6425936520955]
9	[0.6425293531665,	0.6425373968087]
10	[0.6425331717500,	0.6425335694949]
11	[0.6425333631204,	0.6425333776870]
12	[0.6425333701936,	0.6425333705975]
13	[0.6425333703873,	0.6425333704038]

The iteration was stopped as soon as the half-widths of all the including intervals were smaller than 10^{-10}. The method (3) on the other hand requires 81 iteration steps in order to achieve the same accuracy.

Simple considerations show that the number of arithmetic operations is about the same for one iteration step in both the above iterations for this example. In (3) one requires 81 function evaluations and interval evaluations of the derivative, however, compared to the first calculation, where only 13 such calculations are needed. This is reflected in the time required to calculate the example. The total computational time is 6–7 times longer for method (3) for this example.

Remarks: The methods given here are intimately connected with the corresponding methods from Ortega and Rheinboldt [2]. In Ortega and

Rheinboldt [2] convexity was always assumed in order that monotonic sequences including the solution could be calculated. Simple examples show the necessity of convexity. The results in this chapter were proved in Alefeld [1]. Similar statements for Newton-like relaxation methods are also found in the same paper.

If one makes assumptions corresponding to those of Chapter 22 for the given system, then one may also renounce the explicit use of interval computations in the execution of the iteration procedure. Interval computations are then used only for the calculation of the matrix $\mathscr{B}(x^{(k)}, y^{(k)})$ (see Chapter 21). We shall not pursue this further.

If one carries out the iteration (1) completely at each step, then one obtains a sequence of inclusions for y that – as it may be shown – converges under the assumptions of Theorem 1 to $\dot{y}$. The sequence of widths converge quadratically to zero. For systems satisfying the assumptions of Theorem 1, this method therefore shows a large saving in computational effort compared to systems having to use the method investigated in Chapter 19, Theorem 2. In that method one has to invert a point matrix as well as iteratively include the inverse of an interval matrix at each step.

In Cornelius [3] a modification of (3) was considered in which $m(x^{(k)})$ was replaced by an arbitrary point vector from $x^{(k)}$. By choosing this point vector appropriately one can get methods that converge much faster than the sequence $\{m(x^{(k)})\}_{k=1}^{\infty}$ of the centers of the interval vectors $x^{(k)}$ computed by (3). In Schwandt [4] a series of methods similar to methods (2) and (3) are introduced and investigated. An interval arithmetic version of the well-known Buneman algorithm is also introduced and applied to systems of nonlinear equations that have their origin in certain nonlinear boundary value problems. The numerical results are impressive.

REFERENCES

[1] G. Alefeld, Über die Existenz einer eindeutigen Lösung bei einer Klasse nichtlinearer Gleichungssysteme und deren Berechnung mit Iterationsverfahren. *Apl. Mat.* **17**, 267–294 (1972).

[2] J. M. Ortega and W. C. Rheinboldt, "Iterative Solution of Nonlinear Equations in Several Variables." Academic Press, New York, 1970.

Added in proof:

[3] H. Cornelius, Untersuchungen zu einem intervall-arithmetischen Iterationsverfahren mit Anwendungen auf ein Klasse nichtlinearer Gleichungssysteme. Ph.D. Thesis. Techn. Univ. Berlin, Fachbereich Mathematik (1981).

[4] H. Schwandt, Schnelle fast global konvergente Verfahren für die Fünf-Punkte-Diskretisierung der Poissongleichung mit Dirichletschen Randbedingungen auf Rechtecksgebieten. Ph.D. Thesis. Techn. Univ. Berlin, Fachbereich Mathematik (1981).

Appendix A / THE ORDER OF CONVERGENCE OF ITERATION METHODS IN $V_n(I(\mathbb{C}))$ AND $M_{mn}(I(\mathbb{C}))$

Let I be an iteration method generating sequences $\{x^{(k)}\}$ of complex interval vectors $x^{(k)} \in V_n(I(\mathbb{C}))$, $k \geq 0$. Let each such sequence converge to the limit $x^* \in V_n(I(\mathbb{C}))$. Since many of the actual iteration methods are used for the inclusion of specific solutions, we shall assume that

$$(1) \qquad x^* \subseteq x^{(k)}, \qquad k \geq 0.$$

We have two possibilities for measuring the deviation of an element $x^{(k)}$ of the sequence from the limit x^*. These are

$$(2) \qquad e^{(k)} = q(x^{(k)}, x^*)$$

and

$$(3) \qquad h^{(k)} = d(x^{(k)}) - d(x^*).$$

Both are nonnegative real vectors, that is, $e^{(k)}, h^{(k)} \geq o$, and they satisfy

$$(4) \qquad e^{(k)} = o \Leftrightarrow h^{(k)} = o \Leftrightarrow x^{(k)} = x^*.$$

These measures are also monotonic maps with respect to inclusion; that is, they satisfy

$$(5) \qquad x^{(m)} \subseteq x^{(k)} \Rightarrow e^{(m)} \leq e^{(k)}, \qquad h^{(m)} \leq h^{(k)}.$$

This follows immediately from the following property of the metric q

$$A \subseteq B \subseteq C \Rightarrow q(B, A) \leqslant q(C, A), \qquad A, B, C \in I(\mathbb{C}),$$

as well as the property (2.9) that is also trivially true in $I(\mathbb{C})$. A smaller inclusion of $\mathscr{x}^*$ therefore also means a smaller measure for the deviation from $\mathscr{x}^*$.

It is frequently simpler to express the deviation of an element $\mathscr{x}^{(k)}$ from $\mathscr{x}^*$ by a nonnegative real number. This can be achieved by using a monotonic vector norm $\|\cdot\|$ on $\mathscr{e}^{(k)}$ (resp., $\mathscr{h}^{(k)}$) and one obtains

$$e^{(k)} = \|q(\mathscr{x}^{(k)}, \mathscr{x}^*)\|$$

$$(\text{resp.} \quad h^{(k)} = \|d(\mathscr{x}^{(k)}) - d(\mathscr{x}^*)\|).$$

We again have

$$e^{(k)} = 0 \Leftrightarrow h^{(k)} = 0 \Leftrightarrow \mathscr{x}^{(k)} = \mathscr{x}^*,$$

as well as the monotonic property

$$\mathscr{x}^{(m)} \subseteq \mathscr{x}^{(k)} \Rightarrow e^{(m)} \leqslant e^{(k)}, \qquad h^{(m)} \leqslant h^{(k)}.$$

We can give a simple interpretation for the interval space $I(\mathbb{R})$ when $X^* = [x^*, x^*]$. In this case we have that

$$h^{(k)} = d(X^{(k)})$$

is the distance between the including bounds of x^* and

$$e^{(k)} = q(X^{(k)}, x^*) = \max\{|x_1^{(k)} - x^*|, |x_2^{(k)} - x^*|\}$$

is the maximum deviation of an element x in $X^{(k)}$ from x^*.

Now consider the sequences $\{\mathscr{x}^{(k)}\}$ generated by the iteration procedure I converging to $\mathscr{x}^*$ and satisfying (1). To each such sequence there are null sequences $\{\mathscr{e}^{(k)}\}$ and $\{\mathscr{h}^{(k)}\}$. The well-known concept of order of convergence from Ortega and Rheinboldt [2] may now be applied to these.

In order to do this we first define the R factor for such sequences as

$$R_p\{\mathscr{u}^{(k)}\} = \begin{cases} \limsup\limits_{k\to\infty} \|\mathscr{u}^{(k)}\|^{1/k}, & p = 1 \\ \limsup\limits_{k\to\infty} \|\mathscr{u}^{(k)}\|^{1/p^k}, & p > 1, \end{cases}$$

where the sequence $\{\mathscr{u}^{(k)}\}$ may be either of the sequences $\{\mathscr{e}^{(k)}\}$ and $\{\mathscr{h}^{(k)}\}$. This definition of R_p is independent of the norm $\|\cdot\|$, as proven in Ortega and Rheinboldt [2] using the theorem of norm equivalence. The following lemma is also valid.

Lemma 1: Let $\{\mathscr{x}^{(k)}\}$ be a sequence generated by I, converging to $\mathscr{x}^*$ and having the property (1). The null sequences $\{\mathscr{e}^{(k)}\}$ and $\{\mathscr{h}^{(k)}\}$ formed from $\{\mathscr{x}^{(k)}\}$ then satisfy

$$R_p\{\mathscr{e}^{(k)}\} = R_p\{\mathscr{h}^{(k)}\}, \qquad p \geqslant 1.$$

Proof: Using (10.27) we get for $k \geqslant 0$ that

$$\tfrac{1}{2}\mathscr{h}^{(k)} \leqslant \mathscr{e}^{(k)} \leqslant \mathscr{h}^{(k)}.$$

Using a monotonic vector norm we therefore get

$$\tfrac{1}{2}\|\mathscr{h}^{(k)}\| \leqslant \|\mathscr{e}^{(k)}\| \leqslant \|\mathscr{h}^{(k)}\|.$$

Since

$$\lim_{k\to\infty} (\tfrac{1}{2})^{1/p^k} = 1, \qquad p > 1,$$

is clearly valid it follows that

$$\begin{aligned}\lim_{k\to\infty} \sup\|\mathscr{h}^{(k)}\|^{1/p^k} &\geqslant \lim_{k\to\infty} \sup\|\mathscr{e}^{(k)}\|^{1/p^k}\\ &\geqslant \lim_{k\to\infty} \sup(\tfrac{1}{2})^{1/p^k}\|\mathscr{h}^{(k)}\|^{1/p^k}\\ &= \lim_{k\to\infty} \sup\|\mathscr{h}^{(k)}\|^{1/p^k}.\end{aligned}$$

The proof is analogous for $p = 1$, and it will not be presented here. ■

Because of Lemma 1 it follows that in the further considerations it is irrelevant which of the null sequences $\{\mathscr{e}^{(k)}\}$ or $\{\mathscr{h}^{(k)}\}$ belonging to $\{\mathscr{x}^{(k)}\}$ are considered. This is not necessarily true if (1) does not hold. The following example shows this.

Example: Let us consider the sequence

$$\{[0, 1] + (\tfrac{1}{2})^k\}_{k=0}^{\infty}$$

having the limit $[0, 1]$. Clearly

$$e^{(k)} = (\tfrac{1}{2})^k, \qquad k \geqslant 0,$$

and therefore for $p = 1$

$$R_1\{e^{(k)}\} = \lim_{k\to\infty} \sup((\tfrac{1}{2})^k)^{1/k} = \tfrac{1}{2}.$$

On the other hand, $h^{(k)} = 0$, $k \geqslant 0$, and therefore we have that

$$R_1\{h^{(k)}\} = 0. \quad ■$$

The set $\mathfrak{C}(I, \mathscr{x}^*)$ is now defined to be the set of sequences $\{\mathscr{x}^{(k)}\}$ generated by the iteration procedure I for which

$$\lim_{k\to\infty} \mathscr{x}^{(k)} = \mathscr{x}^* \qquad \text{and} \qquad \mathscr{x}^* \subseteq \mathscr{x}^{(k)}, \qquad k \geqslant 0,$$

is valid. The R factor of I is now defined to be

$$R_p(I, \mathscr{x}^*) = \sup\{R_p\{\mathscr{e}^{(k)}\} \mid \{\mathscr{x}^{(k)}\} \in \mathfrak{C}(I, \mathscr{x}^*)\}, \qquad p \geqslant 1.$$

From what was previously proved it follows that this R factor is independent of the norm used, and it could equally well have been defined by the use of the sequence $\{\underset{\sim}{h}^{(k)}\}$.

We may now define the R order of the iteration I as

$$(6) \qquad O_R(I, x^*) = \begin{cases} +\infty & \text{if} \quad R_p(I, x^*) = 0 \quad \text{for } p \geqslant 1 \\ \inf\{p \mid p \in [1, \infty), R_p(I, x^*) = 1\} & \text{otherwise.} \end{cases}$$

This definition was given for vector sequences in $V_n(\mathbb{R})$ by Ortega and Rheinboldt [2].

We now wish to repeat a few useful statements for the evaluation (resp., the estimation) of the R order (compare with Ortega and Rheinboldt [2, p. 297]).

Theorem 2: Let I be an iteration procedure with the limit x^*, and let $\mathfrak{C}(I, x^*)$ be the set of all sequences generated by I having the properties that $\lim_{k\to\infty} x^{(k)} = x^*$ and $x^* \subseteq x^{(k)}$, $k \geqslant 0$. If there exists a $p \geqslant 1$ and a constant γ_2 such that for all $\{x^{(k)}\} \in \mathfrak{C}(I, x^*)$ and for a norm $\|\cdot\|$, it holds that

$$\|\underset{\sim}{e}^{(k+1)}\| \leqslant \gamma_2 \|\underset{\sim}{e}^{(k)}\|^p, \qquad k \geqslant k(\{x^{(k)}\}),$$

or

$$\|\underset{\sim}{h}^{(k+1)}\| \leqslant \gamma_2 \|\underset{\sim}{h}^{(k)}\|^p, \qquad k \geqslant k(\{x^{(k)}\}),$$

then it follows that the R order of I satisfies the inequality

$$O_R(I, x^*) \geqslant p.$$

If, on the other hand, there exists a constant $\gamma_1 > 0$ and a sequence $\{x^{(k)}\} \in \mathfrak{C}(I, x^*)$ such that

$$\|\underset{\sim}{e}^{(k+1)}\| \geqslant \gamma_1 \|\underset{\sim}{e}^{(k)}\|^p > 0, \qquad k \geqslant k_0,$$

or

$$\|\underset{\sim}{h}^{(k+1)}\| \geqslant \gamma_1 \|\underset{\sim}{h}^{(k)}\|^p > 0, \qquad k \geqslant k_0,$$

then it follows that the R order of I satisfies

$$O_R(I, x^*) \leqslant p.$$

If both the above requirements are satisfied with constants γ_1 and γ_2, then it follows that

$$O_R(I, x^*) = p. \quad \blacksquare$$

The above proof of this theorem follows analogously to the corresponding proof in Ortega and Rheinboldt [2].

Theorem 3: Let I be an iteration procedure with the limit x^*, and let $\mathfrak{C}(I, x^*)$ be the set of all sequences generated by I having the properties that $\lim_{k\to\infty} x^{(k)} = x^*$ and $x^* \subseteq x^{(k)}$, $k \geqslant 0$. Let $r = \sum_{i=0}^{n} m_i > 0$ with $n + 1$ nonnegative integers $m_i \geqslant 0$, $0 \leqslant i \leqslant n$.

If there exists a constant $\gamma \geqslant 0$, so that for all $\{x^{(k)}\} \in \mathfrak{C}(I, x^*)$ one of the inequalities

$$\|e^{(k+1)}\| \leqslant \gamma \prod_{i=0}^{n} \|e^{(k-i)}\|^{m_i}, \qquad k \geqslant k(\{x^{(k)}\}),$$

or

$$\|h^{(k+1)}\| \leqslant \gamma \prod_{i=0}^{n} \|h^{(k-i)}\|^{m_i}, \qquad k \geqslant k(\{x^{(k)}\}),$$

holds, then the R order of I satisfies the inequality

$$O_R(I, x^*) \geqslant s,$$

where s is the unique positive solution of the equation

$$s^{n+1} - \sum_{i=0}^{n} m_i s^{n-i} = 0. \quad \blacksquare$$

The proof of Theorem 3 may be carried out using the ideas in Schmidt [3, Sect. 4].

It is often possible to verify one or the other of the inequalities in Theorem 2 or 3 for an iteration and therefore to obtain a bound on the R order for the iteration.

The same considerations as in this appendix may also be carried out in $M_{mn}(I(\mathbb{C}))$. In particular Lemma 1, Theorem 2, and Theorem 3 are still valid.

REFERENCES

[1] G. Alefeld and J. Herzberger, Über die Konvergenzordnung intervallarithmetischer Iterationsverfahren. *Rev. Anal. Numer. Theorie Approximation* **3**, 117–124 (1974).

[2] J. M. Ortega and W. C. Rheinboldt, "Iterative Solution of Nonlinear Equations in Several Variables." Academic Press, New York, 1970.

[3] J. W. Schmidt, Eine Übertragung der Regula Falsi auf Gleichungen in Banachräumen, II. *Z. Angew. Math. Mech.* **43**, 97–110 (1963).

Appendix B / REALIZATIONS OF MACHINE INTERVAL ARITHMETICS IN ALGOL 60

We now wish to expand on the material in Chapter 4. In this appendix we shall take a closer look at the realization of a machine interval arithmetic. We shall essentially follow the work of Christ [3] while also referring to the other literature such as Dewar [4] and Good and London [5].

Formulas (1.2) and Definition 4.1 together give the framework for the realization of machine interval operations:

> a logical part which determines the bounds of the result of combining operands which can be calculated according to (1.2),

and

> an arithmetic part where the pair of bounds of the resulting interval is calculated with the help of a directed rounded machine interval arithmetic as for example in formulas (4.8) and (4.9).

The first part may be easily formulated in ALGOL 60, and it therefore needs no further explanation. The second part, however, requires some more work since directed rounding is not usually available in ALGOL 60 implementations. We shall now generate such a machine arithmetic using a downward-directed rounding $\downarrow$. This will be done in procedure LOW using the usual machine operations. The same considerations hold as in Chapter 4 if we assume that $*$ represents a suitable machine operation. In particular we assume the following:

(a) The set of machine numbers $\mathbb{R}_M$ consists of normalized floating point numbers

$$x = mb^e$$

having the mantissa m, the base b, and the exponent e. $\mathbb{R}_M$ is symmetric; that is,

$$\mathbb{R}_M = -\mathbb{R}_M.$$

(b) The floating point operations $* \in \{+, -, \cdot, /\}$ are optimal: They are generated using a rounding *fl* [with the property (4.3)] on the result of the real operations as was demonstrated in (4.9).

As described in Chapter 4 one obtains the upward-directed rounding $\uparrow$ from $\downarrow$ by using formula (4.5). Since the same sign rules hold for the machine operations $*$ as for real numbers one may also describe the upward rounding as follows:

$$\uparrow(x * y) = \begin{cases} -\downarrow((-x) * (-y)) & \text{for} \quad * \in \{+, -\} \\ -\downarrow((-x) * y) & \text{for} \quad * \in \{\cdot, /\}. \end{cases}$$

The notation $\downarrow$ denotes the effect of the procedure LOW and $*$ describes a machine operation.

We shall now formulate the procedure LOW in ALGOL 60. The result of the procedure LOW acting on the result of a floating point operation is that in every case a lower bound for the real result is computed. The procedure head is given as follows:

```
'REAL' 'PROCEDURE' LOW(X);
'VALUE' X; 'REAL' X;
```

The body of the procedure is dependent on some further properties of the machine operations $* \in \{+, -, \cdot, /\}$. The procedure body is defined dependent on the following properties:

(i) The type of rounding *fl* used to generate the floating point operation from the real operations. We distinguish between

(1) chopping of digits in excess of wordlength to the nearest lower machine number and
(2) rounding to the nearest floating point number.

(ii) Behavior in the neighborhood of 0. If the result of a real operation is smaller in absolute value than the smallest positive machine number, then we distinguish between

(3) setting the result to 0 and
(4) setting the result to the smallest machine number in absolute value with the sign of the real result.

(iii) Behavior when the range of the machine numbers is exceeded. A sensible extension of machine interval operations is only possible in exceptional cases. When the range has been exceeded one should therefore always stop the calculations.

In the procedure body of LOW one can use, for example, the following stop by division

```
'IF' X 'NOTGREATER' ⟨lowerbound⟩ 'THEN'
    'BEGIN' X := ∅; X := 1/X 'END'
```

Here ⟨lowerbound⟩ should be set equal to the smallest floating point number $\min\{x \mid x \in \mathbb{R}_\mathrm{M}\}$.

The following abbreviations are used in the bodies of the procedure LOW. We set k equal to the number of places in the mantissa and $b = 2$, respectively, $b = 10$, is the basis of the floating point numbers. The following abbreviations are then valid:

$$\langle \text{minpos} \rangle := \min\{x \mid x \in \mathbb{R}_\mathrm{M}, x > 0\},$$

$$\langle \text{factor } A \rangle := 1 - b^{1-k},$$

$$\langle \text{factor } B \rangle := 1 - \tfrac{1}{2} b^{1-k}.$$

On binary machines the last two constants should be generated as the quotient of integers; that is,

$$\langle \text{factor } A \rangle := (2^{k-1} - 1)/2^{k-1},$$

$$\langle \text{factor } B \rangle := (2^{k} - 1)/2^{k}.$$

According to the properties of the floating point arithmetic we see that (i) and (ii) give rise to four possible combinations (1)–(3), (1)–(4), (2)–(3), and (2)–(4). We therefore get four possible procedure bodies for the procedure LOW:

Arithmetic (1), (3)

```
'BEGIN'
    'IF' X 'NOTGREATER' ∅ 'THEN'
      'BEGIN'
         X := X/⟨factor A⟩;
         'IF' X 'GREATER' – ⟨minpos⟩ 'THEN' X := – ⟨minpos⟩
      'END';
    LOW := X
'END'
```

Arithmetic (1), (4)

```
'BEGIN'
    'IF' X 'NOTLESS' Ø 'THEN'
      'BEGIN'
        'IF' X 'NOTGREATER' ⟨minpos⟩ 'THEN' X:= Ø
      'END'
    'ELSE'
      'BEGIN'
        X:= X/⟨factor A⟩;
        'IF' X 'GREATER' – ⟨minpos⟩ 'THEN' X:= – ⟨minpos⟩
      'END';
    LOW:= X
'END'
```

Arithmetic (2), (3)

```
'BEGIN'
    'IF' X 'GREATER' Ø 'THEN'
      'BEGIN'
        X:= X × ⟨factor B⟩;
        'IF' X 'NOTGREATER' ⟨minpos⟩ 'THEN' X:= Ø
      'END'
    'ELSE'
      'BEGIN'
        X:= X/⟨factor B⟩;
        'IF' X 'GREATER' – ⟨minpos⟩ 'THEN' X:= – ⟨minpos⟩
      'END';
    LOW:= X
'END'
```

Arithmetic (2), (4)

```
'BEGIN'
    'IF' X 'NOTLESS' Ø 'THEN'
      'BEGIN'
        X:= X × ⟨factor B⟩;
        'IF' X 'NOTGREATER' ⟨minpos⟩ 'THEN' X:= Ø
      'END'
    'ELSE'
      'BEGIN'
        X:= X/⟨factor B⟩;
        'IF' X 'GREATER' – ⟨minpos⟩ 'THEN' X:= – ⟨minpos⟩
      'END';
    LOW:= X
'END'
```

The proof that these procedure bodies generate a rounding $\downarrow$ when the input X is the result of an ordinary machine operation is found in Christ [3].

Using the function procedure LOW and the rounded machine arithmetic generated by this procedure we may now formulate the individual machine interval operations in a simple manner. An interval $A = [a_1, a_2]$ is then declared as an 'ARRAY' A where $A[1]$ is assumed to contain the value a_1 of the lower bound and $A[2]$ is assumed to contain the value a_2 of the upper bound. Furthermore, the arrays A and B contain the operand intervals and C the result interval in the declarations of the procedures.

Addition $C = A + B$:

```
'PROCEDURE'ADD(A, B, C); 'ARRAY'A, B, C;
   'BEGIN'C[1]:= LOW(A[1] + B[1]);
          C[2]:= - LOW(- A[2] - B[2]) 'END'
```

Subtraction $C = A - B$:

```
'PROCEDURE'SUB(A, B, C); 'ARRAY'A, B, C;
   'BEGIN'C[1]:= LOW(A[1] - B[2]);
          C[2]:= - LOW(B[1] - A[2]) 'END'
```

Multiplication $C = A \times B$:

```
'PROCEDURE'MUL(A, B, C); 'ARRAY'A, B, C;
   'BEGIN' 'REAL'A1, A2, B1, B2, C1, C2, P; 'BOOLEAN'BA, BB;
   A2:= - A[2]; B2:= - B[2];
   BA:= A2 'LESS' Ø; BB:= B2 'LESS' Ø;
   'IF'BA  'THEN'A1:= A[1]
          'ELSE' 'BEGIN'A1:= A2; A2:= A[1] 'END';
   'IF'BB  'THEN'B1:= B[1]
          'ELSE' 'BEGIN'B1:= B2; B2:= B[1] 'END';
   C2:= LOW(- A2 × B2);
   'IF'B1  'LESS' Ø
   'THEN' 'BEGIN'C1:= LOW(- A2 × B1);
                'IF'A1  'LESS' Ø
                'THEN' 'BEGIN'P:= LOW(- A1 × B2);
                              'IF'P 'LESS'C1 'THEN'C1:= P;
                              P:= LOW(-A1 × B1);
                              'IF'P 'LESS'C2 'THEN'C2:= P
                       'END'
          'END'
   'ELSE'C1:= 'IF'A1 'LESS' Ø 'THEN' LOW(- A1 × B2)
                             'ELSE' LOW(A1 × B1);
   'IF'BA 'EQUIV'BB
   'THEN' 'BEGIN'C[1]:= C1; C[2]:= - C2 'END'
   'ELSE' 'BEGIN'C[1]:= C2; C[2]:= - C1 'END'
   'END'
```

Division $C = A/B$:

```
'PROCEDURE'DIV(A, B, C); 'ARRAY'A, B, C;
   'BEGIN' 'REAL'A1, A2, B1, B2, C1, C2; 'BOOLEAN'BA, BB;
   A2:= - A[2]; B2:= - B[2];
   BA:= A2 'LESS' Ø; BB:= B2 'LESS' Ø;
   'IF'BA  'THEN'A1:= A[1]
          'ELSE' 'BEGIN'A1:= A2; A2:= A[1] 'END';
   'IF'BB 'THEN'B1:= B[1]
          'ELSE' 'BEGIN'B1:= B2; B2:= B[1] 'END';
```

```
'IF'B1'NOTGREATER' Ø
'THEN' 'BEGIN' 'COMMENT' DIVISIONSSTOP;
                              C1 := Ø; C1 := 1/C1 'END'
'ELSE' 'BEGIN'C2 := LOW(A2/B1);
               C1 := 'IF'A1 'LESS' Ø 'THEN' LOW(A1/B1)
                                       'ELSE'  LOW(- A1/B2)
        'END';
'IF'BA 'EQUIV'BB
'THEN' 'BEGIN'C[1] := C1; C[2] := - C2 'END'
'ELSE' 'BEGIN'C[1] := C2; C[2] := - C1 'END'
'END'
```

We shall now give a short discussion of the reading in of intervals. When one reads in fractional numbers then there is an error committed in the conversion to internal representation. The size of this error is not generally known. In order to read in intervals it is then recommended that the bounds be scaled up by 10^n in order that the bounds become integers. These integers may then be read in without errors. Internally these bounds are then scaled down to their proper values where the application of the procedure LOW guarantees the correct rounding. We present a procedure in which $\langle 10^n \rangle$ denotes a power of 10 that has to be chosen in an appropriate manner.

Reading in A:

```
'PROCEDURE' READIN(A); 'ARRAY'A;
   'BEGIN' 'REAL'A1, A2;
          READ(A1, A2); A[1] := LOW(A1/<10^n>);
                        A[2] := - LOW(- A2/<10^n>)
   'END'
```

We now give a small example for the programming of interval operations. We use the above procedures in the form used on the computer EL X8 at the Computing Center of the University of Karlsruhe by H. Christ. In order to simplify the example it was designed so that there was no need for reading or printing intervals.

Example: The polynomial $p(x)$ is given in the form

$$p(x) = (((\cdots(a_{10}x + a_9)x + a_8)\cdots + a_1)x + a_0.$$

This expression has to be evaluated in the interval sense for the coefficient intervals

$$A_{10} = 10, \qquad A_9 = 9, \ldots, \qquad A_1 = 1, \qquad \text{and} \qquad A_0 = [-0.02, -0.01]$$

and for $X = [-0.1, -0.02]$. We wish to determine whether $0 \in p(X)$ or not.

```
'BEGIN'
'REAL' 'PROCEDURE'LOW(A); 'VALUE'A; 'REAL'A;
'BEGIN'
'IF' A 'NOTLESS' Ø
'THEN' 'BEGIN'A := A × Ø.999999999999; 'IF' A 'NOTGREATER' 10 - 6Ø4
  'THEN'A := Ø 'END'
'ELSE' 'BEGIN'A := A/Ø.999999999999; 'IF' A 'NOTLESS' - 10 - 604
  'THEN'A := - 10 - 6Ø4 'END';
LOW := A;
'IF'A'NOTGREATER' - 10625'THEN''BEGIN'WRITE("OUTOFRANGE"); A := 0;
  A := 1/A 'END'
'END';
'PROCEDURE'ADD(A, B, C); 'ARRAY' A, B, C;
'BEGIN'C[1] := LOW(A[1] + B[1]); C[2] := - LOW(- A[2] - B[2]) 'END';
'PROCEDURE'MUL(A, B, C); 'ARRAY'A, B, C;
  'BEGIN' 'REAL'A1, A2, B1, B2, C1, C2, P; 'BOOLEAN'BA, BB;
A2 := - A[2]; B2 := - B[2]; BA := A2 'LESS' Ø; BB := B2 'LESS' Ø;
'IF'BA'THEN'A1 := A[1] 'ELSE' 'BEGIN' A1 := A2; A2 := A[1] 'END';
'IF'BB 'THEN'B1 := B[1] 'ELSE' 'BEGIN' B1 := B2; B2 := B[1] 'END';
C2 := LOW(- A2 × B2);
'IF' B1 'LESS' Ø
'THEN' 'BEGIN' C1 := LOW(- A2 × B1);
               'IF' A1 'LESS' Ø
               'THEN' 'BEGIN' P := LOW(- A1 × B2); 'IF' P 'LESS' C1
                                'THEN'C1 := P;
                              P := LOW(- A1 × B1); 'IF' P 'LESS' C2
                                'THEN'C2 := P
                   'END'
       'END'
'ELSE'C1 := 'IF' A1 'LESS' Ø 'THEN'LOW(- A1 × B2) 'ELSE'LOW(A1 × B1);
'IF' BA 'EQUIV'BB
'THEN' 'BEGIN'C[1] := C1; C[2] := - C2 'END'
'ELSE' 'BEGIN'C[1] := C2; C[2] := - C1 'END'
'END';
'INTEGER'I; 'BOOLEAN'B; 'ARRAY'A[Ø : 1Ø, 1 : 2], X, F, H[1 : 2];
'FOR'I := 1 'STEP' 1 'UNTIL' 1Ø 'DO'A[I, 1] := A[I, 2] := I;
A[Ø, 1] := LOW(- 2/1ØØ); A[Ø, 2] := - LOW(1/1ØØ);
X[1] := LOW(- 1/1Ø); X[2] := - LOW(1/5Ø);
F[1] := A[1Ø, 1]; F[2] := A[1Ø, 2];
'FOR'I := 9 'STEP' - 1 'UNTIL' Ø 'DO'
              'BEGIN'
                  MUL(F, X, F);
                  H[1] := A[I, 1]; H[2] := A[I, 2];
                  ADD(F, H, F)
              'END';
B := F[1] 'NOTGREATER' Ø 'AND' F[2] 'NOTLESS' Ø;
PRINT(B)
'END'
```

Remarks: A modern treatment of the realization of machine interval arithmetics, which takes into account more recent insights, is treated in Kulisch and Miranker [9].

REFERENCES

[1] G. Alefeld, J. Herzberger, and O. Mayer, Einführung in das Programmieren mit ALGOL 60. Bibliograph. Inst. Mannheim, Hochschultaschenbuch 777 (1972).
[2] N. Apostolatos *et al.*, The Algorithmic Language Triplex-Algol 60. *Numer. Math.* **11**, 175–180 (1968).
[3] H. Christ, Realisierung einer Maschinenintervallarithmetik mit beliebigen ALGOL 60 Compilern. *Elektron. Rech.* **10**, 217–222 (1968).
[4] J. K. S. Dewar, Procedures for interval arithmetic. *Comput. J.* **14**, 447–450 (1970).
[5] D. I. Good and R. L. London, Computer interval arithmetic: Definition and proof of correct implementation. *J. Assoc. Comput. Mach.* **17**, 603–612 (1970).
[6] J. Herzberger, Intervallmäßige Auswertung von Standardfunktionen in ALGOL 60. *Computing* **5**, 377–384 (1970).
[7] U. Kulisch, Implementation and formalization of floating-point arithmetic. *Caratheodory Symp., Athen* (1973).
[8] B. Rothmaier, Die Berechnung der elementaren Funktionen mit beliebiger Genauigkeit. Ph.D. Thesis, Univ. Karlsruhe, Karlsruhe, West Germany (1971).

Added in proof:

[9] U. Kulisch and W. L. Miranker, "Computer Arithmetic in Theory and Practice." Academic Press, New York, 1981.

Appendix C / ALGOL PROCEDURES

A. INCLUSION OF EIGENVALUES

We now describe and give a program listing for an ALGOL procedure for the iterative improvement of inclusions of eigenvalues for real symmetric tridiagonal matrices. A simultaneous method of the form (8.5) is applied. A description of the problem is given at the end of Chapter 8. We denote, as in Chapter 8, the characteristic polynomial of the symmetric tridiagonal matrix $\mathscr{A}$ by $p(x)$ and the eigenvalues of $\mathscr{A}$ by $\lambda_1, \ldots, \lambda_n$. The values of the polynomial can be calculated using (8.10).

When we implement the method (8.5) care must be taken that all the operations occurring in the method are replaced by appropriate machine interval operations. This is done using the ALGOL procedures from Appendix B.

One must also note that all real operations in (8.5) have to be executed as point interval operations. Only in this manner can one be assured that a guaranteed inclusion for the eigenvalues is obtained. For example, if $p([x^{(k,i)}, x^{(k,i)}])$ is evaluated as an interval expression then the result is an inclusion for the real value $p(x^{(k,i)})$. In (8.5) one has to select cases using the function

$$\operatorname{sign}(p([x^{(k,i)}, x^{(k,i)}])),$$

whose values were explained in Chapter 8. The elements of the matrix appearing in the procedure are also given as intervals in the program for the same reasons.

If method (8.5) is executed in the above manner, then one obtains a sequence of intervals

$$X^{(0,i)} \supset X^{(1,i)} \supset X^{(2,i)} \supset \cdots \supset X^{(k(i),i)} = X^{(k(i)+1,i)} = \cdots,$$
$$1 \leqslant i \leqslant n,$$

that do not change after a finite number of steps and that in each case contains an eigenvalue λ_i. The inclusions for the eigenvalues λ_i are therefore no longer computed in the procedure EIGIMP when

$$X^{(k,i)} = X^{(k+1,i)}$$

is satisfied for some k.

The formal parameters for the procedure EIGIMP are

N This is an integer parameter giving the dimension of the matrix under consideration.

A This parameter represents the diagonal elements of the matrix. The individual components $A[I,1], A[I,2]$, represent the interval diagonal elements $[a_{i1}, a_{i2}]$, $1 \leqslant i \leqslant n$.

B The parameter B represents the vector of the superdiagonal elements. The individual components $B[I,1]$, $B[I,2]$ represent the interval elements of the superdiagonal $[b_{i1}, b_{i2}]$, $1 \leqslant i \leqslant n-1$.

X The formal parameter X with the components $X[I,1]$, $X[I,2]$ represents the vector of including intervals $[x_{i,1}, x_{i,2}] \ni \lambda_i$, $1 \leqslant i \leqslant n$. This vector contains the initial including approximations $X^{(0,i)}$, $1 \leqslant i \leqslant n$, upon the call to the procedure. After the procedure call the vector contains the improved inclusions obtained by using (8.5).

BO This parameter represents a Boolean vector having components $BO[I]$, $1 \leqslant i \leqslant n$. $BO[I]$ has the value 'TRUE' if a new inclusion has to be computed for λ_i in the next iteration step. Otherwise the value of $BO[I]$ is 'FALSE'. If two succeeding including intervals are equal then the corresponding component of *BO* is set to 'FALSE'. When calling EIGIMP all those components of *BO* must be set to 'TRUE' for which the eigenvalue inclusions are to be improved according to (8.5). The remaining components are set to 'FALSE'. At the exit from EIGIMP all the values of *BO* are 'FALSE'.

EPS The parameter EPS represents a real constant $\varepsilon > 0$. If the width of the including interval satisfies

$$d(X^{(k,i)}) = s(X^{(k,i)}) - i(X^{(k,i)}) < \varepsilon \max\{|s(X^{(k,i)})|, |i(X^{(k,i)})|\}$$

then the Boolean $BO[I]$ is set to 'FALSE'; that is, the inclusion is no longer improved. If the bound ε cannot be satisfied (as, for example, when $\varepsilon = 0$), then the criterion given in the description of *BO* is used to terminate the iteration.

EMPTY This is an exit that is triggered if the forming of the intersection in (8.5) leads to an empty set for some index i. This will occur if $\lambda_i \notin X^{(0,i)}$ for some i.

EXIT This is an exit which is branched to if any component of the interval inputs A, B, or X has the property that lower bound > upper bound. The same will also happen if the initial including intervals are not pairwise disjoint as required for the method (8.5). It is assumed that the eigenvalues are numbered in increasing value. If for numerical reasons another order is desired, then one should change the segment

```
'FOR'I:= 1 'STEP' 1 'UNTIL' N−1 'DO'
'IF'(BO[I] 'OR'BO[I + 1]) 'AND'X[I, 2] 'NOTLESS'X[I + 1, 1]
'THEN' 'GOTO'EXIT;
```

appropriately.

Global quantities: The procedure uses the procedures LOW and MUL as global quantities

```
'PROCEDURE'EIGIMP (N,A,B,X,BO,EPS,EMPTY,EXIT);
              'VALUE'N,EPS;
              'INTEGER'N;
              'REAL'EPS;
              'ARRAY'A,B,X;
              'BOOLEAN' 'ARRAY'BO;
              'LABEL'EMPTY,EXIT;
'BEGIN'
    'INTEGER'I,J;
    'REAL'N1,N2,U,V,Y,Z;
    'BOOLEAN'BIT;
    'ARRAY'BB,F[1:N,1:2],XM,D[1:N],H,L,M,F1,F2[1:2];

'COMMENT'INPUT DATA;
'FOR'I:= 1 'STEP' 1 'UNTIL' N 'DO'
'BEGIN'
    'IF'A[I,2] 'LESS'A[I,1] 'OR'B[I,2] 'LESS'B[I,1]
    'THEN' 'GOTO'EXIT;
    H[1]:= B[I,1];
    H[2]:= B[I,2];
    MUL(H,H,H);
    BB[I,1]:= H[1];
    BB[I,2]:= H[2]
'END'I;

'FOR'I:= 1 'STEP' 1 'UNTIL' N − 1 'DO'
'IF'(BO[I] 'OR'BO[I+1]) 'AND'X[I,2] 'NOTLESS' X[I+1,1]
'THEN' 'GOTO'EXIT;
```

```
'COMMENT'STARTING VALUES;
'FOR'I := 1 'STEP' 1 'UNTIL' N 'DO'
'IF'BO[I] 'THEN'
'BEGIN'
      XM[I] := (X[I,1] + X[I,2])/2;
      D[I] := -LOW(X[I,1] - X[I,2]);
      F1[1] := F1[2] := 1;
      F2[1] := LOW(XM[I] - A[1,2]);
      F2[2] := -LOW(A[1,1] - XM[I]);
      'FOR'J := 2 'STEP' 1 'UNTIL' N 'DO'
      'BEGIN'
            L[1] := BB[J-1,1];
            L[2] := BB[J-1,2];
            MUL(L,F1,M);
            L[1] := LOW(XM[I]-A[J,2]);
            L[2] := -LOW(A[J,1]-XM[I]);
            MUL(L,F2,L);
            F1[1] := F2[1];
            F1[2] := F2[2];
            F2[1] := LOW(L[1]-M[2]);
            F2[2] := -LOW(M[1]-L[2])
      'END'J;
      F[I,1] := F2[1];
      F[I,2] := F2[2]
'END'I;



'COMMENT'ITERATION;
REPEAT:
BIT := 'FALSE';
'FOR' I := 1 'STEP' 1 'UNTIL' N 'DO'
'IF'BO[I] 'THEN'
'BEGIN'
      'COMMENT' DENOMINATOR;
      H[1] := H[2] := 1;
      'FOR'J := 1 'STEP' 1 'UNTIL'I-1,I+1 'STEP' 1 'UNTIL' N 'DO'
      'BEGIN'
            L[1] := LOW(XM[I]-X[J,2]);
            L[2] := -LOW(X[J,1]-XM[I]);
            MUL(H,L,H)
      'END'J;



      'COMMENT'NEWTON;
      N1 := 'IF'H[1] 'GREATER' Ø
            'THEN'
            ('IF'F[I,1] 'LESS' Ø 'THEN'LOW(F[I,1]/H[1])
                                 'ELSE'LOW(F[I,1]/H[2]))
            'ELSE'
            'IF'F[I,2] 'GREATER' Ø 'THEN'LOW(F[I,2]/H[2])
                                      'ELSE'LOW(F[I,2]/H[1]);
```

```
N2:= 'IF'H[1] 'GREATER' ∅
       'THEN'
       ('IF'F[I,2] 'GREATER' ∅ 'THEN' - LOW( - F[I,2]/H[1])
                                'ELSE' - LOW( - F[I,2]/H[2]))
       'ELSE'
       'IF'F[I,1] 'LESS' ∅ 'THEN' - LOW( - F[I,1]/H[2])
                          'ELSE' - LOW( - F[I,1]/H[1]);
Y:= N1;
N1:= LOW(XM[I] - N2);
N2:= - LOW(Y - XM[I]);
'IF'N1 'LESS'X[I,1] 'THEN'N1:= X[I,1];
'IF'N2 'GREATER'X[I,2] 'THEN'N2:= X[I,2];
'IF'N2 'LESS' N1 'THEN' 'GOTO'EMPTY;

'COMMENT'HALVING;
XM[I]:= (N1 + N2)/2;
F1[1]:= F1[2]:= 1;
F2[1]:= LOW(XM[I] - A[1,2]);
F2[2]:= - LOW(A[1,1] - XM[I]);
'FOR'J:= 2 'STEP' 1 'UNTIL' N 'DO'
'BEGIN'
      L[1]:= BB[J - 1,1];
      L[2]:= BB[J - 1,2];
      MUL(L,F1,M);
      L[1]:= LOW(XM[I] - A[J,2]);
      L[2]:= - LOW(A[J,1] - XM[I]);
      MUL(L,F2,L);
      F1[1]:= F2[1];
      F1[2]:= F2[2];
      F2[1]:= LOW(L[1] - M[2]);
      F2[2]:= - LOW(M[1] - L[2])
'END'J;
F[I,1]:= F2[1];
F[I,2]:= F2[2];
'IF'F[I,1] 'GREATER' ∅
'THEN'
      'BEGIN'
            'IF'H[1] 'GREATER' ∅
            'THEN'
                  'BEGIN'
                        X[I,2]:= XM[I];
                        X[I,1]:= N1
                  'END'
            'ELSE'
                  'BEGIN'
                        X[I,1]:= XM[I];
                        X[I,2]:= N2
                  'END'
      'END'
'ELSE'
'IF'F[I,2] 'LESS' ∅
'THEN'
```

```
            'BEGIN'
                'IF'H[1] 'GREATER' ∅
                'THEN'
                    'BEGIN'
                        X[I,1]:= XM[I];
                        X[I,2]:= N2
                    'END'
                'ELSE'
                    'BEGIN'
                        X[I,2]:= XM[I];
                        X[I,1]:= N1
                    'END'
            'END'
        'ELSE'
            'BEGIN'
                X[I,1]:= N1;
                X[I,2]:= N2
            'END';

        'COMMENT'TEST;
        Z:= -LOW(X[I,1]-X[I,2]);
        U:= ABS(X[I,1]);
        V:= ABS(X[I,2]);
        'IF' V 'LESS' U 'THEN' V:= U;
        'IF' Z 'NOTLESS'D[I] 'OR'Z 'NOTGREATER'EPS × V
        'THEN'BO[I]:= 'FALSE' 'ELSE'BIT:= 'TRUE';
        D[I]:= Z
    'END'ITERATION;
    'IF'BIT 'THEN' 'GOTO'REPEAT
    'END'PROCEDURE
```

B. SINGLE STEP METHOD

We now describe and give a program listing for an ALGOL-procedure ITERATION. This procedure implements the single step method taking intersection after every component (SIC) for the determination of the fixed point x^* of the equation $x = \mathscr{A}x + \ell$ provided that it satisfies the strong row or column criterion. (This method was discussed in Chapter 14). The elements of $\mathscr{A}$ and ℓ are real intervals. In this case it is possible – as described in Alefeld [1] – to determine the required including vector $x^{(0)}$ for x^* very easily. The determination of an including set for a system of equations not amenable to iteration of this type is discussed in Alefeld and Herzberger [2].

The formal parameters are

N This is an integer parameter giving the dimension of the system of equations.

MAT This parameter represents the $n \times n$ interval matrix $\mathscr{A}$. The individual components $MAT[I, J, 1]$, (resp., $MAT[I, J, 2]$) represent the lower (resp., upper) bounds of the element A_{ij} of the matrix.

VEC This parameter represents the interval vector $\mathscr{b}$ with n components. The individual components $VEC[I, 1]$, $VEC[I, 2]$ represent the bounds b_{i1} and b_{i2} of the vector component B_i.

EXIT This label is reached if the matrix $\mathscr{A}$ does not satisfy the row-sum or the column-sum criterion.

X The fixed point x^* is returned in this parameter.

The procedure uses as global quantities the procedures LOW, ADD, and MUL described in Appendix B.

```
'PROCEDURE'ITERATION(MAT,VEC,N,EXIT)RESULT:(X);
'VALUE'MAT,VEC,N; 'ARRAY'MAT,VEC,X; 'INTEGER'N; 'LABEL'EXIT;
'BEGIN' 'INTEGER'I,J; 'REAL'MAX,HMAX,NORM,HNORM; 'BOOLEAN'BV;
    'ARRAY'Y[1:N],IK,HMAT,HX[1:2];
    'REAL' 'PROCEDURE'UP(A); 'VALUE'A; 'REAL'A; UP:= -LOW(-A);
    'REAL' 'PROCEDURE'ABSVAL(LBD,VBD); 'REAL'LBD,UBD;
        'IF'ABS(LBD) 'LESS'ABS(UBD) 'THEN'ABSVAL:= ABS(UBD) 'ELSE'
        ABSVAL:= ABS(LBD);
    MAX:= ∅;
    'COMMENT'ESTIMATE USING THE ROW SUM NORM;
    'FOR'I:= 1 'STEP' 1 'UNTIL' N 'DO'
        'BEGIN'Y[I]:= ∅;
            'FOR'J:= 1 'STEP' 1 'UNTIL' N 'DO'
            Y[I]:= UP(Y[I] + ABSVAL (MAT[I,J,1],MAT[I,J,2]));
            'IF'Y[I] 'NOTLESS' 1 'THEN' 'GOTO'M1
        'END';
    'COMMENT'DETERMINATION OF STARTING VECTOR WHEN ROW SUM
        CRITERION SATISFIED;
    'FOR'I:= 1 'STEP' 1 'UNTIL' N 'DO'
    'BEGIN'HMAX:= ∅;
        'FOR'J:= 1 'STEP' 1 'UNTIL' N 'DO'
        HMAX:= UP(HMAX+UP(ABSVAL(MAT[I,J,1],MAT[I,J,2]) ×
        ABSVAL(VEC[J,1],VEC[J,2])));
        HMAX:= UP(HMAX/LOW(1-Y[I]));
        'IF'MAX 'LESS'HMAX 'THEN'MAX:= HMAX
    'END'; 'GOTO'M2;
    'COMMENT'ESTIMATION USING COLUMN SUM NORM;
    M1 : NORM:= ∅;
    'FOR'J:= 1 'STEP' 1 'UNTIL' N 'DO'
    'BEGIN'HNORM:= ∅;
        'FOR'I:= 1 'STEP' 1 'UNTIL' N 'DO'
        HNORM:= UP(HNORM + ABSVAL(MAT[I,J,1],MAT[I,J,2]));
        'IF'HNORM 'NOTLESS' 1 'THEN' 'GOTO'EXIT;
        'IF'HNORM 'GREATER'NORM 'THEN'NORM:= HNORM
    'END';
```

```
'COMMENT'DETERMINATION OF STARTING VECTOR WHEN COLUMN
    SUM CRITERION SATISFIED;
'FOR'I:= 1 'STEP' 1 'UNTIL' N 'DO'
'FOR'J:= 1 'STEP' 1 'UNTIL' N 'DO'
MAX:= UP(MAX+UP(ABSVAL(MAT[I,J,1],MAT[I,J,2]) ×
ABSVAL(VEC[J,1],VEC[J,2])));
MAX:= UP(MAX/LOW(1-NORM));
M2 : 'FOR'I:= 1 'STEP' 1 'UNTIL' N 'DO'
    'BEGIN'X[I,1]:= LOW(VEC[I,1]-MAX);
    X[I,2]:= UP(VEC[I,2]+MAX) 'END';
'COMMENT'START OF THE ITERATION;
M3 : BV:= 'TRUE';
'FOR'I:= 1 'STEP' 1 'UNTIL' N 'DO'
'BEGIN'IK[1]:= VEC[I,1]; IK[2]:= VEC[I,2];
    'FOR'J:= 1 'STEP' 1 'UNTIL' N 'DO'
    'BEGIN'HMAT[1]:= MAT[I,J,1]; HMAT[2]:= MAT[I,J,2];
          HX[1]:= X[J,1]; HX[2]:= X[J,2];
          MUL(HMAT,HX,HX); ADD(HX,IK,IK)
    'END';
    HX[1]:= X[I,1]; HX[2]:= X[I,2];
    'IF'IK[1] 'LESS'HX[1] 'THEN'IK[1]:= HX[1];
    'IF'IK[2] 'GREATER'HX[2] 'THEN'IK[2]:= HX[2];
    'IF'BV 'THEN' 'BEGIN' 'IF'IK[1] 'NOTEQUAL'X[I,1] 'OR'IK[2]
      'NOTEQUAL'
    X[I,2] 'THEN'BV:= 'FALSE' 'END';
    X[I,1]:= IK[1]; X[I,2]:= IK[2]
'END';
'IF' 'NOT'BV 'THEN' 'GOTO'M3
'END'
```

C. MATRIX INVERSION

We now describe and give a program listing for an ALGOL procedure that implements the iteration method described in Chapter 18 for the inclusion of the inverse of a matrix $\mathscr{A}$. In order to preserve the inclusion of the inverse also when calculating with a machine interval arithmetic we must execute all real operations as point interval operations. Only in this manner are we able to account for all the roundoff errors. Because of this and also because of the unavoidable input conversion errors we shall consider all constants and real matrices as interval matrices. In order to generate a directed rounding for the machine interval operations we use the procedures described in Appendix B.

We first display the procedure MATINV that has the following formal parameters:

N This integer parameter represents the dimension n of the matrix.

A This parameter represents the $n \times n$ interval matrix $\mathscr{A}$. The individual components $A[I, J, 1]$ and $A[I, J, 2]$ represent the lower and upper bounds of the matrix elements A_{ij}. If A_{ij} can be represented as a point interval, then of course both these bounds are equal. The actual parameter is a representation of the matrix $\mathscr{A}$ to be inverted.

X This parameter represents an $n \times n$ interval matrix, and the components $X[I, J, 1]$ and $X[I, J, 2]$ are the lower and upper bounds for the matrix elements X_{ij}. This parameter is used for both input and output to the procedure. When the procedure is called then this parameter is to be replaced by an actual parameter $\mathscr{X}^{(0)}$ for which

$$\{\mathscr{A}^{-1} | \mathscr{A} \in \mathscr{A}\} \in \mathscr{X}^{(0)} \qquad \text{and} \qquad \| |\mathscr{I} - \mathscr{A}m(\mathscr{X}^{(0)})| \| < 1$$

for some matrix norm $\|\cdot\|$. With this $\mathscr{X}^{(0)}$ the iteration (18.5) is repeated (with $r = 2$) until an inclusion matrix satisfying the inequality (18.11) has been computed. The norm used is

$$\|\mathscr{X}\| = n \max_{1 \leqslant i, j \leqslant n} |x_{ij}|.$$

When (18.11) has been satisfied then using $r = 2$ the iteration is continued according to (18.9) until two succeeding including sets are equal. The iteration is then stopped and the last interval matrix is stored in the parameter *X* as an output parameter. A possibility for the determination of the matrix $\mathscr{X}^{(0)}$ is given by the ALGOL-procedure INCLUSIONSET.

The global quantities are the procedures LOW, ADD, and MUL from Appendix B.

```
'PROCEDURE'MATINV(A,N,X);
'VALUE'N;
'INTEGER'N;
'ARRAY'A,X;
'BEGIN' 'INTEGER'I,J,K;
        'REAL'U,V,W,NA,NDX,NDR,H;
        'BOOLEAN'B1,B2;
        'ARRAY'S,H1,H2[1:2],Y[1:N,1:2],M[1:N,1:N],
        Z,R[1:N,1:N,1:2];
U:= LOW(1/N);
B1:= B2:= 'FALSE';

'COMMENT'NORM A;
NA:= Ø;
'FOR'I:= 1 'STEP' 1 'UNTIL' N 'DO'
'FOR'J:= 1 'STEP' 1 'UNTIL' N 'DO'
'BEGIN'V:= ABS(A[I,J,1]); W:= ABS(A[I,J,2]);
    'IF'W 'LESS'V 'THEN'W:= V;
    'IF'W 'GREATER'NA 'THEN'NA:= W;
    Z[I,J,1]:= X[I,J,1]; Z[I,J,2]:= X[I,J,2]
'END'NORM A;
L1:B1:= 'FALSE';
```

```
'COMMENT'CALCULATION OF MIDPOINT;
'FOR'I:= 1 'STEP' 1 'UNTIL' N 'DO'
'FOR'J:= 1 'STEP' 1 'UNTIL' N 'DO'
M[I,J]:= (X[I,J,1] + X[I,J,2])/2;

'COMMENT'RESIDUAL;
'FOR'I:= 1 'STEP' 1 'UNTIL' N 'DO'
'BEGIN' 'FOR'J:= 1 'STEP' 1 'UNTIL' N 'DO'
      'BEGIN'S[1]:= S[2]:= ∅;
           'FOR'K:= 1 'STEP' 1 'UNTIL' N 'DO'
           'BEGIN'H1[1]:= A[I,K,1]; H1[2]:= A[I,K,2];
                    H2[1]:= H2[2]:= M[K,J];
                    MUL(H1,H2,H1); ADD(S,H1,S)
           'END'K;
           R[I,J,1]:= -S[2]; R[I,J,2]:= -S[1]
      'END'J;
      R[I,I,1]:= LOW(1+R[I,I,1]);
      R[I,I,2]:= -LOW(-1-R[I,I,2])
'END' RESIDUAL;

'COMMENT'ITERATION LOOP;
'IF'B2 'THEN' 'FOR'I:= 1 'STEP' 1 'UNTIL' N 'DO'
               'FOR'J:= 1 'STEP' 1 'UNTIL' N 'DO'
               'BEGIN'Z[I,J,1]:= X[I,J,1];
                        Z[I,J,2]:= X[I,J,2]
               'END';
'FOR'I:= 1 'STEP' 1 'UNTIL' N 'DO'
'BEGIN' 'FOR'J:= 1 'STEP' 1 'UNTIL' N 'DO'
      'BEGIN'S[1]:= S[2]:= ∅;
           'FOR'K:= 1 'STEP' 1 'UNTIL' N 'DO'
           'BEGIN'H1[1]:= Z[I,K,1]; H1[2]:= Z[I,K,2];
                    H2[1]:= R[K,J,1]; H2[2]:= R[K,J,2];
                    MUL(H1,H2,H1); ADD(S,H1,S)
           'END'K;
           H1[1]:= H1[2]:= M[I,J]; ADD(S,H1,S);
           Y[J,1]:= S[1]; Y[J,2]:= S[2]
      'END'J;
      'FOR'J:= 1 'STEP' 1 'UNTIL' N 'DO'
      'BEGIN'Z[I,J,1]:= Y[J,1];
               Z[I,J,2]:= Y[J,2]
      'END'J
'END'I;

'COMMENT'INTERSECTION;
'IF'B2 'THEN'
L2 : 'BEGIN' 'FOR'I:= 1 'STEP' 1 'UNTIL' N 'DO'
              'FOR'J:= 1 'STEP' 1 'UNTIL' N 'DO'
              'BEGIN' 'IF'X[I,J,1] 'LESS'Z[I,J,1] 'THEN'
                    'BEGIN'X[I,J,1]:= Z[I,J,1]; B1:= 'TRUE' 'END';
                    'IF'X[I,J,2] 'GREATER'Z[I,J,2] 'THEN'
                    'BEGIN'X[I,J,2]:= Z[I,J,2]; B1:= 'TRUE' 'END'
              'END'I,J;
              'GOTO' 'IF'B1 'THEN'L1 'ELSE'L3
     'END';
```

```
'COMMENT'WIDTH QUESTION;
NDR:= NDX:= ∅;
'FOR'I:= 1 'STEP' 1 'UNTIL' N 'DO'
'FOR'J:= 1 'STEP' 1 'UNTIL' N 'DO'
'BEGIN'V:= ABS(R[I,J,1]); W:= ABS(R[I,J,2]);
      'IF'W 'LESS'V 'THEN'W:= V;
      'IF'W 'GREATER'NDR 'THEN' NDR:= W;
      H:= -LOW(X[I,J,1]-X[I,J,2]);
      'IF' H 'GREATER'NDX 'THEN'NDX:= H
'END'NORM RESIDUAL, NORM WIDTH;
W:= LOW(U-NDR);
U:= LOW(2/N);
W:= LOW(W/NA);
W:= LOW(W × U);
'IF'NDX 'LESS'W 'THEN' 'BEGIN'B2:= 'TRUE';
                              'GOTO'L2
                       'END'
                       'ELSE'
                       'BEGIN' 'FOR'I:= 1 'STEP' 1 'UNTIL' N 'DO'
                              'FOR'J:= 1 'STEP' 1 'UNTIL' N 'DO'
                              'BEGIN'X[I,J,1]:= Z[I,J,1];
                                    X[I,J,2]:= Z[I,J,2]
                              'END';
                              'GOTO'L1
                       'END'WIDTH QUESTION
ITERATION LOOP;
L3 : 'END'PROCEDURE BODY
```

We also describe the ALGOL-procedure INCLUSIONSET that will compute an initial matrix $\mathscr{X}^{(0)}$ to be passed to the procedure MATINV. The formal parameters of this procedure are

N This parameter denotes the degree n of the matrix to be inverted.

B This parameter represents an $n \times n$ interval matrix $\mathscr{B}$ whose components $B[I, J, 1]$ and $B[I, J, 2]$ represent the lower and the upper bound of the matrix elements B_{ij}. The actual parameter for B is a representation of the matrix $\mathscr{A}$ to be inverted.

X This formal parameter represents an $n \times n$ interval matrix $\mathscr{X}$ and the components $X[I, J, 1]$ (resp., $X[I, J, 2]$) represent the lower (resp., upper) bound of the matrix element X_{ij}. After the call of the procedure X represents an interval matrix $\mathscr{X}$ with the property that

$$\{\underset{\cdot}{\mathscr{B}}^{-1} | \underset{\cdot}{\mathscr{B}} \in \mathscr{B}\} \subseteq \mathscr{X}.$$

The calculation of $\mathscr{X}$ is done with the help of the formulas adjoining Theorem 18.2. Either the row-sum or the column-sum norms for $|\mathscr{B}|$ are used.

L This parameter represents a label which is branched to if neither the row-sum nor the column-sum norm of $|\mathscr{B}|$ is smaller than 1. In this case it is not possible to calculate a matrix $\mathscr{X}$ using the given formulas.

The procedure LOW from Appendix B is a global quantity.

```
'PROCEDURE' INCLUSIONSET(B,N,X,L);
        'VALUE'N;
        'INTEGER'N;
        'ARRAY'B,X;
        'LABEL'L;
'BEGIN' 'INTEGER'I,J;
         'REAL'NOR,A1,A2,HI;
         NOR:= Ø;


'COMMENT' ROW NORM;
'FOR' I:= 1 'STEP' 1 'UNTIL' N 'DO'
'BEGIN'HI:= Ø;
     'FOR'J:= 1 'STEP' 1 'UNTIL' N 'DO'
     'BEGIN'A1:= ABS(B[I,J,1]); A2:= ABS(B[I,J,2]);
             HI:= -LOW(-HI-('IF'A1 'GREATER'A2 'THEN'A1 'ELSE'A2))
     'END'J;
     'IF'HI 'GREATER'NOR 'THEN'NOR:= HI;
     'IF'NOR 'NOTLESS' 1 'THEN' 'GOTO'COLNORM
'END'I,ROW NORM;
'GOTO'INCL;
COLNORM : NOR:= Ø;
'FOR'J:= 1 'STEP' 1 'UNTIL' N 'DO'
'BEGIN'HI:= Ø;
     'FOR'I:= 1 'STEP' 1 'UNTIL' N 'DO'
     'BEGIN'A1:= ABS(B[I,J,1]); A2:= ABS(B[I,J,2]);
             HI:= -LOW(-HI-('IF'A1 'GREATER'A2 'THEN'A1 'ELSE'A2))
     'END'I;
     'IF'HI 'GREATER'NOR 'THEN'NOR:= HI;
     'IF'NOR 'NOTLESS' 1 'THEN' 'GOTO'L
'END'J,COLNORM;


INCL :
A1:= LOW(1-NOR);
A1:= -LOW(-1/A1);
'FOR'I:= 1 'STEP' 1 'UNTIL' N 'DO'
'BEGIN' 'FOR'J:= 1 'STEP' 1 'UNTIL'I-1,I+1 'STEP' 1 'UNTIL' N 'DO'
     'BEGIN'X[I,J,1]:= -A1;
             X[I,J,2]:= A1
     'END'J;
     X[I,I,1]:= -A1;
     X[I,I,2]:= -LOW(-2-A1)
'END'I
'END'
```

A detailed description of both procedures as well as a discussion of numerical examples is found in Alefeld and Herzberger [3].

REFERENCES

[1] G. Alefeld, Über die aus monoton zerlegbaren Operatoren gebildeten Iterationsverfahren. *Computing* **6**, 161–172 (1970).

[2] G. Alefeld and J. Herzberger, ALGOL-60 Algorithmen zur Auflösung linearer Gleichungssysteme mit Fehlererfassung. *Computing* **6**, 28–34 (1970).

[3] G. Alefeld and J. Herzberger, Matrizeninvertierung mit Fchlererfassung. *Elektron. Datenverarbeit.* **12**, 410–416 (1970).

[4] G. Alefeld and J. Herzberger, Über die Verbesserung von Schranken für die Eigenwerte symmetrischer Tridiagonalmatrizen. *Angew. Informat.* **16**, 27–35 (1974).

[5] H. Christ, Realisierung einer Maschinenintervallarithmetik mit beliebigen ALGOL-60 Compilern. *Elektron. Rech.* **10**, 217–222 (1968).

[6] G. Zielke, "ALGOL-Katalog Matrizenrechnung." Teubner Verlag, Leipzig, 1972.

BIBLIOGRAPHY

[1] O. Aberth, Iteration methods for finding all zeros of a polynomial simultaneously. *Math. Comp.* **27**, 339–344 (1973).

[2] I. Ahrens, Über Folgen und Reihen von Punktmengen. *Arch. Math.* **23**, 92–103 (1972).

[3] J. Albrecht, Monotone Iterationsfolgen und ihre Verwendung zur Lösung linearer Gleichungssysteme. *Numer. Math.* **3**, 345–358 (1961).

[4] G. Alefeld, Intervallrechnung über den komplexen Zahlen und einige Anwendungen. Ph.D. Thesis, Univ. Karlsruhe (1968).

[5] G. Alefeld, Über die aus monoton zerlegbaren Operatoren gebildeten Iterationsverfahren. *Computing* **6**, 161–172 (1970).

[6] G. Alefeld, Eine Modifikation des Newtonverfahrens zur Bestimmung der reellen Nullstellen einer reellen Funktion. *Z. Angew. Math. Mech.* **50**, T 32–T 33 (1970).

[7] G. Alefeld, Über Eigenschaften und Anwendungsmöglichkeiten einer Intervallarithmetik über den komplexen Zahlen. *Z. Angew. Math. Mech.* **50**, 455–465 (1970).

[8] G. Alefeld, Anwendungen des Fixpunktsatzes für pseudometrische Räume in der Intervallrechnung. *Numer. Math.* **17**, 33–39 (1971).

[9] G. Alefeld, Quadratisch konvergente Einschließung von Lösungen bei nichtkonvexen Gleichungssystemen. *Z. Angew. Math. Mech.* **54**, 335–342 (1974).

[10] G. Alefeld, Über die Existenz einer eindeutigen Lösung bei einer Klasse nichtlinearer Gleichungssysteme und deren Berechnung mit Iterationsverfahren. *Apl. Mat.* **17**, 267–294 (1972).

[11] G. Alefeld and N. Apostolatos, Auflösung nichtlinearer Gleichungssysteme mit zwei Unbekannten. Bericht des Rechenzentrums Inst. Angew. Math., Univ. Karlsruhe (1967).

[12] G. Alefeld and N. Apostolatos, Praktische Anwendung von Abschätzungsformeln bei Iterationsverfahren. *Z. Angew. Math. Mech.* **48**, T 46–T 49 (1968).

[13] G. Alefeld and J. Herzberger, Über das Newton-Verfahren bei nichtlinearen Gleichungssystemen. *Z. Angew. Math. Mech.* **50**, 773–774 (1970).

[14] G. Alefeld and J. Herzberger, ALGOL-60 Algorithmen zur Auflösung linearer Gleichungssysteme mit Fehlererfassung. *Computing* **6**, 28–34 (1970).

[15] G. Alefeld and J. Herzberger, Über die Berechnung der inversen Matrix mit Hilfe der Intervallrechnung. *Elektron. Rech.* **12**, 259–261 (1970).

[16] G. Alefeld and J. Herzberger, Verfahren höherer Ordnung zur iterativen Einschließung der inversen Matrix. *Z. Angew. Math. Phys.* **21**, 819–824 (1970).

[17] G. Alefeld and J. Herzberger, Matrizeninvertierung mit Fehlererfassung. *Elektron. Datenverarbeit.* **12**, 410–416 (1970).

[18] G. Alefeld and J. Herzberger, Über die Verbesserung von Schranken für die Lösung bei linearen Gleichungssystemen. *Angew. Informat.* **13**, 107–112 (1971).

[19] G. Alefeld and J. Herzberger, Zur Invertierung linearer und beschränkter Operatoren. *Math. Z.* **120**, 309–317 (1971).

[20] G. Alefeld and J. Herzberger, Einschließungsverfahren zur Berechnung des inversen Operators. *Z. Angew. Math. Mech.* **52**, T 197–T 198 (1972).

[21] G. Alefeld and J. Herzberger, Nullstelleneinschließung mit dem Newton-Verfahren ohne Invertierung von Intervallmatrizen. *Numer. Math.* **19**, 56–64 (1972).

[22] G. Alefeld and J. Herzberger, Ein Verfahren zur monotonen Einschließung von Lösungen nichtlinearer Gleichungssysteme. *Z. Angew. Math. Mech.* **53**, T 176–T 177 (1973).

[23] G. Alefeld and J. Herzberger, Über die Konvergenzordnung intervallarithmetischer Iterationsverfahren. *Kolloq. Konstrukt. Theorie Funktionen, Conf. Proc., Cluj* (1973).

[24] G. Alefeld and J. Herzberger, Über Simultanverfahren zur Bestimmung reeller Polynomwurzeln. *Z. Angew. Math. Mech.* **54**, 413–420 (1974).

[25] G. Alefeld and J. Herzberger, Über die Verbesserung von Schranken für die Eigenwerte symmetrischer Tridiagonalmatrizen. *Angew. Informat.* **16**, 27–35 (1974).

[26] G. Alefeld and J. Herzberger, On the convergence speed of some algorithms for the simultaneous approximation of polynomial roots. *SIAM J. Numer. Anal.* **11**, 237–243 (1974).

[27] G. Alefeld, J. Herzberger, and O. Mayer, Über neuere Gesichtspunkte beim numerischen Rechnen. *Math. Naturwiss. Unterricht* **24**, 458–467 (1971).

[28] M. Altmann, An optimum cubically convergent iterative method of inverting a linear bounded operator in Hilbert space. *Pacific J. Math.* **10**, 1107–1113 (1960).

[29] N. Apostolatos, Ein Einschließungsverfahren für Nullstellen. *Z. Angew. Math. Mech.* **47**, T 80 (1967).

[30] N. Apostolatos, Nullstellenbestimmung mit Fehlererfassung. Bericht des Rechenzentrums Inst. Angew. Math., Univ. Karlsruhe (1967).

[31] N. Apostolatos, Allgemeine Intervallarithmetiken und Anwendungen. *Bull. Soc. Math., Athens, N. Ser.* **10**, 136–180 (1969).

[32] N. Apostolatos, see G. Alefeld.

[33] N. Apostolatos and U. Kulisch, Grundlagen einer Maschinenintervallarithmetik. *Computing* **2**, 89–104 (1967).

[34] N. Apostolatos and U. Kulisch, Approximation der erweiterten Intervallarithmetik durch die einfache Maschinenintervallarithmetik. *Computing* **2**, 181–194 (1967).

[35] N. Apostolatos and U. Kulisch, Über die Konvergenz des Relaxationsverfahrens bei nichtnegativen und diagonaldominanten Matrizen. *Computing* **2**, 17–24 (1967).

[36] N. Apostolatos and U. Kulisch, Grundzüge einer Intervallrechnung für Matrizen und einige Anwendungen. *Elektron. Rech.* **10**, 73–83 (1968).

[37] N. Apostolatos, U. Kulisch, and K. Nickel: Ein Einschließungsverfahren für Nullstellen. *Computing* **2**, 195–201 (1967).

[38] N. Apostolatos, U. Kulisch, R. Krawczyk, B. Lortz, K. Nickel, and H.-W. Wippermann, The algorithmic language Triplex-Algol 60. *Numer. Math.* **11**, 175–180 (1968).

[39] W. Appelt, Fehlereinschließung bei der numerischen Lösung elliptischer Differentialgleichungen unter Verwendung eines intervallarithmetischen Defektverfahrens. Ph.D. Thesis, Univ. Bonn (1972).

[40] D. W. Arthur, The use of interval arithmetic to bound the zeros of real polynomials. *J. Inst. Math. Appl.* **10**, 231–237 (1972).

[41] J. Avenhaus, Zur numerischen Behandlung des Anfangswertproblems mit exakter Fehlererfassung. Ph.D. Thesis, Univ. Karlsruhe (1970).

[42] J. Avenhaus, Ein Verfahren zur Einschließung der Lösung des Anfangswertproblems. *Computing* **8**, 182–190 (1971).

[43] A. N. Baluev, Zur Methode von Chaplygin (Russian). *Dokl. Akad. Nauk SSSR* **83**, 781–789 (1952).

[44] W. Barth, Nullstellenbestimmung mit der Intervallrechnung. *Computing* **8**, 320–328 (1971).

[45] W. Barth, Ein Algorithmus zur Berechnung aller reellen Nullstellen in einem Intervall. *Computing* **9**, 327–333 (1972).

[46] M. Bauer, Zur Bestimmung der reellen Wurzeln einer algebraischen Gleichung durch Iteration. *Jber. Deutsch. Math.-Verein.* **25**, 294–301 (1917).

[47] H. Beeck, Über intervallanalytische Methoden bei linearen Gleichungssystemen mit Intervallkoeffizienten und Zusammenhänge mit der Fehleranalysis. Ph.D. Thesis, TH München (1971).

[48] H. Beeck, Über die Struktur und Abschätzungen der Lösungsmenge von linearen Gleichungssystemen mit Intervallkoeffizienten. *Computing* **10**, 231–244 (1972).

[49] S. Berti, The solution of an interval equation. *Mathematica* **11**, 189–194 (1969).

[50] S. Berti, The geometry of the relations in the set of real numbers. *Mathematica* **11**, 29–43 (1969).

[51] S. Berti, Intervalele si aritmetica lor in analiza numerica. *Gaz. Mat. A* **75**, 309–313 (1970).

[52] S. Berti, Some relations between interval functions (I). *Mathematica* **14**, 9–26 (1972).

[53] S. Berti, Aritmetica si Analiza Intervalelor. *Rev. Anal. Numer. Teoria Aproximatiei* **1**, 21–39 (1972).

[54] M. Blumhofer, Über einige Algorithmen zur numerischen Einschließung der Lösung bei linearen Gleichungssystemen. Diplomarbeit, Inst. Angew. Math., Univ. Karlsruhe (1972).

[55] R. Boche, Some observations on the economics of interval arithmetic. *Comm. ACM* **8**, 649 (1965).

[56] R. Boche, Complex interval arithmetic with some applications. Lockheed Missiles and Space Company, 4-22-66-1, Sunnyvale, California (1966).

[57] W. Börsch-Supan, A posteriori error bounds for the zeros of polynomials. *Numer. Math.* **5**, 380–398 (1963).

[58] D. Braess, Die Konstruktion monotoner Iterationsfolgen zur Lösungseinschließung bei linearen Gleichungssystemen. *Arch. Rational Mech. Anal.* **9**, 97–106 (1962).

[59] D. Braess, Monotone Iterationsfolgen bei Gleichungssystemen mit fehlerhaften Koeffizienten und Iterationsbeschleunigung. *Numer. Math.* **7**, 32–41 (1965).

[60] D. Braess, Die Berechnung der Fehlergrenzen bei linearen Gleichungssystemen mit fehlerhaften Koeffizienten. *Arch. Rational Mech. Anal.* **19**, 74–80 (1965).

[61] D. Braess and K. P. Hadeler, Simultaneous inclusion of the zeros of a polynomial. *Numer. Math.* **21**, 161–165 (1973).

[62] R. P. Brent, An algorithm with guaranteed convergence for finding a zero of a function. *Comput. J.* **14**, 422–425 (1972).

[63] M. Brockhaus, B. Rothmaier, and P. Schroth, Benutzeranleitung für TRIPLEX-ALGOL im System HYDRA 2. Interner Bericht 11, Inst. f. Informatik, Univ. Karlsruhe (1969).

[64] W. Burmeister, Inversionsfreie Verfahren zur Lösung nichtlinearer Operatorgleichungen. *Z. Angew. Math. Mech.* **52**, 101–110 (1972).

[65] P. Byrnev, see K. Dochev.

[66] W. P. Champagne, On finding roots of polynomials by hook or by crook. TTN-37, Univ. of Texas, Computation Center (1964).

[67] B. A. Chartres, Automatic controlled precision calculations. *J. ACM* **13**, 386–403 (1966).

[68] B. A. Chartres, Computable error bounds for direct solution of linear equations. *J. ACM* **14**, 63–71 (1967).
[69] F. L. Chernous'ko, An optimal algorithm for finding the roots of an approximately computed function. *Ž. Vyčisl. Mat. i Mat. Fiz.* **8**, 705–724 (1968).
[70] H. H. Y. Chien, A multiphase algorithm for single variable equation solving. *J. Inst. Math. Appl.* **9**, 290–298 (1972).
[71] H. Christ, Realisierung einer Maschinenintervallarithmetik auf beliebigen ALGOL-Compilern. *Z. Angew. Math. Mech.* **48**, T 110 (1968).
[72] H. Christ, Realisierung einer Maschinenintervallarithmetik mit beliebigen ALGOL-60 Compilern. *Elektron. Rech.* **10**, 217–222 (1968).
[73] W. Chuba and W. Miller, Quadratic convergence in interval arithmetic, Part I. *BIT* **12**, 284–290 (1972).
[74] L. Collatz, "Funktionalanalysis und numerische Mathematik." Springer-Verlag, Berlin and New York, 1964.
[75] R. H. Dargel, F. R. Loscalzo, and T. H. Witt, Automatic error bounds on real zeros of rational functions. *Comm. ACM* **9**, 806–809 (1966).
[76] D. P. Davey and N. F. Stewart, Guaranteed error bounds for the initial value problem using polytope arithmetic. Publication # 109, Département d'Informatique, Univ. de Montreal (1972).
[77] J. K. S. Dewar, Procedures for interval arithmetic. *Comp. J.* **14**, 447–450 (1970).
[78] K. Dochev and P. Byrnev, Certain modifications of Newton's method for the approximate solution of algebraic equations. *Ž. Vyčisl. Mat. i Mat. Fiz.* **4**, 915–920 (1964).
[79] M. Dowell and P. Jarratt, A modified regula falsi method for computing the root of an equation. *BIT* **11**, 168–174 (1971).
[80] H. Dressel, see J. W. Schmidt.
[81] E. Durand, "Solutions Numeriques des Equations Algébriques," Vol. 1. Masson, Paris, 1960.
[82] R. Dussel, Einschließung des Minimalpunktes einer streng konvexen Funktion auf einem n-dimensionalen Quader. Ph.D. Thesis, Univ. Karlsruhe (1972).
[83] R. Dussel and B. Schmitt, Die Berechnung von Schranken für den Wertebereich eines Polynoms in einem Intervall. *Computing* **6**, 35–60 (1970).
[84] P. S. Dwyer, Matrix inversion with the square root method. *Technometrics* **6**, 197–213 (1964).
[85] H.-P. Dyck, Algorithmen zur numerischen Behandlung von Funktionen mit Fehlererfassung. Diplomarbeit, Inst. Angew. Math., Univ. Karlsruhe (1969).
[86] K. Ecker and H. Ratschek, Intervallarithmetik mit Wahrscheinlichkeitsstruktur. *Angew. Informat.* **14**, 313–320 (1972).
[87] L. W. Ehrlich, A modified Newton method for polynomials. *Comm. ACM* **10**, 107–108 (1967).
[88] H. Ehrmann, Konstruktion und Durchführung von Iterationsverfahren höherer Ordnung. *Arch. Rational Mech. Anal.* **4**, 65–88 (1959).
[89] H. Fischer, Eine Methode zur Einschließung der Nullstellen von Polynomen mit komplexen Koeffizienten. *Computing* **4**, 202–206 (1969).
[90] H. Fischer, Intervall-Arithmetiken für komplexe Zahlen. *Z. Angew. Math. Mech.* **53**, T 190–T 191 (1973).
[91] H. Fischer, Hypernormbälle als abstrakte Schrankenzahlen. *Computing* **12**, 67–73 (1974).
[92] P. Forster, Bemerkungen zum Iterationsverfahren von Schulz zur Bestimmung der Inversen einer Matrix. *Numer. Math.* **12**, 211–214 (1968).
[93] R. Franzen, Die intervallanalytische Behandlung parameterabhängiger Gleichungssysteme. GMD-Bericht No. 47, Bonn (1971).
[94] G. S. Ganshin, Extension of the convergence region of Newton's method. *Ž. Vyčisl. Mat. i Mat. Fiz.* **11**, 1294–1296 (1971).

[95] I. Gargantini and P. Henrici, Circular arithmetic and the determination of polynomial zeros. *Numer. Math.* **18**, 305–320 (1972).

[96] A. Gibb, Algorithm 61. Procedures for range arithmetic. *Comm. ACM* **4**, 319–320 (1961).

[97] A. J. Goldstein and P. L. Richman, A midpoint phenomenon. *J. ACM* **20**, 301–304 (1973).

[98] D. I. Good and R. L. London, Computer interval arithmetic: Definition and proof of correct implementation. *J. ACM* **17**, 603–612 (1970).

[99] S. Gorn, The automatic analysis and control of computing errors. *SIAM J. Appl. Math.* **2**, 69–81 (1954).

[100] J. A. Grant and G. D. Hitchins, The solution of polynomial equations in interval arithmetic. *Comput. J.* **16**, 69–72 (1973).

[101] R. T. Gregory and D. L. Karney, "A Collection of Matrices for Testing Computational Algorithms." Wiley (Interscience), New York, 1969.

[102] W. Gröbner, Matrizenrechnung. Bibliographisches Institut, Mannheim (1966).

[103] N. B. Haaser and J. A. Sullivan, "Real Analysis," Chapter 5, Section 2. Van Nostrand Reinhold, New York, 1971.

[104] K. P. Hadeler, see D. Braess.

[105] W. Hahn, Intervallarithmetik in normierten Räumen und Algebren. Ph.D. Thesis, Graz (1971).

[106] K. Hainer, see F. Stummel.

[107] H. Haller, Externe Zahlendarstellung bei Intervallrechnung. Tagung über Intervallrechnung in Oberwolfach, Tagungsbericht 12/68 (1968).

[108] E. R. Hansen, Interval arithmetic in matrix computations, Part I. *SIAM J. Numer. Anal.* **2**, 308–320 (1965).

[109] E. R. Hansen, On solving systems of equations using interval arithmetic. *Math. Comp.* **22**, 374–384 (1968).

[110] E. R. Hansen, On the solution of linear algebraic equations with interval coefficients. *Linear Algebra Appl.* **2**, 153–165 (1969).

[111] E. R. Hansen, "Topics in Interval Analysis." Oxford Univ. Press (Clarendon), London and New York, 1969.

[112] E. R. Hansen, On linear algebraic equations with interval coefficients. *In* "Topics in Interval Analysis." Oxford Univ. Press (Clarendon), London and New York, 1969.

[113] E. R. Hansen, On solving two-point boundary-value problems using interval arithmetic. *In* "Topics in Interval Analysis." Oxford Univ. Press (Clarendon), London and New York, 1969.

[114] E. R. Hansen, The centered form. In "Topics in Interval Analysis." Oxford Univ. Press (Clarendon), London and New York, 1969.

[115] E. R. Hansen and R. Smith, Interval arithmetic in matrix computations, Part II. *SIAM J. Numer. Anal.* **4**, 1–9 (1967).

[116] R. J. Hanson, Automatic error bounds for real roots of polynomials having interval coefficients. *Comput. J.* **13**, 284–288 (1970).

[117] F. Hausdorff, "Mengenlehre." de Gruyter, Leipzig, 1927.

[118] M. Heidt, Zur numerischen Lösung von gewöhnlichen Differentialgleichungen zweiter Ordnung. Ph.D. Thesis, Univ. Karlsruhe (1971).

[119] P. Henrici, Methods of search for solving polynomial equations. *J. ACM* **17**, 273–283 (1970).

[120] P. Henrici, Circular arithmetic and the determination of polynomial zeros. *Conf. Appl. Numer. Anal.* Lecture Notes in Mathematics 228, pp. 86–92. Springer-Verlag, Berlin and New York, 1971.

[121] P. Henrici, see I. Gargantini.

[122] J. Herzberger, Metrische Eigenschaften von Mengensystemen und einige Anwendungen. Ph.D. Thesis, Univ. Karlsruhe (1969).

[123] J. Herzberger, Definition und Eigenschaften allgemeiner Intervallräume. *Z. Angew. Math. Mech.* **50**, T 50–T 51 (1970).

[124] J. Herzberger, Intervallmäßige Auswertung von Standardfunktionen in ALGOL-60. *Computing* **5**, 377–384 (1970).

[125] J. Herzberger, Zur Konvergenz intervallmäßiger Iterationsverfahren. *Z. Angew. Math. Mech.* **51**, T 56 (1971).

[126] J. Herzberger, Über optimale Verfahren zur numerischen Bestimmung reeller Nullstellen mit Fehlerschranken. Habilitationsschrift, Univ. Karlsruhe (1972).

[127] J. Herzberger, Über ein Verfahren zur Bestimmung reeller Nullstellen mit Anwendung auf Parallelrechnung. *Elektron. Rech.* **14**, 250–254 (1972).

[128] J. Herzberger, Über die Nullstellenbestimmung bei näherungsweise berechneten Funktionen. *Computing* **10**, 23–31 (1972).

[129] J. Herzberger, Bemerkungen zu einem Verfahren von R. E. Moore. *Z. Angew. Math. Mech.* **53**, 356–358 (1973).

[130] J. Herzberger, see G. Alefeld.

[131] G. D. Hitchins, The numerical solution of polynomial equations with special reference to the determination of initial approximations to the roots. Ph.D. Thesis, Univ. of Leeds (1971).

[132] G. D. Hitchins, An interval arithmetic package and some applications. Technical Report 10, Centre for Computer Studies, Univ. of Leeds (1972).

[133] G. D. Hitchins, see J. A. Grant.

[134] A. S. Householder, "The Numerical Treatment of a Single Nonlinear Equation." McGraw-Hill, New York, 1970.

[135] S. Hunger, Intervallanalytische Defektabschätzung bei Anfangswertaufgaben für Systeme gewöhnlicher Differentialgleichungen. GMD-Bericht No. 41, Bonn (1971).

[136] L. W. Jackson, Automatic error analysis for the solution of ordinary differential equations. Technical Report # 28, Department of Computer Science, Univ. of Toronto (1971).

[137] K.-U. Jahn, Aufbau einer 3-wertigen linearen Algebra und affinen Geometrie auf der Grundlage der Intervallarithmetik. Ph.D. Thesis, Univ. Leipzig (1971).

[138] K.-U. Jahn, Eine Theorie der Gleichungssysteme mit Intervallkoeffizienten. *Z. Angew. Math. Mech.* **54**, 405–412 (1974).

[139] K.-U. Jahn, Eine auf den Intervallzahlen fußende 3-wertige lineare Algebra. *Math. Nachr.* **65**, 105–116 (1975).

[140] P. Jarratt, see M. Dowell.

[141] W. M. Kahan: A computable error-bound for linear systems of ordinary differential equations. Abstract, *SIAM Rev.* **8**, 568–569 (1966).

[142] W. M. Kahan, An ellipsoidal error-bound for linear systems of ordinary differential equations. Unpublished manuscript (1967).

[143] W. M. Kahan, A more complete interval arithmetic. Report, Univ. of Toronto (1968).

[144] K. Kansy, Ableitungsverträgliche Verallgemeinerung der Intervallpolynome. GMD-Bericht No. 70, Bonn (1973).

[145] D. L. Karney, see R. T. Gregory.

[146] V. Kartheus, Zur intervallanalytischen Behandlung linearer Gleichungssysteme. GMD-Mitteilungen No. 16, Bonn (1972).

[147] P. Katzan, Faktorzerlegung von Polynomen mit Fehlererfassung. GMD-Bericht No. 18, Bonn (1969).

[148] E. Kaucher, Über metrische und algebraische Eigenschaften einiger beim numerischen Rechnen auftretender Räume. Ph.D. Thesis, Univ. Karlsruhe (1973).

[149] I. O. Kerner, Ein Gesamtschrittverfahren zur Berechnung der Nullstellen von Polynomen. *Numer. Math.* **8**, 290–294 (1966).

[150] R. Klatte, Anwendung einer Verallgemeinerung der Modifikation des Newtonverfahrens. Diplomarbeit, Inst. f. Angew. Math., Univ. Karlsruhe (1969).
[151] D. Klaua, Partiell definierte Mengen. *Mber. Dt. Akad. Wiss.* **10** (1968).
[152] D. Klaua, Partielle Mengen mit mehrwertigen Grundbeziehungen. *Mber. Dt. Akad. Wiss.* **11** (1969).
[153] D. Klaua, Partielle Mengen und Zahlen. *Mber. Dt. Akad. Wiss.* **11**, 585–599 (1969).
[154] H.-O. Klein, Eine intervallanalytische Methode zur Lösung von Integralgleichungen mit mehrparametrigen Funktionen. GMD-Bericht No. 69, Bonn (1973).
[155] D. E. Knuth, "The Art of Computer Programming," Vol. 2. Addison-Wesley, Reading, Massachusetts, 1969.
[156] M. Kracht and G. Schröder, Eine Einführung in die Theorie der quasilinearen Räume mit Anwendung auf die in der Intervallrechnung auftretenden Räume. *Math.-Phys. Semesterber.* **20**, 226–242 (1973).
[157] M. Kracht and G. Schröder, Zur Intervallrechnung in linearen Räumen. *Computing* **11**, 73–79 (1973).
[158] R. Krawczyk, Iterative Verbesserung von Schranken für Eigenwerte und Eigenvektoren reeller Matrizen. *Z. Angew. Math. Mech.* **48**, T 80–T 83 (1968).
[159] R. Krawczyk, Fehlerabschätzung reeller Eigenwerte und Eigenvektoren von Matrizen. *Computing* **4**, 281–293 (1969).
[160] R. Krawczyk, Newton-Algorithmen zur Bestimmung von Nullstellen mit Fehlerschranken. *Computing* **4**, 187–201 (1969).
[161] R. Krawczyk, Verbesserung von Schranken für Eigenwerte und Eigenvektoren von Matrizen. *Computing* **5**, 200–206 (1970).
[162] R. Krawczyk, Einschließung von Nullstellen mit Hilfe einer Intervallarithmetik. *Computing* **5**, 356–370 (1970).
[163] R. Krawczyk, Gleichungen in halbgeordneten Räumen. *Abh. Math. Sem., Univ. Hamburg* **36**, 150–165 (1971).
[164] R. Krawczyk, Zur Einschließung inverser linearer Operatoren. *Methoden Verfahren Math. Phys.* **7**, 1–19 (1972).
[165] R. Krawczyk, see N. Apostolatos.
[166] T. Kreifelts, Über vollautomatische Erfassung und Abschätzung von Rundungsfehlern in arithmetischen Prozessen. GMD-Bericht No. 62, Bonn (1972).
[167] O. Kress, Über Fehlerabschätzungen für Eigenwerte und Eigenvektoren reellsymmetrischer Matrizen. Ph.D. Thesis, Univ. Karlsruhe (1973).
[168] N. Krier, Komplexe Kreisarithmetik. Ph.D. Thesis, Univ. Karlsruhe (1973).
[169] F. Krückeberg, Numerische Intervallrechnung und deren Anwendung. Bonn (1966).
[170] F. Krückeberg, Inversion von Matrizen mit Fehlererfassung. *Z. Angew. Math. Mech.* **46**, T 69–T 71 (1966).
[171] F. Krückeberg, Ordinary differential equations. *In* "Topics in Interval Analysis." Oxford Univ. Press (Clarendon), London and New York, 1969.
[172] F. Krückeberg, Bemerkungen zur Intervallanalysis. *Apl. Mat.* **13**, 152–153 (1968).
[173] F. Krückeberg, Partial differential equations. *In* "Topics in Interval Analysis," Oxford Univ. Press (Clarendon), London and New York, 1969.
[174] U. Kulisch, Grundlagen einer Maschinenintervallarithmetik. *Z. Angew. Math. Mech.* **47**, T 81 (1967).
[175] U. Kulisch, Grundzüge der Intervallrechnung. *In* "Überblicke Mathematik," Vol. 2. Bibliograph. Inst. Mannheim, 1969.
[176] U. Kulisch, Rounding invariant structures. MRC Technical Summary Rep. 1103, Univ. of Wisconsin, Madison, Wisconsin (1970).
[177] U. Kulisch, Interval arithmetic over completely ordered ringoids. MRC Technical Summary Rep. 1105, Univ. of Wisconsin, Madison, Wisconsin (1970).

[178] U. Kulisch, An axiomatic approach to rounded computations. *Numer. Math.* **18**, 1–17 (1971).
[179] U. Kulisch, Axiomatik des numerischen Rechnens. *Z. Angew. Math. Mech.* **52**, T 211 (1972).
[180] U. Kulisch, On the concept of a screen. *Z. Angew. Math. Mech.* **53**, 115–119 (1973).
[181] U. Kulisch, Implementation and formalization of floating-point arithmetic. *Caratheodory Symp., Athen* (1973).
[182] U. Kulisch, see N. Apostolatos.
[183] I. B. Kupermann, Approximate linear algebraic equations and rounding error estimation. Ph.D. Thesis, Dept. of Applied Math., Univ. of Witwatersrand, Johannesburg, South Africa (1967).
[184] I. B. Kupermann, "Approximate Linear Algebraic Equations," Van Nostrand Reinhold, New York, 1971.
[185] I. Kupka, Simulation reeller Arithmetik und reeller Funktionen in endlichen Mengen. *Numer. Math.* **17**, 143–152 (1971).
[186] P. Lancaster, see J. Rokne.
[187] D. Leder, see J. W. Schmidt.
[188] H. Leonhardt, see J. W. Schmidt.
[189] R. L. London, see D. J. Good.
[190] B. Lortz, Eine Langzahlarithmetik mit optimaler einseitiger Rundung. Ph.D. Thesis, Univ. Karlsruhe (1971).
[191] B. Lortz, see N. Apostolatos.
[192] F. R. Loscalzo, see R. H. Dargel.
[193] B. Machost, Numerische Behandlung des Simplexverfahrens mit intervallanalytischen Methoden. GMD-Bericht No. 30, Bonn (1970).
[194] K. Madsen, On the solution of nonlinear equations in interval arithmetic. *BIT* **13**, 428–433 (1973).
[195] O. Mayer, Über die in der Intervallrechnung auftretenden Räume und einige Anwendungen. Ph.D. Thesis, Univ. Karlsruhe (1968).
[196] O. Mayer, Über Eigenschaften von Intervallprozessen. *Z. Angew. Math. Mech.* **48**, T 86–T 87 (1968).
[197] O. Mayer, Algebraische und metrische Strukturen in der Intervallrechnung und einige Anwendungen. *Computing* **5**, 144–162 (1970).
[198] O. Mayer, Über die Bestimmung von Einschließungsmengen für die Lösungen linearer Gleichungssysteme mit fehlerbehafteten Koeffizienten. *Elektron. Datenverarbeit.* **12**, 164–167 (1970).
[199] O. Mayer, Über die intervallmäßige Durchführung einiger Iterationsverfahren. *Z. Angew. Math. Mech.* **50**, T 65–T 66 (1970).
[200] O. Mayer, Über eine Klasse komplexer Intervallgleichungssysteme mit iterationsfähiger Gestalt. *Computing* **6**, 104–106 (1970).
[201] O. Mayer, Über intervallmäßige Iterationsverfahren bei linearen Gleichungssystemen und allgemeineren Intervallgleichungssystemen. *Z. Angew. Math. Mech.* **51**, 117–124 (1971).
[202] O. Mayer, see G. Alefeld.
[203] D. Mehler, Ein Programmiersystem mit Intervallarithmetik für die IBM 7090. Bericht des Instituts für Instrumentelle Mathematik 47, Bonn (1968).
[204] J. Meinguet, On the estimation of significance. *In* "Topics in Interval Analysis." Oxford Univ. Press (Clarendon), London and New York, 1969.
[205] G. Meyer, Zur numerischen Behandlung von linearen Gleichungssystemen mit Intervallkoeffizienten. Diplomarbeit, Inst. Angew. Math., Univ. Karlsruhe (1971).
[206] W. Miller, On an interval-arithmetic matrix method. *BIT* **12**, 213–219 (1972).
[207] W. Miller, Quadratic convergence in interval arithmetic, Part II. *BIT* **12**, 291–298 (1972).
[208] W. Miller, More on quadratic convergence in interval arithmetic. *BIT* **13**, 76–83 (1973).

[209] W. Miller, Automatic verification of numerical stability. *Comm. ACM* (to appear).

[210] W. Miller, see W. Chuba.

[211] W. Mönch, Monotone Einschließung von positiven Inversen. *Z. Angew. Math. Mech.* **53**, 207–208 (1973).

[212] R. E. Moore, Interval arithmetic and automatic error analysis in digital computing. Tech. Rep. 25, Stanford Univ. Applied Mathematics and Statistics Laboratories (1962).

[213] R. E. Moore, The automatic analysis and control of error in digital computation based on the use of interval numbers. *In* "Error in Digital Computation." Wiley, New York, 1965

[214] R. E. Moore, Automatic local coordinate transformations to reduce the growth of error bounds in interval computation of solutions of ordinary differential equations. In "Error in Digital Computation." Wiley, New York, 1965.

[215] R. E. Moore, "Interval Analysis." Prentice-Hall, Englewood Cliffs, New Jersey, 1966.

[216] R. E. Moore, Practical aspects of interval computation. *Apl. Mat.* **13**, 52–92 (1968).

[217] R. E. Moore, Functional analysis for computers. *ISNM* **12**, 113–126 (1968).

[218] R. E. Moore, Introduction to algebraic problems. *In* "Topics in Interval Analysis." Oxford Univ. Press (Clarendon), London and New York, 1969.

[219] R. E. Moore, Introduction to continuous problems. *In* "Topics in Interval Analysis." Oxford Univ. Press (Clarendon), London and New York, 1969.

[220] S. B. Nadler, Multi-valued contraction mappings. *Pacific J. Math.* **30**, 475–488 (1969).

[221] W. Neuland, Die numerische Behandlung von Integralgleichungen mit intervallanalytischen Methoden. GMD-Bericht No. 21, Bonn (1969).

[222] K. Nickel, Über die Notwendigkeit einer Fehlerschrankenarithmetik für Rechenautomaten. *Numer. Math.* **9**, 69–79 (1966).

[223] K. Nickel, Die vollautomatische Berechnung einer einfachen Nullstelle von $F(t) = 0$ einschließlich einer Fehlerabschätzung. *Computing* **2**, 232–245 (1967).

[224] K. Nickel, Quadraturverfahren mit Fehlerschranken. *Computing* **3**, 47–64 (1968).

[225] K. Nickel, TRIPLEX-ALGOL and applications. *In* "Topics in Interval Analysis." Oxford Univ. Press (Clarendon), London and New York, 1969.

[226] K. Nickel, Zeros of polynomials and other topics. *In* "Topics in Interval Analysis." Oxford Univ. Press (Clarendon), London and New York, 1969.

[227] K. Nickel, Das Kahan-Babuska'sche Summierungsverfahren in TRIPLEX-ALGOL 60. *Z. Angew. Math. Mech.* **50**, 369–373 (1970).

[228] K. Nickel, Fehlerschranken zu Näherungswerten von Polynomwurzeln. *Computing* **6**, 9–27 (1970).

[229] K. Nickel, Numerische Mathematik I. Lecture notes, Univ. Karlsruhe (1971).

[230] K. Nickel, On the Newton method in interval analysis. MRC Technical Summary Rep. 1136, Univ. of Wisconsin, Madison, Wisconsin (1971).

[231] K. Nickel, The prae-Euler summation method. MRC Technical Summary Rep. 1072, Univ. of Wisconsin, Madison, Wisconsin (1971).

[232] K. Nickel, The contraction mapping fixpoint theorem in interval analysis. MRC Technical Summary Rep. 1334, Univ. of Wisconsin, Madison, Wisconsin (1973).

[233] K. Nickel, see N. Apostolatos.

[234] K. Nickel, see S. M. Robinson.

[235] K. Nickel and K. Ritter, Termination criterion and numerical convergence. *SIAM J. Numer. Anal.* **9**, 277–283 (1972).

[236] E. Nuding, Intervallarithmetik. Lecture notes, Rechenzentrum der Univ. Heidelberg (1972).

[237] W. Oettli, On the solution set of a linear system with inaccurate coefficients. *SIAM J. Numer. Anal.* **2**, 115–118 (1965).

[238] W. Oettli and W. Prager, Compatibility of approximate solution of linear equations with given error bounds for coefficients and right-hand sides. *Numer. Math.* **6**, 405–409 (1964).

[239] W. Oettli, W. Prager, and J. H. Wilkinson, Admissible solutions of linear system with not sharply defined coefficients. *SIAM J. Numer. Anal.* **2**, 291–299 (1965).

[240] J. M. Ortega, "Numerical Analysis: A Second Course." Academic Press, New York, 1972.

[241] J. M. Ortega and W. C. Rheinboldt, Monotone iterations for nonlinear equations with application to Gauss–Seidel methods. *SIAM J. Numer. Anal.* **4**, 171–190 (1967).

[242] J. M. Ortega and W. C. Rheinboldt, "Iterative Solution of Nonlinear Equations in Several Variables." Academic Press, New York, 1970.

[243] J. M. Ortega and W. C. Rheinboldt, Numerical solution of nonlinear problems. *Stud. Numer. Anal.* **2**, 122–143 (1970).

[244] H.-J. Ortolf, Eine Verallgemeinerung der Intervallarithmetik. GMD-Bericht No. 11, Bonn (1969).

[245] A. Ostrowski, "Solution of Equations and Systems of Equations." Academic Press, New York, 1960.

[246] G. Peters and J. H. Wilkinson, Practical problems arising in the solution of polynomial equations. *J. Inst. Math. Appl.* **8**, 16–35 (1971).

[247] L. Platzöder, Über optimale Iterationsverfahren bei komplexen Gleichungssystemen. Diplomarbeit, Inst. Angew. Math., Univ. Karlsruhe (1974).

[248] T. Popoviciu, Remarques sur le maximum d'un déterminant dont tous les éléments sont non-négatifs. *Mathematica* **13**, 242–256 (1937).

[249] W. Prager, see W. Oettli.

[250] L. R. Rall, "Computational Solution of Nonlinear Operator Equations." Wiley, New York, 1969.

[251] H. Ratschek, Über einige intervallarithmetische Grundbegriffe. *Computing* **4**, 43–55 (1969).

[252] H. Ratschek, Al IV-lea congres al matematicienilor de expresie latina. Rezumatele lucranilor. *Bucaresti-Brasov* 32–33 (1969).

[253] H. Ratschek, Die binären Systeme der Intervallarithmetik. *Computing* **6**, 295–308 (1970).

[254] H. Ratschek, Die Subdistributivität in der Intervallarithmetik. *Z. Angew. Math. Mech.* **51**, 189–192 (1971).

[255] H. Ratschek, Teilbarkeitsgesetze der Intervallarithmetik. *Z. Angew. Math. Mech.* **51**, T 70–T 71 (1971).

[256] H. Ratschek, Teilbarkeitskriterien der Intervallarithmetik. *J. Reine Angew. Math.* **252**, 128–138 (1972).

[257] H. Ratschek, Ergebnisse einer Untersuchung über die Struktur von Intervallpolynomen. GMD-Bericht No. 52, Bonn (1972).

[258] H. Ratschek, Gleichheit von Produkt und Formalprodukt bei Intervallpolynomen. *Computing* **10**, 245–254 (1972).

[259] H. Ratschek, Intervallarithmetik – mit Zirkel und Lineal. *Elem. Math.* **28**, 93–96 (1973).

[260] H. Ratschek, Mittelwertsatz für Intervallfunktionen. Kurzvortrag auf der GAMM-Jahrestagung, München (1973).

[261] H. Ratschek, see K. Ecker.

[262] H. Ratschek and G. Schröder, Über die Ableitung von intervallwertigen Funktionen. *Computing* **7**, 172–187 (1971).

[263] H. Ratschek and G. Schröder, Über die Differentiation von Intervallfunktionen. *Z. Angew. Math. Mech.* **52**, T 219–T 220 (1972).

[264] P. Rechenberg, "Grundzüge digitaler Rechenautomaten." Oldenbourg Verlag, Munich 1964.

[265] W. C. Rheinboldt, see J. M. Ortega.

[266] P. L. Richman, Automatic error analysis for determining precision. *Comm. ACM* **15**, 813–817 (1972).

[267] P. L. Richman, see A. J. Goldstein.

[268] F. N. Ris, Interval analysis and applications to linear algebra. Ph.D. Thesis, Oxford Univ. (1972).

[269] K. Ritter, see K. Nickel.

[270] S. M. Robinson and K. Nickel, Computation of the Perron root and vector of a non-negative matrix. MRC Technical Summary Rep. 1100, Univ. of Wisconsin, Madison, Wisconsin (1970).

[271] J. Rokne, Fehlererfassung bei Eigenwertproblemen von Matrizen. *Computing* **7**, 145–152 (1971).

[272] J. Rokne, Explicit calculation of the Lagrangian interval interpolating polynomial. *Computing* **9**, 149–157 (1972).

[273] J. Rokne, Automatic errorbounds for simple zeros of analytic functions. *Comm. ACM* **16**, 101–104 (1973).

[274] J. Rokne and P. Lancaster, Automatic error bounds for the approximate solutions of equations. *Computing* **4**, 294–303 (1969).

[275] J. Rokne and P. Lancaster, Complex interval arithmetic. *Comm. ACM* **14**, 111–112 (1971).

[276] J. Rokne and P. Lancaster, Complex interval arithmetic. Algorithm 86. *Comput. J.* **18**, 83–86 (1975).

[277] B. Rothmaier, Die Berechnung der elementaren Funktionen mit beliebiger Genauigkeit. Ph.D. Thesis, Univ. Karlsruhe (1971).

[278] B. Rothmaier, see M. Brockhaus.

[279] H. Rutishauser, Versuch einer Axiomatik des numerischen Rechnens. Kurzvortrag auf der GAMM-Jahrestagung 1969 in Aachen.

[280] H. Rutishauser, see H. R. Schwarz.

[281] B. Schicht, Verfahren höherer Ordnung zur iterativen Einschließung der Inversen einer komplexen Matrix. Diplomarbeit, Inst. Angew. Math., Univ. Karlsruhe, 1974.

[282] J. W. Schmidt, Monotone Einschließung mit der Regula falsi bei konvexen Funktionen. *Z. Angew. Math. Mech.* **50**, 640–643 (1970).

[283] J. W. Schmidt, Eingrenzung von Lösungen nichtlinearer Gleichungen durch Verfahren mit höherer Konvergenzgeschwindigkeit. *Computing* **8**, 208–215 (1971).

[284] J. W. Schmidt, Einschließung von Nullstellen bei Operatoren mit monoton zerlegbarer Steigung durch überlinear konvergente Iterationsverfahren. *Ann. Acad. Sci. Fenn.* 502 (1972).

[285] J. W. Schmidt and H. Dressel, Fehlerabschätzungen bei Polynomgleichungen mit dem Fixpunktsatz von Brouwer. *Numer. Math.* **10**, 42–50 (1967).

[286] J. W. Schmidt and D. Leder, Ableitungsfreie Verfahren ohne Auflösung linearer Gleichungen. *Computing* **5**, 71–81 (1970).

[287] J. W. Schmidt and M. Leonhardt, Eingrenzung von Lösungen mit Hilfe der Regula falsi. *Computing* **6**, 318–329 (1970).

[288] G. Schmitgen, Tschebyscheff-Approximation von Intervallfunktionen durch verallgemeinerte Intervallpolynome. GMD-Bericht No. 26, Bonn (1970).

[289] B. Schmitt, see R. Dussel.

[290] G. Schröder, Differentiationsverfahren für Intervallfunktionen. Ph.D. Thesis, Univ. Düsseldorf (1971).

[291] G. Schröder, Differentiation of interval functions. *Proc. Amer. Math. Soc.* **36**, 485–490 (1972).

[292] G. Schröder, Charakterisierung des quasilinearen Raumes $I(\mathbb{R})$ und Klassifizierung der quasilinearen Räume der Dimension 1 und 2. *Computing* **10**, 111–120 (1972).

[293] G. Schröder, see M. Kracht.

[294] G. Schröder, see H. Ratschek.

[295] J. Schröder, Das Iterationsverfahren bei allgemeinem Abstandsbegriff. *Math. Z.* **66**, 111–116 (1956).

[296] J. Schröder, Fehlerabschätzungen bei linearen Gleichungssystemen mit dem Brouwerschen Fixpunktsatz. *Arch. Rational Mech. Anal.* **3**, 28–44 (1959).
[297] P. Schroth, see M. Brockhaus.
[298] G. Schulz, Iterative Berechnung der reziproken Matrix. *Z. Angew. Math. Mech.* **13**, 57–59 (1933).
[299] P. Schwanenberg, Intervallanalytische Methoden zur Lösung von Randwertaufgaben bei gewöhnlichen Differentialgleichungen. GMD-Bericht No. 35, Bonn (1970).
[300] H. R. Schwarz, H. Rutishauser, and E. Stiefel, "Matrizen-Numerik." Teubner Verlag, Stuttgart, 1968.
[301] W. D. Schwill, Fehlerabschätzung für die gewöhnliche Differentialgleichung 1. Ordnung unter Berücksichtigung der Rundungsfehler. Ph.D. Thesis, Univ. Karlsruhe (1972).
[302] L. B. Smith, Interval arithmetic determinant evaluation and its use in testing for a Chebyshev system. *Comm. ACM* **12**, 89–93 (1969).
[303] R. Smith, see E. R. Hansen.
[304] O. Spaniol, Die Distributivität in der Intervallarithmetik. *Computing* **5**, 6–16 (1970).
[305] P. T. Speck, Kreisarithmetik, Realisation und Anwendung auf einer Rechenanlage. Diplomarbeit, Inst. Angew. Math., Univ. Karlsruhe (1974).
[306] H. Stetter, Numerische Mathematik (1. Teil), Lecture notes, 3. Auflage, Technische Hochschule Wien (1969).
[307] N. F. Stewart, Guaranteed local error bound for the Adams method. Unpublished manuscript (1970).
[308] N. F. Stewart, A heuristic to reduce the wrapping effect in the numerical solution of $x' = f(t, x)$. *BIT* **11**, 328–337 (1971).
[309] N. F. Stewart, Centrally symmetric convex polyhedra to bound the error in $x' = f(t, x)$. Publication # 11, Dept. Informatique, Univ. de Montreal (1971).
[310] N. F. Stewart, Interval arithmetic for guaranteed bounds in linear programming. *J. Optimization Theory Appl.* **12**, 1–5 (1973).
[311] N. F. Stewart, see D. P. Davey.
[312] E. Stiefel, see H. R. Schwarz.
[313] J. Stoer, Einführung in die Numerische Mathematik I. Heidelberger Taschenbücher 105. Springer-Verlag, Berlin and New York, 1972.
[314] F. Stummel and K. Hainer, "Praktische Mathematik." Teubner Verlag, Stuttgart, 1971.
[315] J. A. Sullivan, see N. B. Haaser.
[316] T. Sunaga, Theory of an interval algebra and its application to numerical analysis. RAAG Memoirs 2 (1958).
[317] T. D. Talbot, Guaranteed error bounds for computed solutions of nonlinear two-point boundary value problems. MRC Technical Summary Rep. 875, Univ. of Wisconsin, Madison, Wisconsin (1968).
[318] W. J. Thron, "Topological Structures." Holt, New York, 1966.
[319] R. Tost, Lösung der 1. Randwertaufgabe der Laplace-Gleichung im Rechteck mit intervallanalytischen Methoden. GMD-Bericht No. 28, Bonn (1970).
[320] J. F. Traub, "Iterative Methods for the Solution of Equations." Prentice-Hall, Englewood Cliffs, New Jersey, 1964.
[321] Ch. Ullrich, Algorithmen zur intervallmäßigen Behandlung von linearen Gleichungssystemen. Diplomarbeit, Inst. Angew. Math., Univ. Karlsruhe (1969).
[322] Ch. Ullrich, Rundungsinvariante Strukturen mit äußeren Verknüpfungen. Ph.D. Thesis, Univ. Karlsruhe (1972).
[323] S. Ulm, Über Iterationsverfahren mit sukzessiver Approximation des inversen Operators (Russian). *Izv. Akad. Nauk Est. SSR* **16**, 403–411 (1967).
[324] J. S. Vandergraft, Newton's method for convex operators in partially ordered spaces. *SIAM J. Numer. Anal.* **4**, 406–432 (1967).

[325] R. S. Varga, "Matrix Iterative Analysis." Prentice-Hall, Englewood Cliffs, New Jersey, 1962.

[326] A. Walzel, Fehlerabschätzungen bei Anfangswertaufgaben von gewöhnlichen Differentialgleichungen. Ph.D. Thesis, Univ. Köln (1969).

[327] U. Wauschkuhn, Intervallanalytische Methoden zum Existenznachweis und zur Konstruktion der Lösung von Periodizitätsproblemen bei gewöhnlichen Differentialgleichungen. Ph.D. Thesis, Univ. Bonn (1973).

[328] J. H. Wilkinson, Error analysis of direct methods of matrix inversion. *J. ACM* **8**, 281–330 (1961).

[329] J. H. Wilkinson, "Rounding Errors in Algebraic Processes." H. M. Stationery Office, London, 1968.

[330] J. H. Wilkinson, see W. Oettli.

[331] J. H. Wilkinson, see G. Peters.

[332] H.-W. Wippermann, Realisierung einer Intervallarithmetik in einem ALGOL-60 System. *Elektron. Rech.* **9**, 224–233 (1967).

[333] H.-W. Wippermann, Implementierung eines ALGOL-Systems mit Schrankenzahlen. *Elektron. Datenverarbeit.* **10**, 189–194 (1968).

[334] H.-W. Wippermann, Definition von Schrankenzahlen in TRIPLEX-ALGOL. *Computing* **3**, 99–109 (1968).

[335] H.-W. Wippermann, see N. Apostolatos.

[336] P. Wißkirchen, Ein Steuerungsprinzip der Intervallrechnung und dessen Anwendung auf den Gaußschen Algorithmus. GMD-Bericht No. 20, Bonn (1969).

[337] T. H. Witt, see R. H. Dargel.

[338] J. M. Yohe, Interval bounds for square roots and cube roots. *Computing* **11**, 51–57 (1973).

[339] R. C. Young, The algebra of many-valued quantities. *Math. Ann.* **104**, 260–290 (1931).

[340] G. Zielke, "ALGOL-Katalog Matrizenrechnung." Oldenbourg Verlag, Munich and Vienna, 1972.

[341] R. Zurmühl, "Praktische Mathematik." Springer-Verlag, Berlin and New York, 1962.

SUPPLEMENT TO THE ENGLISH-LANGUAGE EDITION

[342] O. Aberth, A precise numerical analysis program. *Comm. ACM* **17**, 509–513 (1974).

[343] E. Adams and H. Spreuer, On the construction of an interval containing the set of solutions of non-inverse isotone linear problems with intervals admitted for both data and coefficients. Center for Applied Mathematics, The Univ. of Georgia, Athens (1978).

[344] E. Adams and W. F. Ames, On contracting interval iteration for nonlinear problems in $\mathbb{R}^n$: part II-applications. *Nonlinear Anal., Theory, Methods and Appl.* **5** (5), 525–542 (1981).

[345] A. C. Aitken, Studies in practical mathematics V. On the iterative solution of linear equations. *Proc. Roy. Soc. Edinburgh Sec. A* **63**, 52–60 (1960).

[346] J. Albrecht, Monotone Iterationsfolgen und ihre Verwendung zur Lösung linearer Gleichungssysteme. *Numer. Math.* **3**, 345–358 (1961).

[347] G. Alefeld, An exclusion theorem for the solutions of operator equations. *Beiträge Numer. Math.* **6**, 7–10 (1977).

[348] G. Alefeld, Das symmetrische Einzelschrittverfahren bei linearen Gleichungen mit Intervallen als Koeffizienten. *Computing* **18**, 329–340 (1977).

[349] G. Alefeld, Intervallanalytische Methoden bei nichtlinearen Gleichungen. *In* "Jahrbuch Überblicke Mathematik 1979," pp. 63–78. Bibliograph. Inst. Mannheim (1979).

[350] G. Alefeld and J. Rokne, On the evaluation of rational functions in interval arithmetic *SIAM J. Numer. Anal.* **18**, 862–870 (1981).

[351] G. Alefeld and J. G. Rokne, On the iterative improvement of approximate triangular factorizations. *Beiträge Numer. Math.* (to be published).

[352] W. F. Ames, see E. Adams.

[353] H. J. Anders and E. Zemlin, Zur Erzeugung einer Maschinenintervallarithmetik auf den Rechenanlagen C 8205 und ICL 1905E. *Wiss. Z. Päd. Hochschule "Karl Liebknecht" Potsdam* **19**, 159–163 (1975).

[354] N. Apostolatos, Eine Methode zur Nullstellenbestimmung von Funktionen. *Computing* **15**, 1–10 (1975).

[355] N. Apostolatos, Eine allgemeine Betrachtung von numerischen Algorithmen. Institut für Angewandte Mathematik. TU München. Institutsbericht 119 (July 1975).

[356] N. Apostolatos, Nullstellenbestimmung mit Fehlererfassung. IV. IKM, VEB Verlag Bd. 2 (1967).

[357] N. Apostolatos and G. Karambatzos, Mengenfunktionen und Anwendungen. Preprint, Athens (1980).

[358] N. Apostolatos, H. Christ, H. Santo, and H. Wippermann, Rounding Control and the Algorithmic Language ALGOL 68. Univ. Karlsruhe, Rechenzentrum, Rep. 6824 (June 1968).

[359] W. Appelt, Ein intervallarithmetisches Abbruchkriterium. *Z. Angew. Math. Mech.* **58**, T 404–T 405 (1978).

[360] U. Baginski, Die Lösung nichtlinearer Gleichungssysteme mit intervallarithmetischen Methoden. Diplomarbeit, Univ. München (1972).

[361] W. Barth and E. Nuding, Optimale Lösung von Intervallgleichungssystemen. *Computing* **12**, 117–125 (1974).

[362] H. Bauch, Zur Konvergenz monotoner Intervallfolgen und monotoner Folgen intervallwertiger Funktionen. *Z. Angew. Math. Mech.* **55**, 605–606 (1975).

[363] H. Bauch, Zur intervallanalytischen Lösungseinschließung bei charakteristischen Anfangswertproblemen mit hyperbolischer Differentialgleichung $Z_{st} = f(s, t, z)$. *Z. Angew. Math. Mech.* **57**, 543–547 (1977).

[364] H. Bauch, Zur Lösungseinschließung bei Anfangswertaufgaben gewöhnlicher Differentialgleichungen nach der Defektmethode. *Z. Angew. Math. Mech.* **57**, 387–396 (1977).

[365] C. Bendzulla, Intervall-Fortran-Präcompiler. *Rechentech./Datenverarbeit.* **11** (1980).

[366] A. Berman and R. J. Plemmons, "Nonnegative Matrices in the Mathematical Sciences." Academic Press, New York, 1979.

[367] S. N. Berti, "Arithmetic of Intervals." Editura Academiei Republicii Socialiste, Bucharest, 1977.

[368] G. P. Bhattacharjee and K. L. Majumder, On the computation of Hermite interval interpolating polynomials. *Computing* **19**, 73–83 (1977).

[369] G. P. Bhattacharjee and K. L. Majumder, Multivariate interval interpolation. *J. Comput. Appl. Math.* **5** (1978).

[370] F. Bierbaum, Intervall-Mathematik, eine Literaturübersicht. Interner Bericht No. 74/2. *Inst. Prakt. Math. Univ. Karlsruhe* (1974).

[371] F. Bierbaum and K. P. Schwiertz, A bibliography on interval mathematics. *J. Comput. Appl. Math.* **4**, 59–86 (1978).

[372] W. Burmeister, Optimal interval enclosing of certain sets of matrices, with application to monotone enclosing of square roots. *Computing* **25**, 283–295 (1980).

[373] O. Caprani and K. Madsen, Mean value forms in interval analysis. *Computing* **25**, 147–154 (1980).

[374] O. Caprani and K. Madsen, Contraction mappings in interval analysis. *BIT* **15**, 362–366 (1975).

[375] O. Caprani and K. Madsen, Iterative methods for interval inclusion of fixed points. *BIT* **18**, 42–51 (1978).

[376] O. Caprani and K. Madsen, Interval Contractions for the Solution of Integral Equations. Preprint (1980).

[377] O. Caprani, K. Madsen, and L. B. Rall, Integration of interval functions. *SIAM J. Math. Anal.* (to appear).

[378] F. L. Chernous'ko, Guaranteed estimates of undetermined quantities by means of ellipsoids. *Sov. Math. Dokl.* **21** (2) (1980).

[379] H. Christ, Prozeduren für Intervallrechnung in ALGOL X8. Univ. Karlsruhe, Rechenzentrum (September 1967).

[380] H. Christ, see also N. Apostolatos.

[381] G. E. Collins, Infallible calculation of polynomial zeros to specified precision. *In* "Mathematical Software III" (*Proc. Symp. Math. Res. Center*) (J. R. Rice, ed.), Publ. No. 39 of the Mathematics Research Center. Academic Press, New York, 1977.

[382] H. Cornelius, Inverse Interpolation zur globalkonvergenten Nullstellenberechnung. *Beiträge Numer. Math.* **10**, 23–35 (1981).

[383] H. Cornelius, Intervallrechnung und Nullstellenbestimmung. Diplomarbeit TU Berlin (1979) (not available).

[384] H. Cornelius, Ableitungsfreie Iterationsverfahren zur globalkonvergenten Nullstelleneinschließung. *Beiträge Numer. Math.* (to appear).

[385] P. J. Davis, see I. Najfeld.

[386] H. Fischer and F. Ost, Gittereigenschaften von Toleranz-Umgebungen. *Z. Angew. Math. Mech.* **57**, T 275–T 277 (1977).

[387] H. Florian and J. Püngel, Elementare Betrachtungen zur Intervallrechnung. Seminarbericht. Institut für Angewandte Mathematik, Graz-Austria (1980).

[388] L. Fox and M. R. Valenca, Some experiments with interval methods for two-point boundary-value problems in ordinary differential equations. *BIT* **20**, 67–82 (1980).

[389] I. Gargantini, Further applications of circular arithmetic: Schroeder-like algorithms with error bounds for finding zeros of polynomials. *SIAM J. Numer. Anal.* **15**, 497–510 (1978).

[390] J. Garloff and K.-P. Schwierz, A bibliography on interval mathematics. *J. Comput. Appl. Math.* **6**, 67–78 (1980).

[391] M. Ginsberg, Monitoring floating-point error propagation in scientific computation. *Comput. Maths. Appl.* **6**, 23–43 (1980).

[392] E. Grassmann and J. Rokne, The range of values of a circular complex polynomial over a circular complex interval. *Computing* **23**, 139–169 (1979).

[393] H. W. Griffith, Preliminary Investigation Using Interval Arithmetic in the Numerical Evaluation of Polynomials. Doctoral Dissertation. Univ. of Austin, Austin, Texas (1971).

[394] J. Grützmann, see W. Wallisch.

[395] E. Hansen, Interval forms of Newton's method. *Computing* **20**, 153–163 (1978).

[396] E. Hansen, Global optimization using interval analysis – the multidimensional case. *Numer. Math.* **34**, 247–270 (1980).

[397] E. Hansen, Global optimization using interval analysis: The one-dimensional case. *J. Optim. Theory Appl.* **29**, 331–344 (1979).

[398] E. Hansen, A globally convergent interval method for computing and bounding real roots. *BIT* **18**, 415–424 (1978).

[399] D. J. Hartfiel, Concerning the solution of $Ax = b$ where $P \leqslant A \leqslant Q$ and $p \leqslant b \leqslant q$. *Numer. Math.* **35**, 355–359 (1980).

[400] M. Hauenschild, Arithmetiken für komplexe Kreise. *Computing* **13**, 299–312 (1974).

[401] M. Hebgen, Eine scaling-invariante Pivotsuche für Intervallmatrizen. *Computing* **12**, 99–106 (1974).

[402] M. Hebgen, Ein Iterationsverfahren, welches die optimale Intervalleinschließung des Inversen eines M-Matrixintervalles liefert. *Computing* **12**, 107–115 (1974).

[403] J. Herzberger, Some recent results in constructing monotone methods for automatic computation. *Proc. Conf. Informat. Sci. Syst.* pp. 444–449. Johns Hopkins Univ. Baltimore, Maryland (1976).
[404] J. Herzberger, Some multipoint-iteration methods for bracketing a zero with application to parallel computation. *In* "Parallel Computers – Parallel Mathematics" (M. Feilmeier, ed.), pp. 231–234. North-Holland Publ., Amsterdam, 1977.
[405] J. Herzberger, Zur Approximation des Wertebereiches reeller Funktionen durch Intervallausdrücke. *In* "Computing Supplementum 1" (R. Albrecht and U. Kulisch, eds.), pp. 57–64. Springer-Verlag, Berlin and New York, 1977.
[406] J. Herzberger, Note on a bounding technique for polynomial functions. *SIAM J. Appl. Math.* **34**, 685–686 (1978).
[407] J. Herzberger, Global konvergente Interpolationsmethoden zur Nullstellenbestimmung. *Beiträge Numer. Math.* **7**, 65–74 (1979).
[408] J. Herzberger, Über einige intervallmäßige Iterationsverfahren. *Anal. Numér. Théor. Approx.* **8**, 53–57 (1979).
[409] J. Herzberger, Optimale Verfahren zur numerischen Einschließung reeller Nullstellen. (submitted for publication).
[410] J. Herzberger, Multipoint-Iterationsformeln hoher Ordnung zur Einschließung von reellen Nullstellen. *Z. Angew. Math. Mech.* **62** (to appear).
[411] L. W. Jackson, Interval arithmetic error-bounding algorithms. *SIAM J. Numer. Anal.* **12**, 223–238 (1975).
[412] K. U. Jahn, Punktkonvergenz in der Intervall-Rechnung. *Z. Angew. Math. Mech.* **55**, 606–608 (1975).
[413] Z. H. Juldasev, see S. A. Kalmykov.
[414] S. A. Kalmykov, Ju. I. Sokin, and Z. H. Juldasev, Über die intervall-analytische Methode zweiter Ordnung für gewöhnliche Differentialgleichungen. *Izv. Akad. Nauk UdSSR, Ser. Fiz.-Mat. Nauk* (3), 28–30 (1976).
[415] L. V. Kantorovich and G. P. Akilov, "Functional Analysis in Normed Spaces" (English translation). Pergamon, Oxford, 1964.
[416] G. Karambatzos, Ein Beitrag zur Intervallarithmetik mit einer Anwendung für die monotone numerische Einschließung von Integralen. Ph.D. Thesis, Athens (1979).
[417] E. Kaucher, Über eine Überlaufarithmetik auf Rechenanlagen und deren Anwendungsmöglichkeiten. *Z. Angew. Math. Mech.* **57**, T 286–T 287 (1977).
[418] E. Kaucher, Allgemeine Einbettungssätze algebraischer Strukturen unter Erhaltung von verträglichen Ordnungs- und Verbandsstrukturen mit Anwendung in der Intervallrechnung. *Z. Angew. Math. Mech.* **56**, T 296–T 297 (1976).
[419] E. Kaucher, Ein Fixpunktsatz für Iterationsverfahren mit isoton gerundeter Iterationsfunktion in bedingt vollständigen Verbänden. Preprint.
[420] E. Kaucher, Algebraische Erweiterungen der Intervallrechnung unter Erhaltung der Ordnungs- und Verbandsstrukturen. *In* "Computing Supplementum 1" (R. Albrecht and U. Kulisch, eds.), pp. 68–69. Springer-Verlag, Berlin and New York, 1977.
[421] E. Kaucher and S. M. Rump, Small bounds for the solution of systems of linear equations. *In* "Computing Supplementum 2" (G. Alefeld and R. D. Grigorieff, eds.), pp. 33–49. Springer-Verlag, Berlin and New York, 1980.
[422] E. Kaucher and S. M. Rump, Zum numerischen Nachweis der Existenz und Eindeutigkeit der Lösung $x \in X$ der Operator-Fixpunkt-Gleichung $f(x) = x$ sowie von $\rho(f') < 1$ mit Anwendungen. Manuscript.
[423] J. B. Kioustelidis, see R. E. Moore.
[424] R. Klatte, Zyklisches Enden bei Iterationsverfahren. *Z. Angew. Math. Mech.* **56**, T 298–T 300 (1976).
[425] R. Krawczyk, Interval extensions and interval iterations. *Computing* **24**, 119–129 (1980).

[426] R. Krawczyk and F. Selsmark, Orderconvergence and iterative interval me
Anal. Appl. **73**, 1–23 (1980).
[427] U. Kulisch, Editorial, *Elektron. Rech.* 203–204 (1968).
[428] U. Kulisch, Zur Anwendbarkeit der naiven Intervallrechnung. M
[429] U. Kulisch, Numerisches Rechnen – Wie es ist und wie es sein
[430] U. Kulisch, see also W. L. Miranker.
[431] P. Lancaster, see J. Rokne.
[432] M. Landsberg and W. Schirotzek, Zur konvexen Optimier
Wiss. Z. TU Dresden **26**, 331–335 (1977).
[433] D. P. Laurie, INTFORT-Interval Arithmetic Interpret
Use with FORTRAN. *Comput. Programs Sci. Tech*
[434] N. J. Lehmann, Über Probleme und Bedeutung
Akad. Wiss. DDR **11**, 48–50 (1973).
[435] K. Madsen, see O. Caprani.
[436] K. L. Majunder, see G. P. Bhattacharjee.
[437] L. J. Mancini and D. J. Wilde, Interva ... al
programming. *J. Optim. Theory Appl.* **20**, 277–
[438] L. J. Mancini and G. P. McCormick, Bounding glo ... metic.
Oper. Res. **27** (4) (1979).
[439] U. Marcowitz, Fehlerabschätzung bei Anfangswertaufgaben ... n gewöhnlichen Differentialgleichungen mit Anwendung auf das Problem d ... eintritts eines Raumfahrzeugs in die Lufthülle der Erde. Dissertation, Univ. K ... (1973).
[440] S. M. Markov, Existence and uniqueness of solutions of the interval differential equation $X' = F(t, X)$. *C. R. Acad. Bulg. Sci.* **31** (12) (1978).
[441] G. P. McCormick, see L. J. Mancini.
[442] M. Mihelcic, Eine Modifikation des Halbierungsverfahrens zur Bestimmung aller reeller Nullstellen einer Funktion mit Hilfe der Intervall-Arithmetik. *Angew. Informat.* **1** (1975).
[443] W. L. Miranker and U. Kulisch, Computer Arithmetic in Theory and Practice. Research Report RC 7776 (33658), July 24, Mathematics, IBM Thomas J. Watson Research Center, Yorktown Heights. (1979).
[444] M. Müller, Einige Probleme der Matrizenrechnung in der Intervall-Arithmetik. Diplomarbeit. TU Berlin (1970).
[445] R. E. Moore, Interval Arithmetic and Automatic Error Analysis in Digital Computing. Ph.D. Thesis. Mathematics Department, Stanford Univ. (October 1962).
[446] R. E. Moore, On computing the range of a rational function of *n* variables over a bounded region. *Computing* **16**, 1–15 (1976).
[447] R. E. Moore, "Elements of Scientific Computing." Holt, New York, 1975.
[448] R. E. Moore, A test for existence of solutions to non-linear systems. *SIAM J. Numer. Anal.* **14**, 611–615 (1977).
[449] R. E. Moore, A computational test for convergence of iterative methods for non-linear systems. *SIAM J. Numer. Anal.* **15**, 1194–1196 (1978).
[450] R. E. Moore, Bounding sets in function spaces with applications to nonlinear operator equations. *SIAM Rev.* **20**, 492–512 (1978).
[451] R. E. Moore, "Methods and Applications of Interval Analysis," SIAM Studies in Applied Mathematics. SIAM, Philadelphia, Pennsylvania, 1979.
[452] R. E. Moore, Interval Methods for Nonlinear Systems. *In* "Computing Supplementum 2" (G. Alefeld and R. D. Grigorieff, eds.), pp. 113–120. Springer-Verlag, Berlin and New York, 1980.
[453] R. E. Moore and J. B. Kioustelidis, A simple test for accuracy of approximate solutions to nonlinear (or linear) systems. *SIAM J. Numer. Anal.* **17**, 521–529 (1980).
[454] I. Najfeld, R. A. Vitale, and P. J. Davis, Minkowski iteration of sets. *Linear Algebra and Its Appl.* **29**, 259–291 (1980).

[455] W. Niethammer, Relaxation bei komplexen Matrizen. *Math. Z.* **86**, 34–40 (1964).
[456] K. Nickel, Die Überschätzung des Wertebereiches einer Funktion in der Intervallrechnung mit Anwendungen auf lineare Gleichungssysteme. *Computing* **18**, 15–36 (1977).
[457] K. Nickel, A globally convergent ball Newton method. *SIAM J. Numer. Anal.* **18**, 988–1003 (1981).
[458] E. Nuding, Schrankentreue Berechnung der Exponentialfunktion wesentlich-nichtnegativer Matrizen. *Computing* **26**, 57–66 (1981).
[459] E. Nuding, see also W. Barth.
[460] D. Oelschlägel and H. Süsse, Fehlerabschätzung bei einem speziellen quadratischen Optimierungsproblem. *Z. Angew. Math. Mech.* **59**, 482–483 (1979).
[461] D. Oelschlägel, see also H. Süsse.
[462] F. Ost, see H. Fischer.
[463] F. A. Oliveira, Error analysis and computer arithmetic. Summary report FAO-2 Mathematical Institute. Univ. de Coimbra (1971).
[464] F. A. Oliveira, Inversion of strongly diagonally dominant matrices with control of errors. *Univ. Lisboa Rev. Fac. Ci., A (2), Mat.* **13** (1970–71).
[465] F. A. Oliveira, Interval analysis and two-point boundary value problems. *SIAM J. Numer. Anal.* **11**, 382–391 (1974).
[466] F. A. Oliveira, Alguns aspectos do problema dos errors em Análise Numérica. *Gaz. Mat. (Lisbon)*, 109–112 (1968).
[467] F. W. J. Olver, A new approach to error arithmetic. *SIAM J. Numer. Anal.* **15**, 368–393 (1978).
[468] M. S. Petkovic, On an interval Newton's method derived from exponential curve fitting. *Z. Angew. Math. Mech.* **61**, 117–119 (1981).
[469] L. Platzöder, Unpublished manuscript (1979).
[470] R. J. Plemmons, see A. Bermann.
[471] J. Püngel, see H. Florian.
[472] L. Rall, A comparison of the existence theorems of Kantorovich and Moore. MRC Technical Summary Report #1944 Mathematics Research Center, Madison, Wisconsin (1979).
[473] L. Rall, Representations of intervals and optimal error bounds. Preprint NI-80-09. Institute for Numerical Analysis, Technical Univ. of Denmark, 1980.
[474] L. Rall, Application of interval integration to the solution of integral equations. Preprint NI-80-07. Institute for Numerical Analysis, Technical Univ. of Denmark.
[475] L. Rall, see also O. Caprani.
[476] H. Ratschek, Über einige Wesenszüge der Intervallarithmetik. *Math.-Phys. Semesterber.* 67–79 (April 1974).
[477] H. Ratschek, Über das Produkt von Intervallpolynomen. *Computing* **13**, 313–325 (1974).
[478] H. Ratschek, Mittelwertsätze für Intervallfunktionen. *Beiträge Num. Math.* **6**, 133–144 (1977).
[479] H. Ratschek, Die subdirekte Unzerlegbarkeit der Intervallrechnung. *J. Reine Angew. Math.* **299/300**, 287–293 (1978).
[480] H. Ratschek, Zentrische Formen. *Z. Angew. Math. Mech.* **58**, T 434–T 436 (1978).
[481] H. Ratschek, Centered forms. *SIAM J. Numer. Anal.* **17**, 656–662 (1980).
[482] H. Ratschek and J. Rokne, About the centered form. *SIAM J. Numer. Anal.* **17**, 333–337 (1980).
[483] H. Ratschek and G. Schröder, Centered forms for functions in several variables. *Math. Anal. Appl.* **82**, 543–552 (1981).
[484] K. Reichmann, Abbruch beim Intervall-Gauß-Algorithmus. *Computing* **22**, 355–361 (1979).

[485] K. Reichmann, Ein hinreichendes Kriterium für die Durchführbarkeit des Intervall-Gauß-Algorithmus bei Intervall-Hessenberg-Matrizen ohne Pivotsuche. *Z. Angew. Math. Mech.* **59**, 373–379 (1975).

[486] P. L. Richmann, Computing a subinterval of the image. *J. ACM* **21**, 454–458 (1974).

[487] J. Rokne, An interval analysis approach to the stock market. Preprint.

[488] J. Rokne, Reducing the degree of an interval polynomial. *Computing* **14**, 5–14 (1975).

[489] J. Rokne, Errorbounds for simple zeros of λ-matrices. *Computing* **16**, 17–27 (1976).

[490] J. Rokne, Bounds for an interval polynomial. *Computing* **18**, 225–240 (1977).

[491] J. Rokne, Polynomial least square interval approximation. *Computing* **20**, 165–176 (1978).

[492] J. Rokne, A note on the Bernstein algorithm for bounds for interval polynomials. *Computing* **21**, 159–170 (1979).

[493] J. Rokne, The range of values of a complex polynomial over a complex interval. *Computing* **22**, 153–169 (1979).

[494] J. Rokne, The centered form for interval polynomials. *Computing* **27**, 339–348 (1981).

[495] J. Rokne, Optimal computation of the Bernstein algorithm for the bound of an interval polynomial. *Freiburger Intervall-Berichte* 81/4 (1981).

[496] J. Rokne, see also G. Alefeld.

[497] J. Rokne, see also E. Grassmann.

[498] J. Rokne, see also H. Ratschek.

[499] J. Rokne and P. Lancaster, Complex interval arithmetic. *Comput. J.* **18**, 83–85 (1975).

[500] J. Rokne and T. Wu, The circular complex centered form. *Computing* **28**, 17–30 (1982).

[501] S. N. Rump, Kleine Fehlerschranken bei Matrixproblemen, Ph.D. Thesis, Univ. Karlsruhe, Fakultät für Mathematik, Karlsruhe (1980).

[502] S. N. Rump, Notiz zur Genauigkeit der Arithmetik in Rechenanlagen. *Elektron. Rech.* **22**, 243–244 (1980).

[503] S. M. Rump, see also E. Kaucher.

[504] H. Santo and H. W. Wippermann, Realisierung eines Rechenwerkes mit Schrankenrundung für die Rechenanlage El X8. Rechenzentrum der Univ. Karlsruhe, Rep. 3969 (1969).

[505] H. Santo, see also N. Apostolatos.

[506] W. Schirotzek, see also M. Landsberg.

[507] J. W. Schmidt, Eine Übertragung der Regula Falsi auf Gleichungen in Banachräume, II. *Z. Angew. Math. Mech.* **43**, 97–110 (1963).

[508] J. W. Schmidt, Ausgangsvektoren für monotone Iterationen bei linearen Gleichungssystemen. *Numer. Math.* **6**, 78–88 (1964).

[509] J. W. Schmidt, Einschließung inverser Elemente durch Fixpunktverfahren. *Numer. Math.* **31**, 313–320 (1978).

[510] J. W. Schmidt and H. Schwetlick, Ableitungsfreie Verfahren mit höherer Konvergenzgeschwindigkeit. *Computing* **3**, 215–226 (1968).

[511] H. Schwetlick, see J. W. Schmidt.

[512] J. Schröder, Computing error bounds in solving linear systems. *Math. Comp.* **16**, 323–337 (1962).

[513] G. Schröder, Zur Struktur des Raumes der Hypernormbälle über einem linearen Raum. *Computing* **15**, 67–70 (1975).

[514] G. Schröder, see also H. Ratschek.

[515] K. P. Schwierz, see F. Bierbaum.

[516] F. Selsmark, see R. Krawczyk.

[517] Bl. Sendov, Segment arithmetic and segment limit. *C. R. Acad. Bulg. Sci.* **30** (7) (1977).

[518] Signum Newsletter, Special Issue October (1979).

[519] S. Skelboe, Computation of rational interval functions. *BIT* **14**, 87–95 (1974).

[520] I. Sokin, see S. A. Kalmykov.

[521] H. Spreuer, see E. Adams.

[522] H. Süsse, Untersuchungen zur Intervallarithmetik. Diplomarbeit. Technische Hochschule "Carl Schorlemmer" Leuna-Merseburg, 1973.
[523] H. Süsse, see also D. Oelschlägel.
[524] H. Süsse and D. Oelschlägel, Nullstellenbestimmung mit der Intervallrechnung. *Wiss. Z. TH Leuna-Merseburg* **17**, 423–432 (1975).
[525] P. Thieler, Eine Anwendung des Brouwerschen Fixpunktsatzes in der Intervallarithmetik der Matrizen. *Computing* **14**, 141–147 (1975).
[526] P. Thieler, On componentwise error estimates for inverse matrices. *Computing* **19**, 303–312 (1978).
[527] R. Tremmel, *Mathematik mit Fehlern* FAZ, 14.7 (1976).
[528] M. R. Valenca, see L. Fox.
[529] R. A. Vitale, see I. Najfeld.
[530] W. Wallisch and J. Grützmann, Intervallarithmetische Fehleranalyse. *Beiträge Num. Math.* **3**, 163–171 (1975).
[531] D. J. Wilde, see L. J. Mancini.
[532] H. W. Wippermann, Definition von Schrankenzahlen in TRIPLEX-ALGOL 60. *Computing* **3**, 99–109 (1968).
[533] H. W. Wippermann, Realisierung einer Intervallarithmetik in einem ALGOL-60 System. *Elektron. Rech.* **9**, 224–233 (1967).
[534] H. Wippermann, see also N. Apostolatos.
[535] H. Wippermann, see also H. Santo.
[536] M. A. Wolfe, A modification of Krawczyk's algorithm. *SIAM J. Numer. Anal.* **17**, 376–379 (1980).
[537] J. M. Yohe, Implementing nonstandard arithmetics. *SIAM Rev.* **21**, 34–56 (1979).
[538] J. M. Yohe, Software for interval arithmetic. A reasonably portable package. *ACM TOMS* **5**, 50–63 (1979).
[539] J. M. Yohe, The interval arithmetic package. MRC Technical Summary Report 1755.
[540] J. Zlamal, A measure for ill-conditioning of matrices in interval arithmetic. *Computing* **19**, 149–155 (1977).
[541] E. Zemlin, see H. J. Anders.

INDEX OF NOTATION

SUBJECT INDEX

I

L

M

N

O

P

Q

R

S